# The **MATH** Olympian

A novel by

Richard Hoshino

Suite 300 - 990 Fort St
Victoria, BC, Canada, V8V 3K2
www.friesenpress.com


**Author photo by:** William Thompson
First Edition — 2015



The five Olympiad problems that form the core of this novel previously appeared on the Canadian Mathematical Olympiad (CMO). Reprinted with the permission of the Canadian Mathematical Society, all five problems: 1995 CMO #1, 1996 CMO #2, 1994 CMO #3, 1996 CMO #4, and 1998 CMO #5, were originally part of the Sun Life Financial Canadian Mathematical Olympiad and appear on the Canadian Mathematical Society Website at http://math.ca/Competitions/CMO/.

**ISBN**
978-1-4602-5872-9 (Hardcover)
978-1-4602-5873-6 (Paperback)
978-1-4602-5874-3 (eBook)

*1. Education, Teaching Methods & Materials, Mathematics*

Distributed to the trade by The Ingram Book Company

"Courage, sacrifice, determination, commitment, toughness, heart, talent, guts. That's what little girls are made of; the heck with sugar and spice."

– Bethany Hamilton, professional surfer

# Contents


Prologue 1

The Canadian Mathematical Olympiad 7

Problem #1: Sum of Exponents 9

Problem #2: Three Equations 103

Problem #3: Circle Voting 183

Problem #4: Measuring Angles 279

Problem #5: Sequence of Integers 373

Press Release 469

Epilogue 471

Acknowledgements 477

Q&A with the Author 479

# Prologue

"I'm ready."

My voice is barely audible. I'm terrified.

Mom wraps me in a tight hug, and we cling to each other, shielding our bodies from the howling wind that's typical for an mid-March morning in Nova Scotia. I glance at my watch.

*8:49 a.m.*

I see a man stare as he walks past us, puzzled by the sight of a broad seventeen-year-old towering over her mother. He, of course, has no idea what's racing through my mind at this moment.

As Mom and I squeeze each other one final time, she knows that I don't need another pep talk or any more words of encouragement. We gently step away from each other, and Mom gets back in the car. Before she closes the door, she turns to give me one last look: a forced, nervous smile.

"I'm ready," I say, as calmly as possible, trying to convince myself that the words I've just spoken are indeed true.

Mom drives off, and I find myself alone, standing a few feet from the main entrance of the Cape Breton regional school board. I take a deep breath, to slow my pounding heart.

*8:50 a.m.*

In ten minutes, I'll be writing the Canadian Mathematical Olympiad, joining forty-nine other high school students from across the country who also qualified to write Canada's toughest math contest.

One exam. Three hours. Five questions.

I'm the last person to qualify, the student closest to the cut-off. But today, that doesn't matter. Whether I'm ranked first or fiftieth, I'm in. And that means I have a chance.

The chance to achieve my childhood dream.

*8:51 a.m.*

I walk up the steps to the school board entrance. As I close the door behind me, I come face-to-face with a slim lady with long black hair, who greets me with a look of intensity. She looks just like Gillian Lowell, but thirty years older.

I flinch and take a step back.

"Bethany MacDonald," says the woman, staring into my eyes. "All of Cape Breton is rooting for you today."

I nod, at a loss for words.

An older well-dressed man comes to the rescue. He introduces himself as Mr. MacKay, the school board superintendent. He asks me to follow him towards the conference room, a large open space he has reserved this morning just for me.

I walk into a room with multiple plaques and pictures hanging on the side walls, with an oval-shaped mahogany desk set right in the centre. I take the seat farthest from the door, where I can see the big clock by looking straight up.

"So, Bethany, how tall are you?"

"Six feet," I reply, knowing that this is the easiest question I'll be asked all morning.

"And yet you're not a basketball player?"

"No. I'm a runner."

"I know," says Mr. MacKay. "You're the captain of the cross-country team at Sydney High School."

I raise my eyebrow. Seeing my reaction, the superintendent smiles.

"Bethany, from what I've heard, you've been breaking stereotypes your whole life."

Mr. MacKay says he'll give me some time to get ready. As the door closes, I'm left all alone, with just my thoughts to keep me company as I prepare myself for the challenge ahead.

*8:54 a.m.*

The International Mathematical Olympiad (IMO) is the world championship of problem-solving for high school students. Nearly one hundred countries are invited to this year's IMO, with each country sending their top six teenage "mathletes".

Ever since my twelfth birthday, I've wanted to be a Math Olympian.

That improbable hope, that one day I'd wear the red and white and represent my country, has sustained me over the past six years. But now I'm in Grade 12, heading to university in the fall, and this is my last chance.

Will I make it? In three hours, I'll know the answer.

Albert, Raju, and Grace are guaranteed to make the team; they're light years beyond the rest of us. Albert Suzuki represented Canada at the IMO the past two years, winning a gold medal both times. Raju Gupta went to the IMO last year, and is a shoo-in again this year. Grace Wong just missed Team Canada by one spot last time, and no one will deny her from making the top six this year.

Because of the four-hour time difference between Nova Scotia and British Columbia, I know that Grace is still sleeping. As the clock ticks ominously above me, I think about my closest friend and reflect on how far we've come since that summer day in Vancouver, when we made our pact.

Since that evening, I've trained non-stop, over thirty hours a week for nearly two years, while juggling all of my responsibilities at school. I've sacrificed so much to make it this far.

And yet, I know the odds are stacked against me.

*8:55 a.m.*

The Canadian IMO team is determined by a secret formula unknown to the fifty of us writing today's competition. All we know is that each math contest is assigned a certain weight, with this final Olympiad exam being the most important. Our scores from all the contests are then added together, from which the top six will be decided.

I'm frustrated and angry at what happened during the previous contests, where my test anxiety flared up at the worst possible time. I know I am much, much better than how I've performed, and this is my last chance to prove it.

Because of how far back I am from the current top six, my only hope is "The Rule": that the winner of today's Canadian Mathematical Olympiad automatically gets a spot on the IMO team.

Even though this is by far the hardest contest we'll write all year, Albert is sure to get a perfect score, just as he did the year before. So I need fifty out of fifty myself, and tie Albert for the top score in Canada. That's the only way I'll make it.

I need to write five complete solutions in just three hours, wowing the judges with an elegant and flawless performance, just like a figure skater at the Olympic trials.

My figure skating analogy triggers a thought – a bad thought – and I wince.

I don't need to be thinking about *that.* Especially not right now.

*8:57 a.m.*

I'm reminded of Gillian Lowell yet again, and I recall the spiteful words she said to me four months ago.

"You risked everything on a stupid dream, trying to be an Olympian in math. In math!"

She spoke loudly enough to be heard by everyone else.

"Bethany, get a life."

I crack a smile, finally realizing that Gillian was right all along. I did get a life.

A life more fulfilling than anything I could have ever imagined.

The door opens. The superintendent walks in, and glances at the clock.

*8:58 a.m.*

"Bethany, shall we begin?"

"Yes," I reply, sitting up straight with my back firmly against the chair.

Mr. MacKay walks over and places in front of me a stack of plain white paper, three blue pens, and five sealed envelopes numbered #1 through #5. He asks me if I have any questions, and I shake my head.

"Okay. As soon as it's nine o'clock, you can start."

Before walking out the door, he pauses and looks back. "Good luck, Bethany. We're all so proud of you."

I stare at the clock above me, and my eyes follow the red second hand moving quickly to the top.

*8:59 a.m.*

At that instant, I know there's an 84.5° angle formed by the hour hand and the minute hand. Recalling the memory of that special day in Halifax, and everything that's happened since, I feel a sense of peace.

*I'm ready.*

I watch the red second hand make another clockwise rotation until it once again points directly north.

*9:00 a.m.*

I rip open the folders, and spend a few minutes studying the five questions.

These problems look hard. Really hard. But this is the Canadian Mathematical Olympiad. Of course they're hard.

As Grace reminded me, most university math professors couldn't solve even one of these five problems. But to be fair, those math professors haven't spent *three thousand hours* training for a moment like this.

I close my eyes and think about the life-changing decision I made on my twelfth birthday, and how this one decision led me to experience hundreds of ups and downs and twists and turns over the past six years.

It's been an amazing ride. In just three hours, this roller-coaster journey will come to an end.

And now I get to write the ending to this story. To my story.

I open my eyes and begin.

# The Canadian Mathematical Olympiad

### Problem #1

Determine the value of:

$$\frac{9^{1/1000}}{9^{1/1000}+3}+\frac{9^{2/1000}}{9^{2/1000}+3}+\frac{9^{3/1000}}{9^{3/1000}+3}+\cdots+\frac{9^{998/1000}}{9^{998/1000}+3}+\frac{9^{999/1000}}{9^{999/1000}+3}$$

### Problem #2

Find all real solutions to the following system of equations.

$$\begin{cases}\dfrac{4x^2}{1+4x^2}=y\\[2ex]\dfrac{4y^2}{1+4y^2}=z\\[2ex]\dfrac{4z^2}{1+4z^2}=x\end{cases}$$

### Problem #3

Twenty-five men sit around a circular table. Every hour there is a vote, and each must respond *yes* or *no*. Each man behaves as follows: on the $n^{\text{th}}$ vote, if his response is the same as the response of at least one of the two people he sits between, then he will respond the same way on the $(n+1)^{\text{th}}$ vote as on the $n^{\text{th}}$ vote; but if his response is different from that of both his neighbours on the $n^{\text{th}}$ vote, then his response on the $(n+1)^{\text{th}}$ vote will be different from his response on the $n^{\text{th}}$ vote. Prove that, however everybody responded on the first vote, there will be a time after which nobody's response will ever change.

### Problem #4

Let $\Delta ABC$ be an isosceles triangle with $AB = AC$. Suppose that the angle bisector of $\angle B$ meets $AC$ at $D$ and that $BC = BD + AD$. Determine $\angle A$.

### Problem #5

Let $m$ be a positive integer. Define the sequence $x_0, x_1, x_2, \ldots$ by $x_0 = 0$, $x_1 = m$, and $x_{n+1} = m^2 x_n - x_{n-1}$ for $n = 1, 2, 3 \ldots$. Prove that an ordered pair $(a, b)$ of non-negative integers, with $a \le b$, gives a solution to the equation $\frac{a^2+b^2}{ab+1} = m^2$ if and only if $(a, b)$ is of the form $(x_n, x_{n+1})$ for some $n \ge 0$.

## The Canadian Mathematical Olympiad, Problem #1

**Determine the value of:**

$$\frac{9^{1/1000}}{9^{1/1000}+3}+\frac{9^{2/1000}}{9^{2/1000}+3}+\frac{9^{3/1000}}{9^{3/1000}+3}+\cdots+\frac{9^{998/1000}}{9^{998/1000}+3}+\frac{9^{999/1000}}{9^{999/1000}+3}$$

## Problem #1: Sum of Exponents

I stare at the first problem, not sure where to start.

I circle the first term in the expression of Problem #1, the one with the ugly exponent $9^{1/1000}$. Am I actually supposed to calculate the 1000th root of 9? Without a calculator, I know that's not possible.

There has to be an insight somewhere. This is an Olympiad problem, and all Olympiad problems have nice solutions that require imagination and creativity rather than a calculator.

I re-read the question yet again, and confirm that I have to determine the following sum:

$$\frac{9^{1/1000}}{9^{1/1000}+3}+\frac{9^{2/1000}}{9^{2/1000}+3}+\frac{9^{3/1000}}{9^{3/1000}+3}+\cdots+\frac{9^{998/1000}}{9^{998/1000}+3}+\frac{9^{999/1000}}{9^{999/1000}+3}$$

There are 999 terms in the sum, and each term is of the form $\frac{9^x}{9^x+3}$. In the first term, $x$ equals $\frac{1}{1000}$; in the second term, $x$ equals $\frac{2}{1000}$; in the third term, $x$ equals $\frac{3}{1000}$; and so on, all the way up to the last term, where $x$ equals $\frac{999}{1000}$.

In the entire expression, there's only one doable calculation, the term right in the middle. I know I can calculate $\frac{9^{500/1000}}{9^{500/1000}+3}$, using the fact that $\frac{500}{1000}=\frac{1}{2}$.

Since raising a quantity to the exponent $\frac{1}{2}$ is the same as taking its square root, I see that:

$$\frac{9^{500/1000}}{9^{500/1000}+3}=\frac{9^{1/2}}{9^{1/2}+3}=\frac{\sqrt{9}}{\sqrt{9}+3}=\frac{3}{3+3}=\frac{3}{6}=\frac{1}{2}$$

But other than this, I'm not sure what to do. Twirling my pen and closing my eyes, I concentrate, hoping for a spark.

One idea comes to mind: setting up a "telescoping series". My mentor, Mr. Collins, introduced me to this beautiful technique years ago at one of our Saturday afternoon sessions at Le Bistro Café. Before explaining the concept to me, Mr. Collins first gave me a simple question of adding five fractions:

Without using a calculator, determine $\frac{1}{2}+\frac{1}{6}+\frac{1}{12}+\frac{1}{20}+\frac{1}{30}$

I solved Mr. Collins' problem by finding the common denominator. In this case, the common denominator is 60, the smallest number that evenly divides into each of 2, 6, 12, 20, and 30. So the answer is:

$$\frac{30}{60}+\frac{10}{60}+\frac{5}{60}+\frac{3}{60}+\frac{2}{60} = \frac{30+10+5+3+2}{60} = \frac{50}{60} = \frac{5}{6}$$

And then I remembered Mr. Collins' smile as he gave me another addition problem:

Determine $\frac{1}{2}+\frac{1}{6}+\frac{1}{12}+\frac{1}{20}+\frac{1}{30}+\frac{1}{42}+\frac{1}{56}+\frac{1}{72}+\frac{1}{90}$

This time, it took me almost fifteen minutes to get the answer. Most of the time was spent trying to figure out the common denominator, which I eventually determined to be 2520. But it was a tedious process of checking and re-checking all of my calculations.

After Mr. Collins congratulated me on getting the right answer, he pointed to the nine fractions on my sheet of paper and asked if there was a pattern. After staring at the numbers for a while, I saw it:

| | | |
|---|---|---|
| **2** = 1 × 2 | **6** = 2 × 3 | **12** = 3 × 4 |
| **20** = 4 × 5 | **30** = 5 × 6 | **42** = 6 × 7 |
| **56** = 7 × 8 | **72** = 8 × 9 | **90** = 9 × 10 |

Mr. Collins suggested I write $\frac{1}{90}$ as the difference of two fractions: $\frac{1}{90}=\frac{1}{9}-\frac{1}{10}$. He then asked whether there were any other terms in this expression that could also be written as the difference of two fractions. I eventually saw that $\frac{1}{2}=\frac{1}{1}-\frac{1}{2}$ and $\frac{1}{6}=\frac{1}{2}-\frac{1}{3}$.

Once I saw the pattern, I discovered this amazing solution, called a "telescoping series":

$\frac{1}{2}+\frac{1}{6}+\frac{1}{12}+\frac{1}{20}+\frac{1}{30}+\frac{1}{42}+\frac{1}{56}+\frac{1}{72}+\frac{1}{90}$ can be re-written as:

$$\left(\frac{1}{1}-\frac{1}{2}\right)+\left(\frac{1}{2}-\frac{1}{3}\right)+\left(\frac{1}{3}-\frac{1}{4}\right)+\left(\frac{1}{4}-\frac{1}{5}\right)+\left(\frac{1}{5}-\frac{1}{6}\right)+\left(\frac{1}{6}-\frac{1}{7}\right)+\left(\frac{1}{7}-\frac{1}{8}\right)+\left(\frac{1}{8}-\frac{1}{9}\right)+\left(\frac{1}{9}-\frac{1}{10}\right)$$

This is just $\frac{1}{1}-\frac{1}{2}+\frac{1}{2}-\frac{1}{3}+\frac{1}{3}-\frac{1}{4}+\frac{1}{4}-\frac{1}{5}+\frac{1}{5}-\frac{1}{6}+\frac{1}{6}-\frac{1}{7}+\frac{1}{7}-\frac{1}{8}+\frac{1}{8}-\frac{1}{9}+\frac{1}{9}-\frac{1}{10}$.

Since one negative fraction cancels a positive fraction with the same value, all the terms in the middle get eliminated:

$$\frac{1}{1}\cancel{-\frac{1}{2}+\frac{1}{2}-\frac{1}{3}+\frac{1}{3}-\frac{1}{4}+\frac{1}{4}-\frac{1}{5}+\frac{1}{5}-\frac{1}{6}+\frac{1}{6}-\frac{1}{7}+\frac{1}{7}-\frac{1}{8}+\frac{1}{8}-\frac{1}{9}+\frac{1}{9}}-\frac{1}{10}$$

Like a giant telescope that collapses down to a small part at the top and a small part at the bottom, this series collapses to the difference $\frac{1}{1}-\frac{1}{10}$, which equals $\frac{9}{10}$. So the answer is $\frac{9}{10}$.

That day, Mr. Collins showed me several problems where the answer can be found using a telescoping series, where a seemingly-tedious calculation can be solved with elegance and beauty.

The key is to represent each term as a difference of the form $x - y$, where $y$ is called the "subtrahend" and $x$ is called the "minuend". From Mr. Collins' examples, I learned that the series telescopes every time the subtrahend of one term equals the minuend of the following term.

As I recall that lesson with Mr. Collins many years ago, I'm hopeful that I can use this technique to solve the first problem of the Canadian Math Olympiad. I look at Problem #1 again, reminding myself of what I need to determine.

$$\frac{9^{1/1000}}{9^{1/1000}+3}+\frac{9^{2/1000}}{9^{2/1000}+3}+\frac{9^{3/1000}}{9^{3/1000}+3}+\cdots+\frac{9^{998/1000}}{9^{998/1000}+3}+\frac{9^{999/1000}}{9^{999/1000}+3}$$

I start with the general expression $\frac{9^x}{9^x+3}$ and try to write it down as the difference of two functions, so that the subtrahend of each term equals the minuend of the following term.

I try a bunch of different combinations to get the difference to work out to $\frac{9^x}{9^x+3}$ such as the expression $\frac{1}{3^x}-\frac{1}{3^x+1}$ which almost works but not quite. I attempt other combinations using every algebraic method I know. All of a sudden, I realize the futility of my approach.

The denominator doesn't factor nicely, so this approach cannot work. Oh no.

*9:19 a.m.*

I feel the first bead of sweat on my forehead, and wonder if I'm going to get another "math contest anxiety attack". I close my eyes and take a deep breath, knowing that if I start to panic and lose focus, my chances of becoming a Math Olympian are over.

*Calm down, Bethany, calm down. There's lots of time left. You can do this.*

I think about the soothing words of Mr. Collins, and am reminded of another important problem-solving strategy I learned from him: simplify the problem by breaking it into smaller and easier parts, in order to find a pattern.

I can do that.

I don't want to deal with the horrible expression given in the problem, a complicated sum of nearly one thousand fractions. I've seen enough contest problems to know that the number 1000 is a distracter, and that it has nothing to do with the question. By making the number big, the problem looks a lot more intimidating than it actually is.

For example, in the addition question that Mr. Collins posed to me that day, as soon as I realize that the series telescopes, it doesn't matter whether there are nine fractions or nine thousand fractions. In the former the answer is $\frac{1}{1} - \frac{1}{10} = \frac{9}{10}$ and in the latter the answer is $\frac{1}{1} - \frac{1}{9001} = \frac{9000}{9001}$. The final answer is different, but at its heart, it's the exact same problem.

I'm sure the same is true with this Olympiad problem. Especially being the first question, I know there has to be a short and elegant solution. Remembering the advice of Mr. Collins, I decide to simplify the problem in order to discover a pattern, which will then allow me to solve the actual problem.

I change the denominator from 1000 to 4, to have just a few terms to play with. So now, instead of the exponents ranging from $\frac{1}{1000}$ to $\frac{999}{1000}$ I only have to consider $\frac{1}{4}$, $\frac{2}{4}$ and $\frac{3}{4}$.

Instead of adding 999 ugly terms as in the actual problem, I only have three terms in the simplified problem. By making the expression easier, I am hopeful that I'll discover something interesting.

So my simplified problem is to determine the value of

$$\frac{9^{1/4}}{9^{1/4}+3}+\frac{9^{2/4}}{9^{2/4}+3}+\frac{9^{3/4}}{9^{3/4}+3}$$

This looks much more reasonable. The middle expression is easy – I figured this out ten minutes earlier.

$$\frac{9^{2/4}}{9^{2/4}+3}=\frac{9^{1/2}}{9^{1/2}+3}=\frac{\sqrt{9}}{\sqrt{9}+3}=\frac{3}{3+3}=\frac{3}{6}=\frac{1}{2}$$

As I ponder how to calculate the values of $\frac{9^{1/4}}{9^{1/4}+3}$ and $\frac{9^{3/4}}{9^{3/4}+3}$, a few ideas occur to me. I scribble some calculations on my notepad, add up the two fractions, and am surprised that the sum is exactly one.

$$\frac{9^{1/4}}{9^{1/4}+3}+\frac{9^{3/4}}{9^{3/4}+3}=1$$

Interestingly, the first and last terms of my simplified problem add up to 1. I have a hunch that this might also be true in the more complicated Olympiad problem with 999 terms.

$$\frac{9^{1/1000}}{9^{1/1000}+3}+\frac{9^{2/1000}}{9^{2/1000}+3}+\frac{9^{3/1000}}{9^{3/1000}+3}+\cdots+\frac{9^{998/1000}}{9^{998/1000}+3}+\frac{9^{999/1000}}{9^{999/1000}+3}$$

To my delight, the hunch is correct.

$$\frac{9^{1/1000}}{9^{1/1000}+3}+\frac{9^{999/1000}}{9^{999/1000}+3}=1$$

I run through the calculations one more time, double-checking that I haven't made any mistakes. Yes, the terms in the numerator perfectly match the terms in the denominator, and the sum is indeed one.

I wonder whether this pattern continues, and am shocked to discover that

$$\frac{9^{2/1000}}{9^{2/1000}+3}+\frac{9^{998/1000}}{9^{998/1000}+3}=1$$

I suddenly feel a lump in my throat. I know how to solve the Olympiad problem.

All I need to do is to apply the technique I discovered in Mrs. Ridley's class seven years ago, when I was in Grade 5. I can't believe it.

It's the Staircase.

# 1

"Two-hundred ten!" I blurted out.

Every head in the classroom turned towards me. Several students stared at me in shock. Mrs. Ridley stood there with her back leaning against the chalkboard and her jaw dropped, and for several uncomfortable seconds that seemed like an eternity, I could hear the sound of my own heart thumping.

One person broke the silence. Of course, it was Gillian.

"Is that the answer?"

My Grade 5 teacher turned towards me and smiled. "Yes, it is."

Michael, the loud boy sitting to my left, began clapping.

"It looks like Gillian finally has some competition."

A few people joined in the applause, which prompted Gillian to turn around from her seat right in the middle of the second row and glare at Michael, who was sitting directly behind her.

"Bethany," said Mrs. Ridley. "Well done. How did you do that so quickly?"

I shrugged and looked at my shoes.

"Let me try that again, Bethany," pressed Mrs. Ridley, taking a couple of steps towards me. "You couldn't have just added up the numbers. Can you show all of us how you got the answer?"

I slouched back into my chair, and stared at my notepad with the picture of the two joining staircases.

Mrs. Ridley knew that public speaking terrified me. Surely she knew that every time I spoke, my baritone voice reduced half the class to giggles.

"I'm sorry," she said. "I shouldn't have asked you to share in front of the entire class. But how did you add up the numbers from one to twenty so quickly? Can you show me what you did, Bethany?"

I shook my head and closed my notepad, afraid that I'd get in trouble if Mrs. Ridley saw all the pictures I had drawn in class that day.

I couldn't tell Mrs. Ridley that I was tired from all the mindless drills we did every day, and that I needed something to do to pass the time. I couldn't explain to my teacher that I was only drawing staircases because I found math so boring.

Even though my head was down, I could feel everyone staring at me.

All of a sudden, I heard a high-pitched cackle from Vanessa, the freckled redhead sitting in front of me. As always, I was sure it was her best friend Gillian who leaned over and whispered something cruel.

Gillian and Vanessa were inseparable, and formed a tight clique with Alice and Amy, the two Chinese twins who sat directly in front of them in the first row. Whenever a teacher wasn't around, the four of them called me names like Big Ugly Bethany. I was so much bigger and taller, yet they were the ones bullying me.

When Mrs. Ridley posed her usual "mental math" question at the halfway mark of the class, it was a crazy coincidence that my doodling provided the spark needed to answer her question of calculating the sum of the first twenty positive integers: 1 + 2 + 3 + 4 + 5 + 6 + 7 + 8 + 9 + 10 + 11 + 12 + 13 + 14 + 15 + 16 + 17 + 18 + 19 + 20.

Or maybe the creative spark came from all those jigsaw puzzles I'd been doing since I was five, and realizing how the two staircases would perfectly fit together.

Mrs. Ridley's mental math questions made me uncomfortable, and I hated how Gillian always made it a personal competition. Even though I understood ideas like fractions and long division, I just wasn't fast at doing calculations. I could do them – they were easy – it just took time.

Gillian's brain worked faster than the rest of ours, and she always got the mental math question first.

Until now.

"Bethany," Mrs. Ridley whispered. "Instead of sharing your solution in front of the class, perhaps you could write it down for me, and I'll present it to the class? Would you do this for me?"

I looked at the clock, and saw that class wouldn't finish for another thirty minutes.

I wanted to get away to a place where I could be safe – where I could be alone.

But unlike Meg, the main character in the Madeleine L'Engle book I finished last night, I couldn't do a "wrinkle in time". I was stuck here, in three dimensions, with no chance of escape.

Sighing, I stared at Mrs. Ridley and nodded. I picked up my pen and opened my notepad.

Mrs. Ridley returned to the front of the classroom and started explaining another tedious drill while I concentrated on writing my "solution".

I wondered whether I should talk about the staircase. If I explained the staircase, would the truth come out that I drew the picture only because I was bored and was just scribbling on my page? Could I demonstrate my solution without the two staircases?

No, I couldn't. I had to draw the staircases.

I ripped out a fresh sheet of paper, and took out a small ruler from my pencil case.

I began writing. About fifteen minutes later, when I was sure everything was just right, I slowly lifted my hand. Mrs. Ridley saw me, walked over and took my sheet of paper.

With the class working on something else, Mrs. Ridley began reading what I had written:

Answer = One staircase

Two staircases joined together = One rectangle

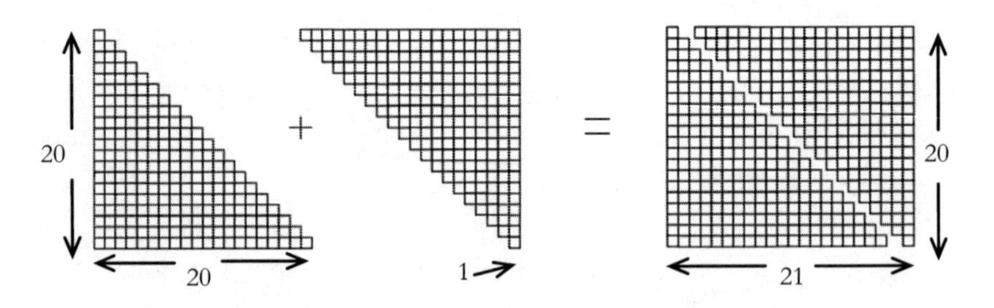


| | |
|---|---|
| Two staircases | $= 21 \times 20$ |
| One staircase | $= 21 \times 10$ |
| Answer | $= 210$ |

I didn't take a breath as I stared at Mrs. Ridley reading over my solution. Her eyes kept moving up and down the page. Finally she glanced up with a confused look on her face.

"Sorry, I don't get it."

My teacher didn't understand. My shoulders sagged. Mrs. Ridley bent down and leaned so close that I could smell the perfume on her face.

"Bethany, can you explain it to me?"

We whispered back and forth, pointing to the diagram, until it all clicked in her mind.

"Oh, wow," said Mrs. Ridley. "I understand. Yes, I understand."

Mrs. Ridley got the attention of the class, and walked up to the front of the classroom, placing my sheet of paper on top of the fancy projector before turning the machine on. After a minute, the projector was ready and my staircase diagram filled the screen at the front of the classroom.

"Class, listen carefully," said Mrs. Ridley. "In the mental math question, I asked you to calculate the sum 1 + 2 + 3 + 4 + 5 + 6 + 7 + 8 + 9 + 10 + 11 + 12 + 13 + 14 + 15 + 16 + 17 + 18 + 19 + 20. I want to show you how Bethany solved the problem."

"Why doesn't Bethany explain it?" interrupted Gillian. "I'm sure she'd be happy to."

Vanessa started giggling. Mrs. Ridley shook her head and pointed to the diagram.

"As I was saying, here is Bethany's solution. Look at the staircase on the left. There's one square in the top row, two squares in the second row, three squares in the third row, all the way down to twenty squares in the last row. So the answer to the mental math question is equal to the number of squares in that *one* staircase."

Mrs. Ridley paused. "Do you all understand Bethany's picture?"

A bunch of heads nodded. I held my breath.

"Now here is Bethany's amazing idea. Make a copy of the staircase and *flip* it," said Mrs. Ridley, pointing to the second staircase.

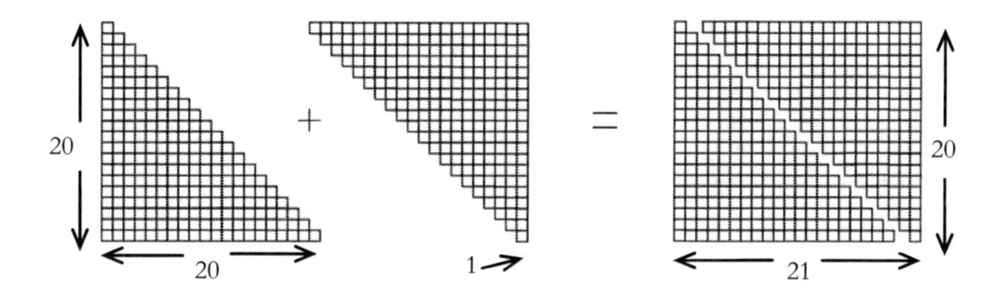

She then pointed to the rectangle on the right.

"And if we take those two staircases and stitch them together, we end up with a rectangle that's twenty squares high and twenty-one squares wide."

She explained how each row in the rectangle would have 21 squares, since the first row has 1+20 = 21 squares, the second row has 2+19 = 21 squares, and the third row has 3+18 = 21 squares, and so on until we get to the twentieth and final row, which has 20+1 = 21 squares.

"Oh, I get it!" exclaimed Michael. "You take two staircase-shaped pieces and join them together – and you get a rectangle. It's like a jigsaw puzzle."

"That's exactly right, Michael. Now, who can explain the rest of Bethany's solution?"

Gillian raised her hand. With her chin lifted up, she spoke in her usual high-pitched tone.

"The answer is the number of squares in one staircase. Twice the answer is the number of squares in two staircases, which is the same as the number of squares in that big rectangle. The number of squares in that big rectangle is twenty-one times twenty, width times height. So the number of squares in *two* staircases is twenty-one times twenty."

"Good," said Mrs. Ridley. "So how many squares are in one staircase?"

"Half of that," said Gillian. "Twenty-one times ten."

"Thank you, Gillian. As always, that's an excellent explanation. Did everyone understand that?"

One student shook his head to indicate that he hadn't. Mrs. Ridley called on Gillian again to explain why the answer had to be half the number of squares in that big 21×20 rectangle, and why this worked out to 21×10=210. This time, the student nodded in understanding.

"Well done, Gillian," said Mrs. Ridley. "And well done, Bethany."

I smiled.

Mrs. Ridley glanced towards the middle of the classroom, beaming at the two of us.

"We should call you the Bethany and Gillian team. The B & G Team."

My smile disappeared.

Gillian raised her hand. "The G should come first. The G & B Team."

Mrs. Ridley ignored the comment and turned to write something on the chalkboard.

After a few seconds, Gillian snickered and looked at Vanessa. She pointed to herself with her index finger and then pointed her thumb directly at me. She lowered her voice.

"G & B. Good and Bad."

Upon hearing Vanessa's cackle, Mrs. Ridley turned back to face us. "What was that?"

"Oh, nothing," replied Vanessa. "I was just laughing about something that happened yesterday."

Mrs. Ridley shook her head. After a slight hesitation, she picked up her chalk and continued writing on the board. Gillian looked at Vanessa again. She pointed to herself and then stuck her finger towards me.

"G & B. Girl and Boy."

This time Vanessa's laugh could be heard by the entire classroom. Mrs. Ridley glared hard at Vanessa.

"I'm sorry, Mrs. Ridley," replied Vanessa. "It won't happen again."

I glanced at the clock. The bell would ring in five minutes. I looked at the clock in desperation.

*Hurry up, clock! Move faster!*

I saw Gillian take out a sheet of paper and two pencils. As she began sketching, I could see her smile getting broader every minute.

I leaned forward to see the image Gillian was creating.

A feeling of nausea came over me. It was a picture of two people. The girl on the left had a big *G* on her shirt. She had long flowing dark hair, tanned skin, and a perfectly-shaped face. The girl on the right had a big *B* on her shirt. She was really tall, with broad shoulders, a pointed nose, frazzled hair, and huge feet. The girl on the left looked like a ballerina, while the girl on the right looked like a deformed tree.

I didn't need to be reminded which one was me. A tear crept out of my eye.

Gillian scribbled some words below the two pictures and showed her masterpiece to Vanessa and the twins sitting in front of them. They burst into laughter.

As Gillian turned the page towards Vanessa, I saw the word *BIGFOOT* written below my portrait, and felt more tears flowing down my cheeks.

Then I saw the eight-letter word *GORJEOUS* written under Gillian's portrait.

Mrs. Ridley angrily walked over and grabbed the picture. When she looked at what Gillian had created, her face flushed.

The bell rang. As people began to pack up their stuff, Mrs. Ridley held up her hand.

"Gillian, Vanessa, Alice, Amy, Bethany. You're not dismissed."

After everyone had left, there were just six of us in the classroom: Mrs. Ridley, me, and Gillian's gang. The twins looked scared knowing that they were in big trouble, while Gillian and Vanessa looked indifferent. I quickly wiped the tears away from my eyes, and sniffled.

"Gillian Lowell," snapped Mrs. Ridley, pointing to her picture. "Did you draw this?"

"Yes," she said, nonchalantly. "But it was just a joke. I didn't mean it."

Mrs. Ridley glared at her. "Gillian, what you did was absolutely unacceptable. It was deeply hurtful, and you should be ashamed of yourself. What do you have to say to Bethany?"

Gillian turned to me and shrugged. "Sorry."

"Apologize like you mean it."

"Sorry," said Gillian, a bit louder but still totally unconvincing.

With a pained expression in her eyes, Mrs. Ridley sighed and looked at me: "Do you accept her apology?"

"Yeah," I replied, and stood up from my seat, taking my books and binder with me.

A strange feeling came over me, just as I was about to exit the classroom. It was a sense of conviction, something I had never felt before.

I grabbed a piece of chalk from the side board and wrote down an eight-letter word in big block letters.

*GORGEOUS*

Standing up straight, I locked eyes with Gillian, and tapped the chalkboard.

"I might not be gorgeous. But at least I can spell it."

# 2

*ELEVATOR OUT OF SERVICE.*

"Not again," I muttered.

I slowly walked up the stairs, carrying my heavy backpack, all the way up to the sixth floor.

Today was June 15. My twelfth birthday. It was a terrible day from start to finish, and the broken elevator was the icing on the cake.

Grade 6 would finish in a couple of weeks. I couldn't wait for school to be over. Then I wouldn't have to see Gillian and Vanessa and anybody else until September, when we'd all start Grade 7 at a new school.

I didn't need to be reminded that there was a party at Gillian's this evening, since June 15 was also Gillian's twelfth birthday.

I couldn't blame people for wanting to go to her place instead of mine. After all, Gillian had the biggest house in Cape Breton, complete with a backyard pool, while Mom and I rented an apartment on the sixth floor of a dumpy old building, where the elevator broke at least once a week.

I slowly opened the door.

"Happy birthday, darling!"

"Thanks, Mom."

"How was school today?"

"Terrible," I said, bursting into tears. Mom turned off the stove and reached out to give me a hug.

I told Mom about what Gillian said to me in the girls' locker room. That I had no friends. That I was the only person in the entire Grade 6 class not invited to her birthday party. That my father left Mom a month before I was born because he knew that the baby would end up being a stupid loser.

Mom sighed and hugged me tighter. She didn't say anything. She didn't need to.

I hated my life.

After a while, I calmed down and I wasn't crying anymore. I just wanted to be alone, and get back to reading *Harry Potter*.

"Can I turn off the TV?"

"Of course, darling. Dinner will be ready in just a few minutes."

While Mom returned to the kitchen, I looked for the remote control in our tiny living room. "Mom, where's the remote?"

Mom didn't look up from the pot she was stirring.

"I don't know, but the power button's just under the TV. Dinner's almost ready."

Just as I was about to hit the power button, I saw a clip of six teenagers wearing matching red and white shirts, writing math equations on a big chalkboard. They were all smiling and laughing, and the camera focussed on the short freckled girl in the middle, with her straight auburn hair falling just below her shoulders.

A deep male voice resonated from the television set.

"Are you a good problem-solver? Think you can match wits with the best in the country? When we come back, you'll meet Rachel Mullen and the rest of Canada's team to the International Math Olympiad."

I stood there in silence as a commercial came on.

*Math Olympiad?*

"Ready for dinner, honey?"

I continued to stare at the screen.

"Bethany?"

"Uh, not ready. Two minutes – just want to see something first."

Confused by my sudden interest in the news, Mom walked over and stood behind the worn-down Lazy-Boy recliner and motioned for me to take her seat. I nodded and sat down, finding the remote control sandwiched between the covers of the recliner.

"What are we watching?" asked Mom.

The commercials were still rolling. "I'm not sure."

The screen returned to the CBC news, and the anchor's face appeared. I turned up the volume.

"When Rachel Mullen was growing up in Brandon, Manitoba, she was an over-enthusiastic child interested in everything, especially mathematics. Even though some of her teachers encouraged her to pursue other interests, she stuck with the subject she enjoyed the most. And today, she is one of our country's brightest young minds, representing Canada at the biggest stage of them all, the International Mathematical Olympiad next month. Here is her story."

I stared at the TV as Rachel stood in front of the classroom, pointing at some equations on the whiteboard and explaining something to a teenage boy who nodded in agreement. In the next shot, Rachel was sitting opposite a middle-aged reporter with the Canadian flag in the background. She laughed at something the reporter said, and smiled at him.

"I was a hyperactive kid who had a hard time concentrating on tasks, always moving from one activity to another. I drove my parents crazy. But my parents were patient with me, and they were so supportive of my dreams and ambitions – even though some of my dreams only lasted a few weeks! Luckily, I met an amazing teacher in Grade 7 who noticed my interest in puzzles, and she introduced me to math contests. She saw my potential even before I did, but eventually I realized it too."

"And what was that?" asked the reporter.

"That I could develop a passion for math and grow to love it. And through writing contests, I could get invited to math camps and meet people from all across Canada. Even though we grew up in different cities and provinces, we have so much in common. They'll be my closest friends for the rest of my life."

"What do you like about math, Rachel?"

"The beauty, the patterns. Everyone thinks math is about memorizing formulas and rules – but it's not. At its heart, math is about problem-solving. I'm not the smartest person at my school, but I'm probably the most creative and imaginative. That's how I developed as I trained for the Math Olympiad."

"Tell me about the Olympiad. Are you looking forward to it?"

"Definitely. I've been wanting to make the team for the past three years, and finally got it on my last try. I'm nervous, but I'm ready."

The news story then switched to an older man whose perfectly-parted thin blond hair appeared almost child-like, and looked so strange on top of his head. As the man began to speak, a small caption appeared on the bottom of the screen: "J. William Graham, Executive Director, Canadian Mathematical Society".

"The six members of our Canadian IMO team were selected from among *two hundred thousand* students from Grades 7 to 12 who participated this year in local, provincial, and national mathematics competitions. They represent the very best of Canada, and will be excellent ambassadors for our country at the Math Olympiad."

"Can you tell me more about the Olympiad competition?" asked the reporter.

"I'd be happy to," said Dr. Graham. "Our six team members will pit their skills against the top math students from over one hundred countries. They will attempt to solve six problems over nine hours, a mathematical 'hexathlon' that requires exceptional problem-solving skills, mathematical understanding, daring, and imagination – the types of skills that we Canadians will require if we're going to be at the forefront of innovation in the twenty-first century."

"Excuse me, Dr. Graham. Did you say six problems over *nine* hours?"

"That's right. Six problems over nine hours. Preparing for a competition like this one requires years of training, just like our sports athletes at the Summer and Winter Olympics. Just as our Canadian athletes amaze us with their physical prowess and push the boundaries of athletic performance, our *mathletes* do the same thing with their intellectual prowess. They are truly an inspiration to Canada."

Rachel's face reappeared in front of the screen. She was sitting next to the Canadian flag.

"Rachel, what advice would you give to a young person who might be watching this?"

She paused. "Find out what gets you excited and passionate. Some of my teachers tried to turn me away from math because I was a girl, and felt I should pursue other hobbies and interests. But I stuck with math because that's what got me excited. That's what inspired my passion. That's what grew my self-esteem and confidence. That's what added meaning and joy to my life. I found my voice, and because I did, I'm now a Math Olympian."

The story ended. I sat there, on the recliner, unable to move.

Mom was saying something to me, but I couldn't hear her.

I had an epiphany of what direction my life could take over the next six years. It was a moment of shocking clarity, a voice from deep within shouting into my head and heart.

I shuddered excitement. Since coming up with the "staircase" insight in Mrs. Ridley's class last year, I had been struck by the beauty of math but never had any teacher help me draw that out. I was never challenged or stretched in class, so I found the subject boring and mindless.

But I felt that there was something more to math. There had to be something more to math – for me.

I was moved by Rachel's words – self-esteem, confidence, meaning, joy – and in that one moment, I knew.

"Mom," I whispered, "I want to be a Math Olympian."

"That's nice, darling," said Mom, walking towards the kitchen. "Ready to eat dinner?"

"Mom!" I shouted. "I'm not kidding! I want to be a Math Olympian."

She turned to face me, and stood in front of the recliner. She saw the intensity in my eyes. Her face froze, and for a brief moment I saw her face flush.

"What's wrong, Mom?"

She hesitated.

"Nothing, Bethany. I'm fine."

She reached over and hugged me, squeezing me a lot tighter than usual. She took the remote from my hand, turned off the TV, and led me towards the dining table.

We chit-chatted a bit, but not as much as we normally did. I ate my chicken noodle soup, silently chewing each bite, daydreaming about the future. Mom looked a bit flustered, but assured me everything was okay.

Once we had finished eating, Mom invited me to unwrap my birthday present, a board game called Scrabble.

"You'll love this, darling. It's perfect for you."

"Thanks, Mom. This looks really fun. You want to play?"

"Of course I do. But I have one final surprise for you."

As I was reading the rules for Scrabble, Mom returned to the kitchen, and came back with a chocolate cupcake with a candle in the middle.

"Now make a wish, birthday girl. Anything you want."

I closed my eyes.

My mind kept coming back to a single thought that gave me goose bumps, and renewed me with a sense of joyful hope.

*I want to be a Math Olympian.*

I smiled and blew out the candle.

# 3

"Oh my God. Rachel won a gold medal!"

Mom looked up from her book. She stood up from her seat at the dining table and slowly walked towards our computer in the living room.

"Check this out," I said, giving Mom my chair.

Looking over her shoulder, I re-read the story I just found on *cbc.ca.*

**Canadian solves her way to Math Olympic Gold**

When Rachel Mullen clears customs at Pearson International Airport this afternoon, she will declare that she is in possession of a bright and shiny object obtained during her recent trip abroad – a gold medal from the International Mathematical Olympiad (IMO).

Ms. Mullen, a seventeen-year old from Brandon, Manitoba, represented Canada at this year's IMO, the world championship of mathematical problem-solving for high school students. Out of one hundred countries, the Canadian team finished in fifteenth place, with one gold medal and five bronze medals amongst its six team members.

"This year's IMO was one of the most difficult in years. The problems were extremely challenging yet all six students performed exceptionally well," said J. William Graham, the executive director of the Canadian Mathematical Society, the organization responsible for the selection and training of Canada's IMO team.

The annual IMO contest is set by an international jury of mathematicians, with one from each participating country. On each day of the contest, three questions had to be solved within a time limit of four-and-a-half hours.

Six hundred students wrote this year's IMO. Gold medals were given to the fifty students with the highest total score. Ms. Mullen correctly solved five of the six problems, placing tenth overall.

"I'm thrilled with our team's result," said Ms. Mullen, who will head to the University of Waterloo in September on a full scholarship. "The problems were so hard this year! I'm still in shock that I won a gold medal. I will treasure this experience for the rest of my life."

I felt goose-bumps on my arms. Rachel had done it – the best in Canada, one of the best in the world.

"I want to be a Math Olympian too."

Mom sighed.

"You said that last month, right after we saw that TV clip. Do you really mean it?"

"Yeah," I replied instantly. "Totally."

Mom shook her head. "Trust me, Bethany. You don't want to pursue this. It's not worth it."

"How would you know?" I asked.

"From personal experience," said Mom, hesitating.

"What personal experience?"

Mom changed the subject. "If you go for this Math Olympiad, you're going to need to make sacrifices and train for thousands of hours. You won't have the time to hang out with friends, play on sports teams, join school clubs, and just enjoy being a teenager."

"But what if this is what I want to do?"

"You don't, Bethany," said Mom, rising up from the chair. "Trust me, you don't."

"Why?" I pressed.

"Because these young Olympians do nothing else but train. They have no social life because there's so much pressure to perform. It's a terrible thing to ask a young person to endure. And I don't want you to get hurt like that."

"I want to do this."

"Remember when we were watching Wimbledon a few weeks ago?"

"You're changing the subject."

"No, I'm not. Remember Wimbledon? Think about all those players we were cheering for. Do you know how they all got so good?"

I didn't answer.

"Their parents enrolled them in private tennis academies when they were little kids. They had personal coaches. They practiced all day, every day. Their parents hired tutors to help them with their school work in the evenings. None of them had a normal life. I want you to have a normal life."

"But what if I don't want a normal life?"

"Look at our circumstances, Bethany! I'm raising you on my own. I'm sure all these Math Olympian kids have parents who are math professors, who teach them all sorts of complicated math, and work with them for hours every night. Or their parents have tons of money and can send their kids to schools where their teachers give them special coaching. I can't do that."

I remembered the part in the TV clip when Rachel spoke so lovingly of her parents, who supported her dreams and encouraged her every step of the way.

"Why can't you be supportive of my dream?"

"Because your dream isn't realistic," she responded. "If you put all of your eggs in one basket and that basket breaks, what happens then? You get shattered. Your life gets completely shattered."

Mom's voice trailed off. She looked away from me. A few seconds later, I heard sniffling.

I stared in disbelief. "Mom, are you . . . crying?"

Mom didn't answer. She walked away from the computer and sat on the recliner, dabbing her eyes with her fingers. I had only ever seen Mom cry once before, years ago, when Grandpa got really sick.

I sat on the floor facing Mom, and looked up at her. Neither of us spoke for several minutes.

"I'm sorry," I said, meaning it.

After a long pause, Mom wiped her eyes with another tissue and put her hand on my arm.

"When I was your age," whispered Mom, "I wanted something big. It took over my life. It robbed me of my childhood. It robbed me of everything. I don't want you to have to go through what I did."

"Go through what, Mom?" I asked.

She was so private about her past – about her childhood, about her adult life, even about my father's identity. I knew Mom grew up in Cape Breton but she didn't hang out with people who knew her well; she was an only child so I had no relatives other than Grandpa since Grandma died before I was born. As for my father, all I knew was that he was a star hockey player and would have made it to the NHL had he not gotten injured, and that he left Mom for a woman in Alberta while Mom was pregnant with me.

Mom looked into my eyes. I could sense she was ready to share. I moved closer.

"You know how I used to do figure skating."

I nodded, remembering what my Geography teacher told me in Grade 4, that she watched Mom skate on TV. From my teacher, I learned that Mom won the provincials three years in a row, between the ages of seventeen and nineteen. Until then, I had no idea that Mom was a star athlete.

"I hated figure skating. Even though I was the provincial champion, the pressure got to me. I made myself vulnerable. I don't want you to open yourself like that, and risk so much for a goal that's so uncertain."

I held Mom's hands and didn't say anything. I wondered why she was so upset about being the best figure skater in Nova Scotia.

"Please, darling. Don't risk your future on trying to be amazing at math. It's better to be well-rounded. It's better to have a normal life."

Mom reached over and gave me a hug. But inside I felt empty.

Normal was so boring.

Whenever I read books like *Harry Potter* or *The Hobbit*, I saw people doing stuff that was interesting. They were pursuing adventures. They were having fun. They were living out their dreams.

All I could think of was Rachel and that TV clip, where she talked about the passion she got from striving for the Math Olympiad and how that pursuit gave her life meaning and joy.

"Are you okay?" asked Mom.

"Yeah, I'm fine," I replied, not meaning it.

Mom put her hand on my shoulder. She held it there for a bit, and then went back to her seat at the dining table and continued reading her book. I went to the kitchen to get myself a glass of orange juice, and stood there in silence.

I wanted life to be interesting.

After standing by the fridge for a few minutes, staring blankly into space, I had an idea. Returning to the computer, I did a Google search on "International Math Olympiad".

How hard could these Olympiad problems be? After all, I was pretty good at math already, and knew that I would get a lot better in the next few years. Maybe I could pursue this dream . . . secretly.

I clicked on the first link, and after a few more clicks, was taken to a site containing all the IMO problems since the annual event began in 1959.

I chose a year at random, and downloaded the English-version of the problems from that year.

My jaw dropped when I saw the first problem. There were no numbers anywhere.

Wasn't this supposed to be a math contest?

**Question #1**:

Determine all functions $f: R \rightarrow R$ such that the equality

$$f(\lfloor x \rfloor y) = f(x) \lfloor f(y) \rfloor$$

holds for all $x, y \in R$.

(Here, $\lfloor z \rfloor$ denotes the greatest integer less than or equal to $z$.)

The three problems on Day 1 seemed like they were written in a foreign language. Scrolling down to the first problem of Day 2, I recognized the word "triangle" but not much else.

**Question #4**:

Let *P* be a point inside triangle *ABC*. The lines *AP*, *BP* and *CP* intersect the circumcircle of triangle *ABC* again at the points *K*, *L* and *M*, respectively. The tangent to the circumcircle at *C* intersects the line *AB* at *S*. Suppose that *SC*=*SP*. Prove that *MK*=*ML*.

What was a "circumcentre"? A "tangent"? And what were all those letters *P*, *K*, *L*, *M* to keep track of?

Glancing at the six problems, and not understanding the meaning of a single one, I realized that the Math Olympiad was so much harder than the calculations and formulas I was used to. I found the page with the solutions to each of these six problems and clicked on the link for Question #4.

There were a bunch of complicated geometrical diagrams accompanying each of the two solutions that were posted, with one mentioning the "Tangent-Chord Theorem", and the other mentioning the "Power-of-a-Point Theorem". As I skimmed the two solutions to try to make sense of them, my head started to hurt.

I closed the web browser and got up from the computer table. Moving to Mom's recliner, I sank down into the seat, in shock.

So much for wanting to be a Math Olympian.

I'd been thinking and dreaming about the Math Olympiad every day for the past month. But I now knew I could never get to that level.

Mom was right. The Math Olympiad was just for super-special people, the natural prodigies with math professors for parents, or gifted teenagers whose parents could afford personal tutors.

The Math Olympiad was for the Rachel Mullens of this world, not for ordinary people like me.

My eyes started to well up.

I heard the phone ring, but couldn't move.

"Hello?" said Mom, picking up the phone.

After a couple of seconds, she spoke again. "Hi, Dad. How are you?"

All of a sudden, she began to cry. It started off slowly but her sniffles turned into heavy sobs. I jolted from my seat and turned to face Mom, wondering what had happened to Grandpa.

I stared at Mom as she continued her call. Her voice was muffled and I could only make out certain words through her tears, but one word was unmistakable.

*Cancer.*

# 4

"Bye, Grandpa. See you in a couple of days."

We closed the door to his room at Cape Breton Regional Hospital, and walked towards the elevator in silence.

We had visited Grandpa at least four times a week since mid-July, when we learned that Grandpa only had months to live. Even though Grade 7 would be starting in two days, I wanted to come to the hospital as often as possible, and told Mom we could go together every night after dinner.

The elevator went down to the main floor of the hospital. We stepped out and headed towards the exit.

"Is that you, Lucy?"

An old man was walking towards us.

Mom was the first to react. "Mr. Collins, what a surprise."

The man gave Mom a warm hug. "I'm so sorry about your father. I'm here to visit him now."

"Thank you," said Mom. "I know he'll be happy to see you."

"Of course," he replied. "Your father and I go back many years."

After an awkward pause, Mom pointed to me. "Mr. Collins, this is my daughter Bethany."

I shook hands with the man, and stared at the thick grey hair on top of his pear-shaped face.

"It's nice to meet you, Bethany," he said, smiling at me. "My name is Taylor Collins, and I was your mother's math teacher at Sydney High School. Your Grandpa and I grew up on the same street."

Mr. Collins invited us to join him in the hospital cafeteria, and offered to buy us any drink we wanted. Mom resisted, but Mr. Collins insisted we join him.

Mr. Collins bought a coffee for Mom and an orange juice for me. We found an empty table, right by the cafeteria door.

He turned to Mom. "How are you, Lucy?"

"Other than the news about Dad, I'm doing well."

"Are you still with the federal government?"

"Same old, same old," she said, nodding.

He chuckled. "I assume that means you really don't like your job."

"No, I do," she said, correcting herself. "The work is good."

I glanced over at Mom, surprised she lied to her former teacher.

I knew Mom hated her job at the Canada Revenue Agency. For the past ten years, she was the administrative assistant to one of the senior directors. The director yelled all the time, especially at Mom. Whenever Mom talked to anyone on the phone, she complained about her boss.

Mr. Collins asked about me. I told him about my boring, ordinary life. He seemed to disagree, and was happy to hear about my favourite books, and my evening Scrabble games with Mom. Unlike Gillian and her friends, Mr. Collins didn't tease me for having a manly voice, and he instantly made me feel comfortable.

"In a couple of days, you'll start Grade 7 at Pinecrest Junior High. Are you excited?"

"Yes. But I'm also nervous."

"That's completely natural, especially when you start at a new school. In a few years, you'll be a student at Sydney High School, where I've been a teacher for the past thirty-nine years. I've had the privilege of working with thousands of talented students throughout my career."

Mom looked at me sheepishly. "He wasn't including me in that list."

Mr. Collins nodded. "Well, you were preoccupied with something else when you were in high school."

Mom forced a smile, and returned to sipping her coffee.

I stood up. "Just need to use the washroom. Be right back."

The ladies' room was just around the corner. As I walked back towards the cafeteria, I could hear Mom and Mr. Collins in a serious conversation.

I stood close to the cafeteria door, where I could hear everything without being seen.

"My granddaughter Ella is really sad. She loves figure skating and wants to compete in the Olympics someday. But her coach got a new job in Toronto and moved out west last month. You know, I feel very strange asking you this, but I was wondering if you'd think about . . ."

"No chance, Mr. Collins."

"But there's no one in Cape Breton more qualified to coach Ella. We would pay whatever you wanted."

"I'm sorry. I'm not interested. After what happened, I have no desire to ever go back to the rink."

"You're right, Lucy. I apologize for asking. Please forgive me for being so insensitive, especially at a time like this."

"It's okay. It's just that I have so many regrets. Because of figure skating, I couldn't go to college or university. And now, years later, I'm stuck in a job I hate."

"Can you start over?"

"It's much too late for that, especially when you're thirty-four years old and only have a high school diploma. Besides, the salary is too good and it's too much of a risk to walk away from a permanent government job, especially when I'm raising Bethany on my own."

"I admire you, Lucy. You've had to deal with so much in your life."

I nervously stepped into the cafeteria and sat back down.

"Is everything okay?" asked Mom.

"Yeah," I replied, patting my stomach. "It must have been something I ate."

Even though Mom looked concerned, I noticed Mr. Collins grinning.

"Bethany," said Mr. Collins. "I can tell you have the ability to think quickly on your feet. From what I've heard, you're an excellent student."

"How do you know?" I asked.

"Cape Breton is a small place, and it turns out that Mrs. Ridley and I are close friends. She told me about your creativity and writing skills, and she thinks the world of you."

"Thank you," I said.

Mr. Collins smiled and sipped his coffee.

"Bethany, from what Mrs. Ridley has told me, I don't think you'll find Grade 7 math either challenging or interesting. The subject is so beautiful, but the curriculum is so boring. Yes, I'll admit it. The curriculum is *boring*.

"Imagine an art class where you spend the entire year practicing brushstrokes and learning how colours combine without actually creating a single painting, or being inspired by how simple techniques could produce an artistic masterpiece revealing elegance and beauty? In art class, you create art and get inspired by beautiful art. But in math class, it's a different story. It's shameful."

"But aren't you changing that?" asked Mom. "I saw your name in the paper about some provincial committee you're on."

"That's right," said Mr. Collins. "I'm chairing the committee that will create the new mathematics curriculum for all of Atlantic Canada. Even though I'm two years from retirement, the next two years will be the most exhausting of my life. But it's a challenge I'm eagerly anticipating."

"Ever since I've known you, you've been involved in a lot of things."

"Yes," he said. "Life is far too busy right now, and I can't possibly take on any more than I've already got on my plate. But it will be worth it when students like Bethany experience the new curriculum."

"When will that happen?" I asked.

"In two years," he replied. "When you start Grade 9."

Mom turned to Mr. Collins.

"In the meantime, what should Bethany do? For the past few years, she's been coming back from school, unexcited. She says school homework is too easy and I can tell she's bored. I'm not confident that Grade 7 will be any different. Is there anything you could recommend for her?"

"Yes," said Mr. Collins, turning to me. "I think you would enjoy writing math contests. Have you ever seen a math contest?"

I nodded, remembering the day we learned about Grandpa.

"I saw a math contest on the internet. Last month."

"What was it called?" he asked.

"The International Mathematical Olympiad."

"Wow, you're ambitious," he said, with a big smile stretched across his face. "You're already solving the problems from the IMO?"

"Oh no," I said, turning red. "I didn't understand the meaning of a single question."

"Well, you're only heading into Grade 7, so that's understandable. From what I've heard about you, I think you have the potential to do well in contests. There's a fun contest open to all Grade 7 students across Canada, and Pinecrest students always participate. Maybe you'll consider participating yourself."

"Definitely," I said.

"How would Bethany get ready for a math contest?" asked Mom.

"Well, it seems like she would already do quite well," replied Mr. Collins. "But if she wanted to excel and get a really high score, she would have to study on her own. Sadly, there aren't any after-school math enrichment programs offered in Cape Breton, unlike bigger cities where students have access to more resources."

Mom tapped her fingers on the table. She looked like she was deep in thought, like at the end of a Scrabble game when the score was really close.

After some chit-chat about the weather, Mom turned to Mr. Collins. "Can I ask you a question?"

"Of course, Lucy."

"Of all the coffee shops you know around town, what's the best place to study?"

"Le Bistro on Charlotte Street," said Mr. Collins. "It's well-lit, the tables are big, and the background music is nice and soft."

Mom took a deep breath.

"That would be perfect. Especially because Le Bistro is so close to the rink."

Mr. Collins looked at Mom in surprise.

"Please tell your granddaughter I will pick her up at her house at one forty-five on Saturday, and will drive her to Centre 200. I will drop her off at Le Bistro around three o'clock, just after she's finished her first session with me, her new figure skating coach."

"Really, Lucy?" said Mr. Collins, beaming. "I'm thrilled to hear this. How much should we pay you?"

"Zero," replied Mom. "I will coach Ella every Saturday. For free."

"I can't tell you how much this means to me. Thank you."

"You're welcome," she said. "Ella and I will go to Le Bistro next Saturday, to meet you – and Bethany."

Mom smiled and continued. "I will drop off Ella at three o'clock, just after Bethany has finished her first session with you, her new math coach."

I stared at Mom in shock.

She looked at Mr. Collins. "After all, there's no one in Cape Breton more qualified to coach Bethany."

Mr. Collins laughed out loud. "That settles it. I accept."

"Mom, you don't have to," I said. "If you don't want to coach . . ."

"It's okay," replied Mom, putting her hand on my shoulder. "I love figure skating. I've always wanted to be a coach."

"So, what do you say, Bethany?" asked Mr. Collins. "Do we have a deal?"

I nodded, at a loss for words.

"Great," he replied. "Well, I should get going and say hello to your Grandpa. It was nice to meet you, Bethany. I'll see you next Saturday."

He turned to Mom, and gave her a hug. "Thank you so much, Lucy."

"No," she whispered. "Thank you."

I watched Mr. Collins walk out of the cafeteria and head towards the elevator. I turned to Mom and saw she had a big smile on her face.

"Thanks, Mom. You're the best."

# 5

"Next stop, Charlotte Street."

I sighed in relief, and hit the yellow button next to my seat. As the bus turned onto downtown Sydney's main waterfront street, I could see my destination on my right.

As soon as the door opened, I got out and sprinted towards Le Bistro Café, aware that I was already ten minutes late for my first session with Mr. Collins.

I ran into the café, and was grateful to see Mr. Collins sitting by the big table in the corner, waving at me.

"I'm so sorry I'm late," I said, walking towards him. "The bus . . ."

"No problem," said Mr. Collins, interrupting me with a smile that immediately put me at ease.

"I know that the buses here don't always come on time. Let's just end ten minutes later that we planned, okay?"

"Yes," I replied, taking off my backpack and jacket. "Thank you."

"Here, let's get something to drink."

We walked to the counter and lined up behind several people. Looking up at the elaborate menu of artisan sandwiches and specialty drinks that were handwritten using coloured chalk, I realized how different Le Bistro was from the two places Mom and I most often went: McDonald's and Tim Hortons.

While we were waiting to place our order, Mr. Collins asked me about my first week at Pinecrest. I excitedly told him that I had made two new friends, Bonnie and Breanna. We met during homeroom on the first day at our new school, and we had an instant connection since we were all brunettes, and were the three tallest girls in our grade.

I didn't bother telling Mr. Collins that Gillian and her friends were in all of my classes too.

After we sat down with our drinks, a medium coffee for Mr. Collins and a small orange juice for me, Mr. Collins got down to business.

"Bethany, I'm delighted to work with you. Every Saturday afternoon, we'll spend an hour learning together, here at Le Bistro. We'll learn the heart of mathematics, a beautiful subject that revolves around deep patterns connected together in unexpected ways."

He was moving his hands, waving them in all directions.

"You'll discover how mathematics develops creativity and problem-solving skills, and makes important connections to everything you see in the world. However, you won't be learning any formulas that you'll need to memorize, nor will you be doing a single drill or calculation."

I nodded, pretending to understand, but confused because I thought math was all about formulas, drills, and calculations.

"Are you ready to do some math, Bethany?"

"Yes," I said, picking up my pen.

Mr. Collins placed a stack of white index cards in the centre of the table, and gave the top card to me. "All right, let's get started. On that card, please spell out your name."

I did what I was told. B-E-T-H-A-N-Y.

"Now, spell your name backwards."

I raised by eyebrow, confused by the simple instructions. Staring at my card, I wrote the seven letters of my name in reverse order, one line below: Y-N-A-H-T-E-B.

"Bethany, from your reaction I can tell you're wondering why I'm asking you to do something so basic. So let's make it more interesting. Put down your pen and card, and spell your last name backwards. And this time spell it out loud."

I hesitated. I said my last name a few times out loud, visualizing it in my head. *MacDonald, MacDonald, MacDonald.*

I quickly got the first two letters: "D-L."

*MacDonald, MacDonald.* Then I got the next letter. "A."

*MacDonald, MacDonald.* Then I got the next letter. "N."

Repeating the process, I finally came to the end. That was a lot harder than I expected.

"Great work," said Mr. Collins. "Now Bethany, I'd like to introduce to you a problem-solving technique that you can apply to almost any situation in life. The technique is to *simplify the problem by breaking it into smaller parts.*

"Let's take our example with backwards spelling. Instead of thinking of your last name as MacDonald, think of it as Mac – Don – Ald. Three blocks, three letters."

Mr. Collins repeated the three syllables, gesturing with three fingers on his left hand. "Now, Bethany, reverse-spell your last name again. But this time, do it block by block, rather than letter by letter."

*Mac – Don – Ald*, I thought to myself while staring at my hand, one finger for each syllable.

"D-L-A – N-O-D – C-A-M."

I got that in just a few seconds.

"Great! Now spell *Cape Breton Nova Scotia*, doing exactly what you just did."

*Cape – Bre – Ton – Nova – Sco – Tia*, I said silently, using both hands to mark the six blocks. I shook my fingers, three times on each hand.

"A-I-T – O-C-S – A-V-O-N – N-O-T – E-R-B – E-P-A-C."

"Excellent. You reverse-spelled twenty letters in less than thirty seconds. See how much simpler that is?"

I nodded.

Mr. Collins gave me more words and phrases to reverse-spell. Each time, I was getting faster and faster.

"Excellent," said Mr. Collins, after about fifteen minutes or so. "Now let's move on to something else. I'm going to write down a ten-letter word and you'll tell me what's special about it."

BOOKKEEPER

"That's another word for librarian?"

"Not quite," said Mr. Collins. "But that's a good guess. A bookkeeper is someone who records business transactions, so it's a synonym of 'accountant'. Perhaps some of your Mom's colleagues at the Canada Revenue Agency are bookkeepers. Do you notice anything unique and special about the spelling of this word?"

"Yes," I replied. "The double letters. O-O then K-K then E-E."

"Very good. It turns out that BOOKKEEPER and SWEETTOOTH are the only two words in the English language that have three consecutive sets of double letters. There aren't any others."

"How about BOOKKEEPERS?" I asked. "And BOOKKEEPING?"

"Good point," said Mr. Collins, laughing out loud. After a short pause, he spoke again.

"Here's a challenge for you, Bethany. Determine a well-known seven-letter word that has two consecutive sets of double letters, where the third and fourth letters are both L."

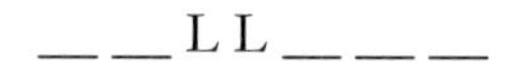


I looked at my math coach, confused that we had now moved from backwards spelling to word puzzles. He nodded silently, clearly indicating he wanted me to focus on his question.

I quickly saw one word that fit the pattern, BALLBOY. But that didn't have two consecutive sets of double letters. I racked my mind to find other seven-letter words that fit that pattern. After a couple minutes of concentration, I came up with TALLEST, which made me think of myself. And moments after that, I came up with BULLIES, which made me think of Gillian and her friends.

But none of these words had two consecutive sets of double letters.

I looked up at Mr. Collins and shrugged. "I don't know."

"No problem. Have you found any seven-letter words that fit the pattern?"

"Just two," I said, hesitating. "BALLBOY and TALLEST."

"Good. There are a few other words that fit the pattern, like CELLARS and CALLING and WILLING, and also my last name, COLLINS. Now you know you're looking for a seven-letter word with two consecutive pairs of double letters. We talked about a powerful problem-solving strategy earlier. What was that strategy?"

"Simplify the problem."

"That's right. How do we simplify the problem?"

"Break it into small parts."

"Excellent. Let's apply that strategy to this problem. How do we do that?"

"The double letters."

"Right. You want to focus on the double letters. You've got one pair of double letters already, and you know there must be another pair of double letters somewhere else. So what's the next step?"

"Figure out where those double letters can go."

I took three index cards from the stack, one for each of the three possible cases. For each case, I scribbled an "x" where the double letters could go.

**x x** L L _ _ _          _ _ L L **x x** _          _ _ L L _ **x x**

Mr. Collins smiled. "You made one small mistake. Remember the statement of the problem? What are we looking for? A seven-letter word with what?"

"Two pairs of double letters."

"Two *consecutive* pairs of double letters."

"Oops," I said, flipping over the third index card. So we were down to two cases.

**x x** L L _ _ _          _ _ L L **x x** _

"Excellent," said Mr. Collins. "Now let's simplify the problem further by breaking it down into even smaller parts. What are the possibilities for these double letters? Look at your first index card, the one where the first two letters are the same. Can the first two letters be a pair of consonants?"

I thought for a few seconds. No, we couldn't have words beginning with BBLL, CCLL, DDLL, FFLL, and so on. That was impossible. The first pair of letters had to be vowels.

Clearly the word couldn't begin with AALL, IILL, or UULL. I thought for a couple of moments and realized that OOLL didn't make any sense either. I racked my brain for seven-letter words beginning with EELL. I smiled.

"Is it EELLIKE?"

"Is it what?"

"You know, EEL-LIKE, like an eel?"

Mr. Collins grinned. "That's a great try, but no, that's not a word."

I turned over the first card. I was convinced that the correct seven-letter word couldn't begin with two consecutive pairs of double letters, since EELLIKE was the only possibility that fit the pattern. I picked up the second index card, the one where I had marked the pattern _ _ L L **x x** _.

I realized that this pair of double letters couldn't be consonants either, since no word would fit a pattern like _ _ L L B B _ or _ _ L L C C _. So this pair of letters had to be vowels.

I was sure that the pattern had to be _ _ L L E E _ or _ _ L L O O _. The other three possibilities didn't make sense.

What could the word be? I figured that the answer would probably be two separate words stuck together to form a single word, like BALL-BOY or MALL-RAT. That made the most sense. But what could the first four letters be? There were plenty of options: BALL, TALL, SELL, FILL, ROLL, PULL, and so on. For each four-letter option, I tried to join it to a three-letter word to form the seven-letter answer.

But what could the last three letters be? EEL was the only possibility that I could think of. What other three-letter words began with EE or OO? BALL-EEL certainly wasn't a word. And the double O's didn't make much sense. Of course, BALL-OOF, BALL-OOM, and BALL-OON weren't words.

And then I saw it.

"BALLOON!"

Several customers turned to stare at me. I put my head down, embarrassed at the attention. But inside I was ecstatic.

Mr. Collins chuckled. "Outstanding. How do you feel?"

I let out a deep breath. "Good. Tired, but good."

Mr. Collins proceeded to give me more word puzzles with missing information. The problems got progressively harder, with the last one requiring me to determine a well-known word with U-F-A in consecutive letters. After fifteen minutes, I finally got it.

"MANUFACTURE!"

"Yes, that's right! Now quick, give me *four* more English words that also have those same three letters appearing consecutively."

"Four more?" I asked. "It took me forever to come up with just one."

"I know you can do it, Bethany. Another powerful strategy is to solve a problem by reducing it to a previously-solved problem. Remember BOOKKEEPERS?"

I laughed, understanding what he was implying. After a slight pause, I came up with four more words that fit the pattern: MANUFACTURES, MANUFACTURING, MANUFACTURED, and MANUFACTURER.

"Well done."

Mr. Collins glanced at his watch. He and I were both surprised that our hour was nearly complete.

"Well, Bethany, it was a pleasure working with you. You're extremely bright, and once you learn a problem-solving strategy, you quickly figure out how to apply it. You must be feeling tired."

I nodded. Although I was drained mentally, I felt alive and exhilarated, a feeling I hadn't experienced since that day in Mrs. Ridley's Grade 5 class nearly eighteen months before.

"Now I want to ask you an important question. How much math do you think we did today?"

Just as I was about to answer, I saw a quirky smile on Mr. Collins' face, as if he was asking me a trick question.

"A little?" I responded. "Simplifying questions into smaller parts. That's math, right?"

"Absolutely. Our context was word puzzles and backwards spelling, but we spent every minute of our time thinking mathematically. The exercises we did are improving your concentration and deductive reasoning skills, and training your mind to make complex associations. Because at the end of the day, math is not about remembering the right formula to get the right answer, but applying your imagination and problem-solving skills to tackle complex problems. Speaking of which, do you use Google?"

"Yes," I said. "All the time."

"Did you know that Google's search algorithm was created by a pair of twenty-three-year-old students in California? Their solution is an ingenious application of Linear Algebra, a field of mathematics you'll first encounter before you graduate from high school. While the Google algorithm is simple, the ability to come up with that solution required extraordinary creativity and innovative thinking. But these are skills that students can develop. These are the skills I want to develop in you."

"Those Google guys must be rich."

"Indeed. Their breakthrough turned them both into billionaires. It still astounds me that you can rank trillions of websites in the correct order in less than a quarter of a second. But that's what you can do with Linear Algebra, and that's how internet search engines work. By building your skills in critical thinking and deductive reasoning, you will become a more confident, creative problem-solver, able to apply your mind to serve society – just as those Google guys did."

"How can I build that?"

"Build what? Build a billion-dollar company, or build your problem-solving skills?"

I blushed. "The second one."

"I'd recommend you spend some time each week on your own, *cross-training* your mind, the same way sports athletes cross-train their body to enhance their physical performance: tennis players run to develop their endurance; football players lift weights to build muscle; figure skaters do yoga for more flexibility and range of motion."

"Figure skaters do yoga?"

"Absolutely."

"Did Mom ever do yoga?"

"I don't know," said Mr. Collins. "Remember I was her math teacher, not her skating coach."

I smiled, trying to picture Mom as a teenager in a class with Mr. Collins.

"Can you tell me about Mom's figure skating career?" I asked. "There's so much I don't know."

"Let's not talk about that," replied Mr. Collins, suddenly turning serious. "It's not that there's anything to hide; there's just no sense talking about the past."

"Okay," I replied, disappointed.

"Bethany, as I was saying, you need to cross-train your mind, just as an athlete cross-trains her body. As a young *mathlete*, I'd recommend you spend a minimum of thirty minutes a day doing mental gymnastics to develop your mind. Our local newspaper, the Cape Breton Post, has a page of puzzles every day, which are just perfect. Have you heard of the Cryptoquote, the Jumble, the Sudoku, and the Kenken?"

"Just the Sudoku," I said.

"Well, I'd encourage you to get the local paper so that you can do the other puzzles too. They're easy to learn, and they're great training for the mind. We can discuss them next week if you'd like. Would you like to do that?"

I nodded.

"Grampy!"

I turned and saw a petite blonde running towards us, with Mom trailing a few steps behind her.

Mr. Collins introduced me to Ella, his eight-year-old granddaughter. She turned to Mr. Collins and started raving about her first lesson with her new coach. Mom had a smile on her face.

"I had so much fun today, Grampy! Coach Lucy says I'm really good."

"She's right," said Mom, looking at Mr. Collins. "Ella is a natural athlete, and she covers the rink so gracefully. Her jumping skills are amazing for someone her age."

After I told Mom about my first session with Mr. Collins, Ella turned to her grandfather.

"Do you think my Mummy will want to do yoga with me?"

"Why do you ask?" asked Mr. Collins.

"Because Coach Lucy says that if I do yoga, I'll get better at spins and turns. Coach Lucy says that yoga will help me skate better."

"Really?" said Mr. Collins. "I had no idea there was a connection between yoga and figure skating."

I snickered.

"What's so funny?" asked Mom.

"Nothing," I replied, smiling at Mr. Collins. "Nothing at all."

# 6

"Orange juice is boring, Bethany. Why not try something different this time?"

"What do you suggest?" I asked, pointing to the beverage menu on the wall, with weird-looking names like Chai Latte, London Fog, and Espresso Macchiato written in bright blue chalk.

"How about the Steaming Hot Chocolate?"

"That would be great. Thank you."

Mr. Collins ordered a specialty coffee for himself and a medium hot chocolate for me, with the total coming out to $6.85. Mr. Collins opened up his wallet and took out a ten-dollar bill. Just as the cashier was pressing the register to get the change, he stopped her.

"Hang on. I've got some change. Here's $12.10," he said, handing over a two-dollar coin and a dime.

The cashier looked at him strangely, and entered the amount. She was surprised that the change came out to exactly $5.25, and wordlessly handed Mr. Collins a five-dollar bill and a quarter.

Turning to me, he smiled. "Useful application of math, eh?"

"Do you always do that?"I asked.

"Yes," he replied. "Instead of getting four extra coins, I got one coin and gave up two, and now my wallet is lighter. I have a simple system that's guaranteed to leave me with at most seven coins in my wallet at any time. Like those people who prefer the status quo, I can't stand *change*."

I groaned at the pun, and took my seat at our usual table. Moments later, a young man came by with our drinks. I took my first sip.

"It's good, eh?" said Mr. Collins. "They make it with real chocolate. No powder or syrup."

"It's amazing," I replied, licking my lips. "Thank you so much."

"So, Bethany, what did you think of today's Jumble?"

From my backpack, I took out the puzzle page from the Cape Breton Post, with my completed answers. The Jumble showed a picture of a well-dressed man yawning with the caption: *Why the school superintendent was always so tired.*

To the left of the picture were six words that needed to be unscrambled, with the underlined letters requiring a final unscrambling to form the correct answer, which was always a visual or verbal pun.

I handed my completed Jumble to Mr. Collins.

| | | |
|---|---|---|
| SNURB | → | **BU**R**N**S |
| AAMMD | → | MA**DA**M |
| OIAPT | → | PA**TI**O |
| WEFRE | → | **F**EW**ER** |
| OOOODV | → | V**OO**D**O**O |
| EEEDSC | → | SE**C**E**DE** |

The underlined letters were **BUNDATIFEROOOCDE**, which unscrambled to the answer of "Why the school superintendent was always so tired".

He was **"BORED" OF EDUCATION**

Mr. Collins laughed. "I'll add that to my list of *pun*-ishing jokes."

I noticed a familiar-looking man a few tables over waving to Mr. Collins. He smiled and waved back at the well-dressed man who looked around forty or forty-five.

"Who's that?" I asked.

"He's the mayor of Sydney, and he's also a former student of mine. His youngest niece is in your class at Pinecrest."

"What's her name?"

"Gillian Lowell."

I tried to act as nonchalant as possible. "I see."

Mr. Collins, perhaps sensing my reaction, decided it was time to start working.

"Before I state today's question, let me ask you if Pinecrest's hallway is still the same as when I was last there, with several hundred lockers on one side, where the lockers are numbered #1, #2, #3, and so on?"

"Yup, it's still like that."

"And what locker do you have?"

"Locker #100."

"Oh, how perfect. By the way, how many students are at the school this year?"

"Miss Carvery said that there were about 450," I replied, recalling my vice-principal's message on the first day of school.

"Great. Here's our problem for today. Pretend that there are exactly 450 students at Pinecrest, and they conspire to play a strange game with all the lockers. Suppose all of the lockers are initially closed. The first student, the one with Locker #1, goes down the hallway and opens each of the lockers. After she's done, the second student, the one with Locker #2, goes down the hallway and closes all the lockers that are multiples of 2. After he's done, the third student, the one with Locker #3, changes the state of all lockers that are multiples of 3, closing the open lockers and opening the closed lockers.

"And this process continues, with each student going down the hallway one after the other, altering the lockers whose numbers are multiples of their locker number. You've got Locker #100, so when it's your turn, you would alter the $100^{th}$, $200^{th}$, $300^{th}$, and $400^{th}$ locker. Does that make sense?"

"I think so," I replied. "So when it's my turn, if Locker #100 is open, I would close it?"

"That's right. And if it's closed, you would open it. You do the same thing when you get to Locker #200, Locker #300, and Locker #400."

"I see."

"So the exercise continues until all 450 students have completed their tour down the hallway. The last student, the one with Locker #450, would just change the state of Locker #450. When she's done, the game ends. Here's my question. At the end, how many lockers are open, and which ones are they?"

Oh boy. There was no way I was going to write down 450 marks on my notepad and figure out one by one which lockers were going to remain open. That would take several hours, at least.

I looked at Mr. Collins and shrugged.

"Since we started working together two months ago, we've used a powerful problem-solving strategy almost every week. What is that strategy?"

"Break down a difficult problem into smaller parts."

"That's right. Last week, you showed that a positive integer is divisible by 99 precisely when it's divisible by both nine and eleven. So in coming up with the rule for divisibility by 99, you split the task into two easier problems, and then put it together to solve the harder problem."

"But I don't see how to do that with this locker question. You can't break it down into smaller parts."

"You're right," he replied. "But you can do something else. What's the scariest part of this problem?"

"The number 450. It's too hard to keep track of that many people."

"If it were a smaller number, would the problem be less scary?"

"Of course," I replied. "But you can't just pretend you've got less people to make the problem easier."

"*Fewer* people, not *less* people. Why can't you pretend you've got fewer people in the problem?"

I stared at him. "Well, you're not allowed. You said there were 450 people and 450 lockers. You can't just change the number of people to something less . . . I mean, fewer."

Mr. Collins grinned. "The great thing about problem-solving is that we can change whatever we feel like, and remove any restriction that we think is placed upon us. By pretending than we have fewer than 450 people, we're simplifying the problem so that we can discover a pattern. Once we figure out what the pattern is, we can see if it holds for the actual problem with 450 students."

"Are we really allowed to do that?"

"Definitely," said Mr. Collins. "My suggestion is for you to simplify the locker problem by pretending that there are only a handful of students at Pinecrest, say exactly twenty-five students. By considering this small scenario, you'll be able to figure out what's going on, enabling you to solve the actual problem."

Mr. Collins handed me a fresh sheet of paper.

I created a square grid, and filled in the numbers from one to twenty-five on the rows and on the columns. I then marked the first three rows of my diagram.

| | 1 | 2 | 3 | 4 | 5 | 6 | 7 | 8 | 9 | 10 | 11 | 12 | 13 | 14 | 15 | 16 | 17 | 18 | 19 | 20 | 21 | 22 | 23 | 24 | 25 |
|---|---|---|---|---|---|---|---|---|---|---|---|---|---|---|---|---|---|---|---|---|---|---|---|---|---|
| **1** | O | O | O | O | O | O | O | O | O | O | O | O | O | O | O | O | O | O | O | O | O | O | O | O | O |
| **2** | | C | | C | | C | | C | | C | | C | | C | | C | | C | | C | | C | | C | |
| **3** | | | C | | | O | | | C | | | O | | | C | | | O | | | C | | | O | |

"That's really good, Bethany. Before you go any further, can you explain what you are doing?"

"Sure," I said. "The rows represent the students, and the columns represent the lockers. The first student opens all the lockers; that's why I marked the entire first row with the letter O. The second student closes every other locker; that's why in the second row, I marked all the even lockers with the letter C. And for the third student, for each multiple of three, every O locker turns into a C, and vice-versa."

"That's an excellent explanation. Now please fill out the rest."

It took just a few minutes to fill out the rest of the grid, especially since every student after the thirteenth person could only alter one locker. I highlighted the lockers that remained open at the end, and recognized the pattern instantly.

"So which lockers remain open?"

"Locker #1, #4, #9, #16, and #25," I said, surprised.

"Do you notice anything interesting about these five numbers?"

"They're all perfect squares," I replied. "1×1, 2×2, 3×3, 4×4, and 5×5. Cool."

I put down my pen, certain that this pattern held for the more difficult problem.

"For 450 lockers, the answer will be all the perfect squares less than 450. We're done."

"Not so fast, Bethany," said Mr. Collins, raising up his hand like a referee. "You've taken a small sample and used it to find a pattern. This nice pattern suggests that the answer is all the perfect squares less than 450, but doesn't actually prove it. You need to add a step in between: first find the pattern, then clearly explain the reason for the pattern, and then use your explanation to determine the answer to the actual problem with 450 students."

"How do I do that?"

"You tell me. Why do we get this nice pattern?"

I thought about it for a few minutes, looking at the rows to see if I could see anything. When that led nowhere, I looked at the columns and noticed that Locker #1, #4, #9, #16, #25 were each altered an odd number of times, which explained why these lockers remained open at the end. For example, Locker #9 was altered three times (Open to Closed to Open), and similarly Locker #16 was altered five times. But it wasn't clear to me why it worked this way, and why the other lockers, the non-perfect squares, had to have been altered an even number of times.

Mr. Collins jumped in. "Let's look at one of the lockers, say #16. How many times was #16 touched?"

"Five times."

"Which five people altered this locker?"

"Student #1, #2, #4, #8, #16," I said, pointing to my grid.

"Now let's look at Locker #12. Which students altered this locker?"

"Student #1, #2, #3, #4, #6, #12."

"Do you see the pattern?"

I nodded. Instead of just focussing on how many times each locker was altered, I need to figure out which students altered each locker.

"Okay, now tell me which students touch Locker #20. But this time, don't look at your table."

"Student #1, #2, #4, #5, #10, #20," I said, listing all the *divisors* of the number twenty. Each student, whose locker number evenly divided into twenty, had to alter Locker #20 when they went down the hallway.

"And how about Locker #17?"

"Since seventeen is a prime number, it's just Student #1 and #17."

"Excellent. You've figured out the pattern. Since the number twenty has six divisors, this locker remains closed at the end, since it's altered by an even number of students. But the number sixteen has five divisors, so this locker remains open at the end, since it's altered by an odd number of students. Bethany, can you explain why perfect squares must have an odd number of divisors and non-perfect squares must have an even number of divisors?"

I hesitated. The pattern made sense, but I couldn't explain it.

"Remember how you solved the staircase problem in Mrs. Ridley's Grade 5 class?"

I looked up in surprise. "You know about that?"

"As you know, Cape Breton is a small place. Sydney is even smaller. And the math teacher community in Sydney is even smaller than that. I ran into Mrs. Ridley shortly after you solved that problem in her class, and she was glowing as she told me about Bethany's Staircase."

"Cool," I said, smiling.

"So the staircase," continued Mr. Collins, re-focussing my attention. "Mrs. Ridley asked you to sum up the integers from one to twenty. How did you do that?"

I drew a sketch of the two staircases and explained how to stick them together to form a rectangle. I explained how each row contained twenty-one squares.

"Excellent. I love that solution. Let's unpack that further. Your brilliant solution works because you paired each number." He pointed to the first row, with the one square on the top row of my first staircase next to the twenty squares from the bottom row of my inverted staircase.

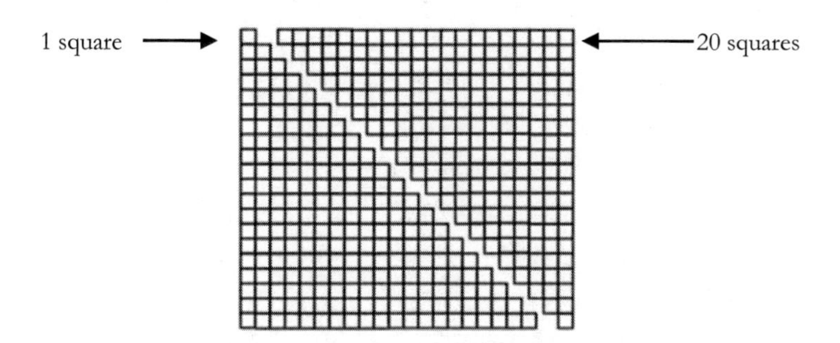


"In the first row, you've got one square in the first staircase and twenty squares in the second, which adds up to twenty-one. In the second row, you've got two squares and nineteen squares, which also adds up to twenty-one. Every number is paired: one is paired with twenty, two is paired with nineteen, and so on."

I nodded, seeing how each pair added up to twenty-one: (1, 20), (2, 19), (3, 18), and so on.

"Let's get back to the locker room problem," said Mr. Collins. "I want you to pair each number. For example, let's look at Locker #20. You said that this locker was altered by six students. What should the three pairs be?"

I wrote the six divisors of twenty on a piece of paper, namely 1, 2, 4, 5, 10, and 20. Like the staircase problem, I paired the numbers from the endpoints, working in. So my pairs were (1, 20), (2, 10), and (4,5).

"Excellent. What do you notice about each pair?"

"They multiply to twenty."

"Right. Locker #20 remains closed at the end because twenty has an even number of divisors. We have an even number of divisors since we can group the divisors into pairs. To express it in another context, think of a wedding party where all the guests are married and each person comes with their spouse. Since each person comes with their partner, there must be an even number of guests at the wedding. Does that make sense?"

I nodded. In the wedding party context, one would be "married" to twenty, and similarly two and ten would be a couple, as would four and five. That explained why an even number of people touched Locker #20.

"Now do the same with Locker #24. What are its pairs?"

I wrote down the eight divisors of twenty-four, seeing that the four pairs were (1, 24), (2, 12), (3, 8), and (4, 6).

"Great. One more. What about Locker #16?"

I wrote down the five divisors of sixteen, namely 1, 2, 4, 8, and 16. I got two pairs (1, 16) and (2, 8). But then one number was left over. The only way I could pair four was if I could pair it to itself, to produce (4, 4). But that was illegal since there was only one Student #4.

I saw that $16 = 4 \times 4$. So that explained why perfect squares behaved differently.

"Good, Bethany. From your reaction, it's clear you've got it. So let's answer the original question posed at the very beginning. You have 450 students and 450 lockers. How many lockers remain open at the end, and which ones are they?"

"All the lockers whose numbers are perfect squares: $1 \times 1$, $2 \times 2$, $3 \times 3$, and so on. Since $21 \times 21$ is 441, and that's just under 450, there are twenty-one lockers that remain open at the end. The lockers that remain open are 1, 4, 9, and so on, all the way up to 441."

"Well done, Bethany. Let me recap what you just did. You learned another powerful problem-solving strategy today, of *simplifying a difficult problem in order to find a pattern*, which you could then apply to solve the original problem. By pretending there were only twenty-five students at Pinecrest instead of 450, you discovered the pattern of the perfect squares. Once you recognized the pattern, you could then figure out how to explain and solve the harder problem.

"Simplification is a great strategy for real-world problem-solving, whether it's building a small model of an airplane before creating the real thing, or making just a couple of pancakes to figure out the correct proportion of flour to baking soda before cooking a huge batch to feed an army.

"Now, Bethany, here's my challenge for you. Sometime this week, write up a full solution to the locker problem with 450 students, clearly describing the process you just followed to arrive at your final answer. We'll go over your proof next week."

"So you want me to write up what I just explained?" I asked.

"Yes," replied Mr. Collins. "You did a great job explaining it orally; what I want now is a written justification. Your math class needs to emphasize proof-writing a lot more than it does now. I argued for years with the textbook writers and math consultants but got nowhere with them."

"But aren't you now in charge?"

"Yes," he said. "Now that I'm heading up the math curriculum, I'll have the chance to re-shift the focus for our students, with more emphasis on mathematical communication. Say, in all your years in math classes, have you ever written up a proof to a problem you solved?"

"Just once," I said. "The staircase."

"Right. But that was something you did on your own, not a deliberate skill that's practiced and developed as part of a school curriculum. This is why students go off to college or university and can get most of the answers, but they can't explain how they arrived at those answers, nor can they rigorously justify why their calculations are correct. Society doesn't benefit if we graduate engineers and programmers and laboratory scientists who can get the right answers, but can't communicate it to anyone who might be analyzing their reports or de-bugging their computer code or reading their research publications."

"So you're training me to become an engineer or programmer or scientist?"

"No, Bethany. I'm helping you develop your problem-solving skills, as well as your oral and written communication skills. This will all come in handy for whatever career you end up choosing for yourself."

"And what is that?"

He smiled. "Anything you want to be."

"Anything? I can have *any* career I want, as long as I put my mind to it and work hard to achieve it?"

"I believe that with all my heart."

"You're saying that I can be a famous soprano singer? Or a world-champion Olympic gymnast?"

Mr. Collins paused. He looked at me for a few seconds and broke out into a wide grin.

"Okay. Almost any career you want."

# 7

"S-H-E-P-H-E-R-D."

"That is correct," said Miss Carvery, our vice-principal. The students and teachers in the auditorium applauded as Gillian strolled back to her seat on stage. She passed by Vanessa and gave her a high-five. As Gillian walked by me, she calmly tossed her hair back, lifted her chin up, and gave me a cold stare.

I looked away.

It was now Vanessa's turn. After hearing her word, Vanessa hesitated before slowly responding.

"C-E-M-E-T-A-R-Y."

A loud bell went off.

"I'm sorry, that is incorrect. The correct spelling is C-E-M-E-T-E-R-Y."

I knew how to spell that word, but for all the wrong reasons.

Grandpa passed away two weeks ago.

I sighed, thinking about Grandpa. I missed him a lot. And of course, Mom did too.

After Vanessa walked off the stage, a boy named Rodney made a mistake on the word "deductible", and he too was eliminated. We were now down to the final two spellers. I needed to re-focus.

I took a deep breath, stood up from my seat, and stepped up to the microphone at the centre of the stage. I could see hundreds of people staring at me.

After Miss Carvery gave me my word, I let out a sigh of relief.

"N-E-C-E-S-S-A-R-Y."

The audience applauded. I sat back down, next to Gillian, with neither of us acknowledging the other.

"We have now completed the seventh round of the inaugural Pinecrest Junior High spelling bee," said Miss Carvery. "Congratulations to all of you who participated today. After starting with fifteen spellers, we are now down to our two finalists, Gillian Lowell and Bethany MacDonald. One of these two Grade 7 students will represent Pinecrest at next month's provincial spelling bee in Halifax. The winner of the provincials will represent Nova Scotia at the CanSpell National Spelling Bee in Ottawa."

She looked at us. "Round eight will begin. Gillian, it's your turn."

I glanced over at Gillian, who nodded confidently and walked up to the microphone. I noticed that my hands were wet, and I could feel the sweat on the back of my neck. My legs were shaking.

Though I read many books and was a strong speller, this Spelling Bee was 100% stress and 0% fun. But as I stared at Gillian's back, I knew that I didn't want to lose – to her.

My thoughts were interrupted as Gillian began spelling her word.

"K-E-R-O-S-E-N-E."

It was my turn. Feeling a bit queasy, I relaxed when I heard my word.

"P-A-R-A-L-L-E-L."

Gillian was next. "R-H-Y-T-H-M."

I paused slightly on my next word, but got it. "K-H-A-K-I."

Gillian didn't miss a beat. "U-K-U-L-E-L-E."

Miss Carvery said a word I had heard before, but wasn't sure how to spell. I asked her to say the word again. She leaned into the microphone and slowly enunciated the four syllables: *ko-rah-lay-t.*

Seeing a crowd of people staring at me, I closed my eyes and paused.

Was it one *R* or two *R*'s? Was it spelled with an *A* or an *E*? I visualized the four possible options for the correct spelling, putting each one on its own index card:

CORALATE, CORRALATE, CORELATE, CORRELATE

"Can you say the word again?" I asked.

Miss Carvery repeated the word twice. I still didn't know whether the ending was RALATE or RELATE.

"Can you use it in a sentence?"

She did, but that didn't help. I had one more lifeline.

"Can I have the definition?"

"To place in, or bring into, mutual or reciprocal relation."

As soon as I heard the word *relation*, I knew the word had to end in "RELATE". I was now certain the correct spelling had to be CORELATE or CORRELATE.

But which one? Was it one *R* or two *R*'s?

CO + RELATE looked too easy for it to be the right spelling. COR + RELATE seemed more likely, especially as I knew many words with two R's, like CORRUPT, CORRECT, and CORRODE.

I leaned into the microphone. "C-O-R-R-E-L-A-T-E."

"That is correct."

I smiled and walked back to my seat.

It was now Round 11. Miss Carvery gave Gillian a word I'd never heard. It sounded like *sigh-key*.

Gillian hesitated and took her time before carefully saying each letter.

"P-S-Y-C-H-E."

"That is correct."

"Yes!" said Gillian, as she pumped her fist and returned to her seat, flipping her hair again.

I walked up to the front of the stage. When I heard my word, I calmly nodded. Just to be sure, I asked Miss Carvery for the definition.

This was an easy word I first learned in Grade 1 when we were studying colours. Just to be sure, I visualized the two possible ways the word could be spelled:

MAROON    MARROON

Was it one *R* or two *R*'s? I pictured similar words that had two pairs of consecutive double letters, like RACCOON and BUFFOON and BASSOON, and of course, BALLOON. Yes, it definitely had two pairs of double letters.

"M-A-R-R-O-O-N."

The loud bell went off. I looked at Miss Carvery in disbelief.

"I'm sorry, that is incorrect. The correct spelling is M-A-R-O-O-N."

Shaking my head, I walked over to the other side of the stage, to join my friends Bonnie and Breanna and the other students who had also been eliminated. I heard the audience clapping for me and felt someone pat me on the back.

*How did I miss that?*

Feeling dazed, I could barely hear Miss Carvery invite Gillian up to the front. I stared blankly at my shoes, annoyed at myself for making such a careless mistake.

"Gillian, this is the championship word. Are you ready?"

"Yes," she replied.

Miss Carvery said the final word. Based on Gillian's excited reaction, I knew she would get it.

"C-A-T-A-C-L-Y-S-M."

"That is correct. Congratulations, Gillian Lowell, you are the winner of Pinecrest's first-ever Spelling Bee!"

The audience stood up and cheered, as Gillian raised both arms in the air. The fourteen other contestants on stage got up from our seats and began to clap.

As Gillian received a shiny trophy and posed for pictures with Miss Carvery, a few students turned towards me.

"Good try, Bethany."

"I was pulling for you."

"Sorry about the last word."

I thanked my classmates, but it was of no comfort.

I turned to see Gillian celebrating with Vanessa and their friends, and exchanging high-fives. A few teachers walked up to Gillian to shake her hand.

After thirty seconds of watching this, I was ready to leave. I turned back to my friends.

"Ready to go?"

"Yeah," said Bonnie. "Let's roll."

Before the Spelling Bee began, the three of us had decided that no matter who won, we were going to treat ourselves to a chocolate sundae at McDonald's. We had lots of time to walk across the street to get our dessert before the school bus came to take us home.

"Wait," I replied. "Just give me one second."

Leaving my friends behind, I walked towards Gillian. As she held up the trophy and celebrated with her clique, I decided I should do the honourable thing. When Gillian saw me, her smile turned into a frown.

"What do you want?"

I held out my hand. "Congratulations. You deserved it."

She looked at my hand and ignored it.

"Maroon?" she snapped. "We learned that in Grade 1. Are you really that dumb?"

Vanessa cackled out loud. She looked at me and rolled her eyes. The Chinese twins had a sympathetic look on their faces, but I knew their allegiances were with Gorgeous Gillian and not with Big Ugly Bethany.

With a clenched jaw, I walked back towards Bonnie and Breanna. "Yeah, I'm ready."

The three of us walked towards the auditorium exit. Just as we got to the door, I felt someone tap me on the shoulder. I turned back and saw Miss Carvery looking up at me.

"Bethany, can I talk with you?"

"Sure," I replied, turning to face the vice-principal.

Miss Carvery looked at Bonnie and Breanna.

"Girls, just head on to wherever you're going. Bethany will meet you there."

"Yes, Miss Carvery."

As Bonnie and Breanna left the auditorium, Miss Carvery led me towards the far end of the auditorium, away from the crowd of teachers and students.

"Please have a seat," she said, pointing towards the seat next to her.

I sat down and looked at my favourite teacher at Pinecrest. She had light black skin and her hair was clipped short, just above the ears. Bonnie's mother told us that Miss Carvery had won a huge scholarship to a famous university in England, and returned back home to Cape Breton to start teaching eight years ago. I couldn't believe that Miss Carvery was only thirty-three years old – a year younger than Mom.

"Bethany, I'm sorry about what happened at the end there."

"Yeah," I sighed. "Such an easy word – I don't know why I spelled it with two *R*'s."

"That's not what I was talking about."

I looked at her in surprise.

"I admire how you handled yourself just now, that even though you were disappointed at finishing second, you had the dignity to walk over and congratulate the winner. I heard what Gillian said to you, and I was disappointed by her actions. But I wanted to commend you for what you did, and the integrity you displayed by calmly walking away. I applaud that."

"Thank you," I replied.

"Other than the last word, did you enjoy the Spelling Bee?"

I swallowed. "Can I be honest with you?"

"Yes, of course."

"I hated it, Miss Carvery. All that pressure, all that tension. It was too competitive. I wish we didn't have to keep going and going until there was only one person left."

"Thank you for your honesty. I agree that the Spelling Bee was competitive, like what happens when hundreds of people are competing for a prize that only one person can win – such as a prestigious entrance scholarship to a university, an executive leadership position in a company, or a spot on the Canadian Olympic team. But I believe today's Spelling Bee served much deeper purposes."

"Such as?"

"To celebrate excellence in academic achievement. To encourage young people to develop their skills in spelling and language through friendly competition. To make people stronger in the face of . . ."

"But it's not friendly competition," I said. "Everyone who didn't win feels like a loser."

"Do you feel like a loser, Bethany? You correctly spelled ten difficult words, without a single error, in an extremely challenging high-pressure environment, with hundreds of people watching you. That's not the mark of a loser. That's the mark of a champion. I know that today's second-place result was hard for you. But you have strong character, and you'll bounce back. I know you will."

"Thank you."

Miss Carvery paused and gently put her hand on my arm.

"I heard that your grandfather passed away. I'm so sorry for your loss. How are you and your mother doing?"

"Okay, I guess. We miss him a lot."

"Of course you miss him. How is your mother doing? I mean, really doing?"

"She's fine," I said, pausing at the question. "Mom hates her job but she isn't unhappy."

Miss Carvery smiled. "I hear she's been coaching figure skating for the past couple of months."

"Yes," I said, surprised she knew. "She coaches an eight-year old on Saturday afternoons."

"The granddaughter of Taylor Collins," said Miss Carvery, nodding. "I know, since I ran into Mr. Collins the other day. He's delighted to be working with you. Both you and Ella are fortunate to have found such excellent coaches."

"Thanks," I said. "I'm very lucky. Ella too."

"Though I admit I was shocked to hear that your mother agreed to come back to the sport as a coach."

"Why?" I asked, confused. "Wasn't she an amazing figure skater?"

"Of course she was," replied Miss Carvery. "Everyone in Cape Breton knew Lucy MacDonald. All of us used to cheer for her."

"But you just said you were shocked that Mom would coach figure skating. Of course she would. Mom was the best figure skater in Nova Scotia three years in a row!"

Miss Carvery looked at me strangely. After a few seconds of uncomfortable silence, she gasped and put a hand to her mouth.

"What's wrong?" I asked.

"Nothing's wrong," said Miss Carvery, quickly moving her hand back on to her lap. She paused and put her hands on my shoulders.

"You should go, Bethany. Your friends are waiting for you."

# 8

"My best score ever," I said, passing a sheet to Mr. Collins.

Mr. Collins smiled as he checked my answers to the twenty-five questions from a previous Grade 7 math contest. The exam had taken me seventy-five minutes to complete, exceeding the one-hour time limit, but this was for fun – and not for competition.

The Grade 7 paper was organized by the University of Waterloo and was called the "Gauss", named after some famous German mathematician. The Gauss contest was written by nearly twenty thousand students each year from all across Canada.

"Yes, Bethany, I think this might be your best result yet," said Mr. Collins, comparing his multiple-choice responses to mine. "Your first nineteen are all correct."

In January, Mr. Collins and I had begun a new routine, where we spent some time at the beginning of each session reviewing the problems from an old Gauss contest. While I decided that I wouldn't be writing the actual contest with other Pinecrest students in May, I really enjoyed these contest problems – they were a lot more interesting than the stuff we covered in Grade 7 Math. And because it wasn't a high-pressure contest environment, I could take as long as I wanted, and not have to rush through the problems. Mom often told me how I had the right perspective: doing math because it was fun, rather than competing or comparing myself with Gillian Lowell or anybody else at Pinecrest.

"Any ideas on #20?" asked Mr. Collins, noticing the one question I had left blank.

"Not really. I got stuck because there were too many words. It was too much stuff to keep track of."

"Hang on one second, Bethany. I'll be right back."

Mr. Collins stood up, walked over to the counter at the front of Le Bistro, and returned a few seconds later with ten packets of sugar.

"What's this?"

"Hold these packets in your hand, so that you have something concrete to work with. It will be much easier to solve problems like this if you use props."

"Isn't that cheating?"

"Not at all," he said, smiling. "It's creative problem-solving."

Taking the ten packets of sugar in my hand, I looked at the one question I left blank.

**20.** Anne, Beth and Chris have ten candies to divide amongst themselves. Anne gets at least three candies, while Beth and Chris each get at least two. If Chris gets at most three, the number of candies that Beth could get is:

(A) 2 (B) 2 or 3 (C) 3 or 4 (D) 2, 3, or 5 (E) 2, 3, 4, or 5

Anne got at least three packets of sugar, while Beth and Chris got at least two. So I put three packets on my left, two in the middle, and two on my right, keeping the remaining three in my hand. I wanted to know how many total packets could be placed in the middle, the pile belonging to Beth. Of the three remaining packets, how could I distribute them?

I suddenly realized that the information about Chris getting at most one extra packet was unimportant. Since Anne and Beth could be given additional packets with no restriction, I could just distribute the remaining three packets between Anne and Beth. So Beth could receive zero, one, two, or three additional packets of sugar, and Anne could take the rest:

| Scenario | Anne | **Beth** | Chris |
|---|---|---|---|
| Beth gets zero additional packets | 6 | **2** | 2 |
| Beth gets one additional packet | 5 | **3** | 2 |
| Beth gets two additional packets | 4 | **4** | 2 |
| Beth gets three additional packets | 3 | **5** | 2 |

Yes, that worked for sure. I looked at the five possible answers on the contest sheet, and saw that option (E) corresponded to Beth getting 2, 3, 4, or 5 candies. I circled (E).

"Correct," replied Mr. Collins. "Question #21 is straightforward, and we've covered problems like that many times. So let's move on to Question #22. Can you show me how you did this one?"

**22.** The total number of squares and rectangles, of all sizes, that appear in a 3×3 square is:

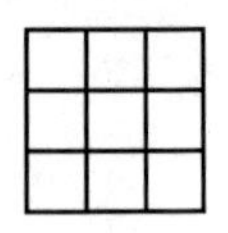

(A) 28 (B) 30 (C) 32 (D) 34 (E) 36

"I broke the problem down into cases, turning a hard question into a bunch of smaller simpler questions. I figured out all the types of squares and rectangles that could appear, you know, 1×1, 1×2, 2×2, 2×3, and so on. I counted each case, and then added it all up."

"Excellent," replied Mr. Collins. "You simplified the problem by figuring out the possible dimensions of the squares and rectangles that could appear in the picture, and then found its total. What did you get?"

I handed him a sheet of paper with all the different possibilities.

There were nine 1×1 squares, since there were nine unit squares in the diagram.

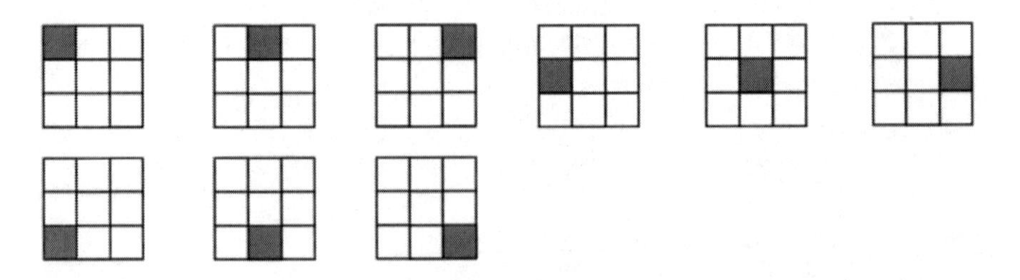

As for 1×2 rectangles, there were twelve of them in the picture: six "horizontal" and six "vertical".

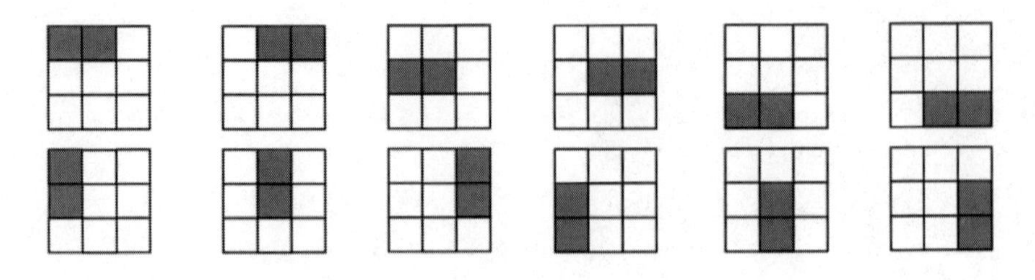

Counting 1×3 rectangles, I found six.

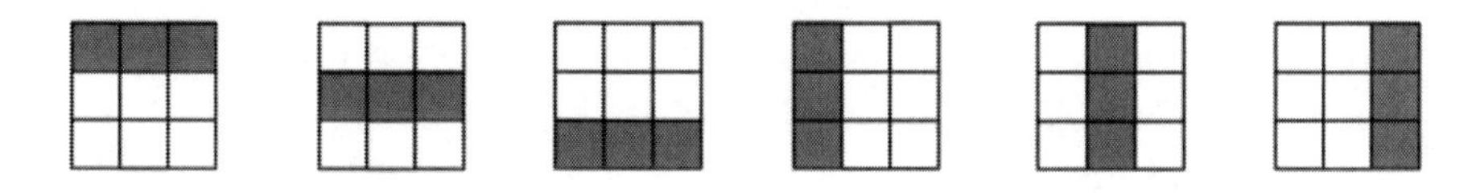

Counting 2×2 squares, I found four.

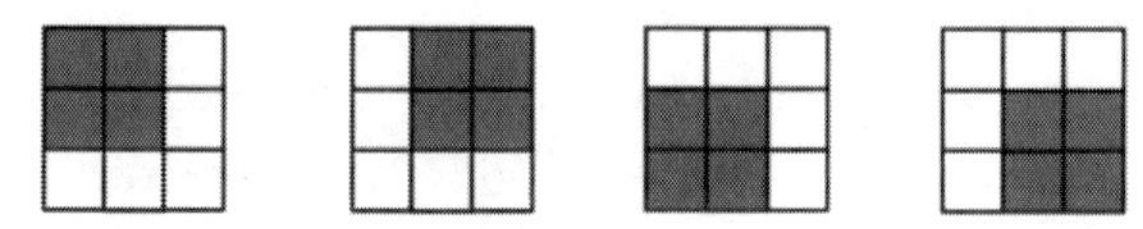

Counting 2×3 rectangles, I found two.

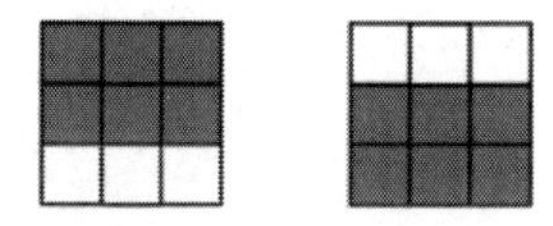

Finally, there was the 3×3 square. Of course, there was only one.

"So I found all the cases, and added them up. I got 9+12+6+4+2+1, which adds up to 34. So that's the answer."

Mr. Collins paused. "Are you sure you considered every case?"

"Yes," I said, before hesitating. "I think so."

Mr. Collins picked up a piece of paper and proceeded to draw a little table.

*Column Dimension*

*Row Dimension*

| | **1** | **2** | **3** |
|---|---|---|---|
| **1** | | | |
| **2** | | | |
| **3** | | | |

"Let's look at the row and column dimensions of each possible rectangle that can appear in our big 3×3 square. I want you to summarize this information in this table. Note that in any rectangle, its row dimension or column dimension can't be more than three, since the big square is only 3×3. Do you agree?"

I nodded. We couldn't have a 4×5 rectangle inside a 3×3 square, nor could we have rectangles with dimensions such as 2×4 or 5×1. Both the row and column dimension could be at most three.

"Bethany, I'd like you to put an **X** mark next to every scenario you've considered. For example, you counted twelve rectangles that were either 1×2 (horizontal) or 2×1 (vertical). Mark both of those entries in my table with an **X**. Do the same with all the other cases."

Looking at my diagrams of shaded squares and rectangles, I put an **X** next to every case I had considered.

*Column Dimension*

*Row Dimension*

| | **1** | **2** | **3** |
|---|---|---|---|
| **1** | X | X | X |
| **2** | X | X | X |
| **3** | X | | X |

Once I completed the table, I realized I had forgotten to consider 3×2 rectangles. It was easy to see that there were two of them.

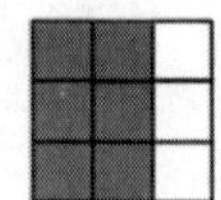

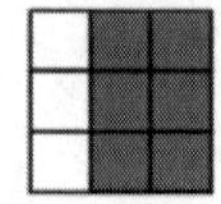

So the total number of squares and rectangles wasn't 34, but 34+2.

"The answer is thirty-six."

"That's right," said Mr. Collins. "To answer this difficult problem, you simplified it into smaller cases, which in this context, was to enumerate the squares and rectangles of all possible dimensions."

"Enumerate?" I asked.

"Sorry. Enumerate is just a fancy word for count."

"I see."

"You broke the problem into different cases and enumerated each case separately. You looked at 1×1 squares, then 1×2 rectangles, then 2×1 rectangles, and so on. But what happened?"

"I forgot a case."

"That's right. You were convinced that you had found all the possible cases, but once you summarized this information in your little table, you realized you had forgotten the 3×2 rectangles. You discovered this only because you presented the information in a highly structured way.

"When you're breaking down a hard problem into smaller simpler cases, you need to figure out how to structure your information so you can be certain that you haven't forgotten anything. Having a clearly-defined structure is the difference between *thinking* you've considered every case, and *knowing* you've considered every case. Say, that reminds me. Can I tell you a story?"

"Yes, please."

"Have you ever heard of the Challenger space shuttle disaster, the one from 1986?"

I shook my head. That was way before I was born.

"It was a terrible tragedy. The American space shuttle Challenger exploded seventy-three seconds after takeoff, killing all seven astronauts on board. NASA had never launched a shuttle with the temperature below ten degrees Celsius, but they let Challenger launch on a day when the temperature was only two degrees. They figured the cold weather wouldn't affect any part of the shuttle. But they overlooked something.

"The rocket boosters of a space shuttle are comprised of several parts, which are fitted together by giant rubber bands called O-rings, to ensure air doesn't escape. In the cold weather, these O-rings didn't seal properly, allowing pressurized hot gas to escape the rocket booster and reach the external fuel tank. That created a deadly combustion effect, and the space shuttle exploded."

"Did anyone know about these O-rings?" I asked.

"Yes, people did know, and that's what makes the disaster even worse. The NASA engineers knew the O-rings would work properly as long as the temperature remained above ten degrees Celsius. But once the temperature dropped to freezing, they predicted the O-ring would lose resilience, that it would be unsafe to launch. The engineers reported the problem to their higher-ups, but they were overruled."

"Overruled?"

"The mission commanders at NASA, the big guns making the decisions, insisted the engineers prove it was *not* safe to launch rather than demonstrate the conditions *were* safe to launch. Notice the difference between the two. The engineers couldn't show that it was not safe to launch as they had never tested the O-rings in such cold temperatures.

"During one of the public hearings in the aftermath of the disaster, a Noble-prize winning physicist took a small O-ring, and placed it in some ice water for a few minutes. When he took out the O-ring from the ice water, he showed how it lost elasticity and became brittle."

"But I don't get it," I said. "Why did they launch if they knew something could go wrong?"

"Well, NASA was under a lot of stress. The space shuttle launch was delayed a bunch of times that previous week, because of bad weather. Everyone was waiting for the launch to occur, and people were getting impatient due to the weather delays. So despite the objections and concerns from the engineers and other scientists, the mission control at NASA decided to lift off, with horrible consequences. It was a sad case of public relations trumping scientific and engineering reality.

"The moral of the story is the importance of performing rigorous scientific analysis. In the real world, a single omission can lead to disaster. Even if you remember ninety-nine out of one hundred items, it's not enough, especially when the one item you forget is the O-ring. Through the process of writing mathematical proofs, you're learning how to present arguments that are air-tight, where every line follows logically from the previous line, with no holes in your reasoning. Bethany, you'll learn never to forget the O-ring as you master a subject that's never bo-ring."

I laughed out loud.

"Let's continue. Of the first twenty-two questions on the practice Gauss contest, you solved twenty of them on your own, and together we figured out the other two. Take me through the last three."

We went through the final three problems, and I was delighted when Mr. Collins found no holes in any of my arguments.

"Wonderful, Bethany. So how many questions did you correctly solve?"

"Twenty-three out of twenty-five."

"What's that as a percentage?"

I multiplied twenty-three by four in my head. "That's ninety-two percent."

"That's right. That's your best score so far. I went online this morning to check the statistics on that contest. Of the twenty thousand Canadian students who wrote the Gauss that year, did you know that less than one percent got ninety-two percent or more? Bethany, you have become an outstanding mathlete!"

I smiled, realizing that one percent of twenty thousand was just two hundred.

"But I can't compare myself to them," I said, remembering that those two hundred students had to write the contest under pressure. "I had all the time I wanted this morning. They only had an hour."

"Good point. But if you were presented with the same circumstances, I bet you would do quite well."

"I don't think so."

"Why do you say that, Bethany?"

"Because I know myself. I don't do well under pressure. Remember the Spelling Bee?"

"Don't be so hard on yourself. As you told me, you were eliminated in the eleventh round. That means you spelled ten out of eleven words correctly."

"But I lost."

"No, you didn't," said Mr. Collins. "You misspelled a single word, and came in second place."

"Second place is the first loser. Someone at Pinecrest had that on their T-shirt."

"Oh, Bethany, don't think of it that way. That T-shirt slogan sends a horrible message, especially to young people. And while you are naturally disappointed in finishing second, I want to encourage you to think of it a different way: it was just you and Miss Carvery that day. She gave you eleven words to spell and you correctly got ten of them. That's over ninety percent. And maybe next year, when you're in Grade 8, you'll get an even higher percentage right."

"I won't be doing the Spelling Bee next year."

"That's fine, and I respect your decision. If the Spelling Bee isn't fun for you, then there's no sense entering it next year. It's the same with this Gauss Contest. If you find the problems interesting and want to do the contest because you find math enjoyable and challenging, then I'd encourage you to participate with the other Pinecrest students in May. But if it only leads to stress and anxiety, and you find yourself competing against Gillian Lowell and feeling bad if she solves more questions than you, then you're right – it's better that you don't participate."

I nodded.

"Bethany, let me change the subject for a minute. I'm sixty-three years old. I was never a star athlete but I always enjoyed running. And about twenty years ago, I started running half-marathons, gradually building up to full marathons."

"You've run a marathon?" I said in surprise.

"Yes. In fact, I've run five of them. There's a big race in Halifax each May, called the Bluenose Marathon, and this May will be my sixth time running that course. My goal each year is to simply beat my personal best time. I'm not interested in how many people I beat, or what place I finish. It's just me versus the clock. Last year I ran it in four hours and two minutes, forty-two kilometres on a cold rainy day."

"That's amazing you can run for over four hours."

"Thank you," replied Mr. Collins. "It's taken years of training, pushing my body to see how far and how fast I can go. My target this May is to run the marathon in just under four hours, which would be a first. And I think I have a good chance, especially if I'm not running in a rainstorm."

"Do you enjoy it?" I asked, wondering how anyone could possibly enjoy long-distance running.

"I love running, especially when I'm with my Sunday morning group along the Sydney Waterfront. Of course, they're all younger and faster, but that doesn't bother me. I feel so alive when I run. I can tell how alive you feel when you do math. What I get from running is what you get from math."

"Can I ask you a question?"

He smiled. "Of course, Bethany."

"I got twenty-three out of twenty-five on this practice Gauss contest, and that's my best score so far. So maybe I should write the actual Gauss contest in May?"

"It depends on your attitude. Tell me why you might want to write the contest."

"To try and beat my personal best of twenty-three questions. To see how much I've improved since September. Just me versus the questions, that's it."

"And what if one of your classmates gets a better score than you?"

I paused.

"Then that's fine. It just means that one of my classmates got a better score. No big deal."

"Are you being honest? If Gillian Lowell solves more questions than you, would you really not care?"

"No," I admitted. "But maybe one day, that wouldn't bother me."

Mr. Collins looked at me and smiled.

"Maybe one day, you and Gillian will be teammates."

# 9

"Twenty-four out of twenty-five. Well done."

"That's ninety-six percent," I replied, satisfied with my score because I had no chance on the final question.

"That ties your personal best," said Mr. Collins, handing me a sheet of paper filled with a bunch of numbers. "With the Gauss contest less than a week away, this is a great sign. You're definitely ready."

"Thanks, Mr. Collins. Can you tell me what these numbers are?"

"You tell me."

I looked at the sheet of paper more carefully.

| | | | | | | | | | |
|---|---|---|---|---|---|---|---|---|---|
| 10 | 11 | 12 | 13 | 14 | 15 | 16 | 17 | 18 | 19 |
| 20 | 21 | 22 | 23 | 24 | 25 | 26 | 27 | 28 | 29 |
| 30 | 31 | 32 | 33 | 34 | 35 | 36 | 37 | 38 | 39 |
| 40 | 41 | 42 | 43 | 44 | 45 | 46 | 47 | 48 | 49 |
| 50 | 51 | 52 | 53 | 54 | 55 | 56 | 57 | 58 | 59 |
| 60 | 61 | 62 | 63 | 64 | 65 | 66 | 67 | 68 | 69 |
| 70 | 71 | 72 | 73 | 74 | 75 | 76 | 77 | 78 | 79 |
| 80 | 81 | 82 | 83 | 84 | 85 | 86 | 87 | 88 | 89 |
| 90 | 91 | 92 | 93 | 94 | 95 | 96 | 97 | 98 | 99 |

"They're all the two-digit numbers," I said.

"That's right. I've given you the set of two-digit positive integers. And how many are there?"

"Ninety," I replied instantly, seeing that the numbers on the card appeared in a $9 \times 10$ table.

"Now suppose you randomly select a number from this table. What's the probability the last digit is one?"

Looking at the second column of the card, I saw there were nine numbers with last digit one, namely the integers 11, 21, 31, 41, 51, 61, 71, 81, and 91. Since there were ninety numbers in all, the answer was 9/90, which reduced to 1/10.

"One-tenth," I responded.

"Good. Now what's the probability the *first* digit is one?"

This was also straightforward. Looking at the first row of the card, I saw there were ten numbers with first digit one, namely the integers 10, 11, 12, 13, 14, 15, 16, 17, 18, and 19. The answer was 10/90.

"One-ninth."

"Excellent. Now suppose I gave you a card that listed all the three-digit integers. And say I got you to pick a number from that card at random. What would be the probability the first digit is one?"

As the two-digit scenario had probability 10/90, I reasoned that the three-digit scenario would have probability 100/900.

"It's the same," I said. "One-ninth."

"Good. Now how about five-digit integers, or fifty-digit integers, or hundred-digit integers?"

"Exact same – it's always one-ninth."

"What can you conclude? Be specific."

"That if you pick any integer at random, no matter how large it is, there's a probability of one-ninth that the first digit of that number is one."

Mr. Collins gave me another sheet of paper. "Now tell me what this is."

| | | | | | |
|---|---|---|---|---|---|
| 1) | China | 1,330,000,000 | 6) | Pakistan | 177,000,000 |
| 2) | India | 1,210,000,000 | 7) | Nigeria | 158,000,000 |
| 3) | USA | 312,000,000 | 8) | Bangladesh | 151,000,000 |
| 4) | Indonesia | 237,000,000 | 9) | Russia | 142,000,000 |
| 5) | Brazil | 190,000,000 | 10) | Japan | 127,000,000 |

I quickly figured out what these numbers represented.

"They're population numbers."

"Correct. China has approximately 1.3 billion people, followed by India with 1.2 billion people, and so on. This is the list of the ten most populated countries. If you were to pick a country at random from this table, what's the probability that its population has first digit one?"

Eight of the ten countries had their population beginning with the digit one, with the exception of the United States and Indonesia. So the probability was 8/10.

"It's eight out of ten – eighty percent."

"But didn't you say that the expected probability was one-ninth? Eighty percent is a lot more than eleven percent."

"Yeah, but these are special numbers," I said.

"What do you mean by *special*?" asked Mr. Collins.

"Well, you just gave me the populations for the top ten countries, so it was just luck that eight of them had first digit one. If you looked at every country's population, the first digits would be evenly spread out."

"Are you saying that if we looked at population data for all the world's countries, it would be equally likely for the first digit to be one as for the first digit to be nine? That the probability would be about one-ninth for each of the nine possible first digits?"

"Yes."

Mr. Collins smiled and took out some papers from his clipboard.

"Well, let's see, shall we?"

I stared at the three-page Wikipedia printout, listing the approximate population numbers for the top two hundred countries, ranging from 1.3 billion for China (#1) to 56,000 for Greenland (#200).

"How did you know I was going to say that?"

Mr. Collins smiled. "Just a lucky guess."

"What do you want me to do?"

"There are about two hundred and twenty-five countries in the world, if you include small dependent territories like the Falkland Islands and the Cayman Islands. I've taken the top two hundred, just to simplify the calculations. What I want you to do is make a table that lists the number of countries, and the percentage of countries, that have populations beginning with each of the digits from one to nine."

I flipped through the handout, tallying up the first digits for each country's population, and then calculating the proportion of countries with that first digit. Since there were two hundred countries in my list, the percentages were easy to calculate.

| **First Digit** | **1** | **2** | **3** | **4** | **5** | **6** | **7** | **8** | **9** |
|---|---|---|---|---|---|---|---|---|---|
| **Countries** | 58 | 28 | 26 | 18 | 17 | 17 | 10 | 17 | 9 |
| **Percentage** | 29% | 14% | 13% | 9% | 8.5% | 8.5% | 5% | 8.5% | 4.5% |

"Great," said Mr. Collins. "What do you notice?"

"Twenty-nine percent of countries have their population beginning with the digit one, while only four and a half percent of countries have their population beginning with the digit nine."

"So these statistics are not evenly spread out. It's not one-ninth for each of the nine possible first digits."

"But why?"

"Yes, that's exactly what I said when I first learned about this phenomenon. Whenever you have a list of *naturally-occurring numbers*, like the lengths of the world's longest rivers, the heights of the world's tallest buildings, the numbers that appear in your local newspaper, or the house numbers of residents in Cape Breton, it is far more likely for the numbers to have a low first digit than a high first digit."

"I don't believe you."

Mr. Collins handed me another card. "At first, I didn't believe it myself. Here's the magic table that states the distribution of first digits in such naturally-occurring numbers."

| **First Digit** | 1 | 2 | 3 | 4 | 5 | 6 | 7 | 8 | 9 |
|---|---|---|---|---|---|---|---|---|---|
| **Percentage** | 30.1% | 17.6% | 12.5% | 9.7% | 7.9% | 6.7% | 5.8% | 5.1% | 4.6% |

"Notice how these numbers are really close to the ones you came up with for country population?"

I nodded, seeing the similarity between his magic table and my calculations for the population numbers: 29%, 14%, 13%, 9%, all the way down to 4.5%.

Mr. Collins continued. "Let me illustrate with another example that I found on the internet yesterday. California has a whole bunch of cities: huge cities like Los Angeles with 3.9 million people and San Diego with 1.3 million; medium-sized cities like Pasadena with 150,000 people; and tiny cities that you've never heard of with just a few thousand people."

He handed me another sheet of paper.

"If you determine the first digit of each city's population, and tally the results, here's what the graph looks like."

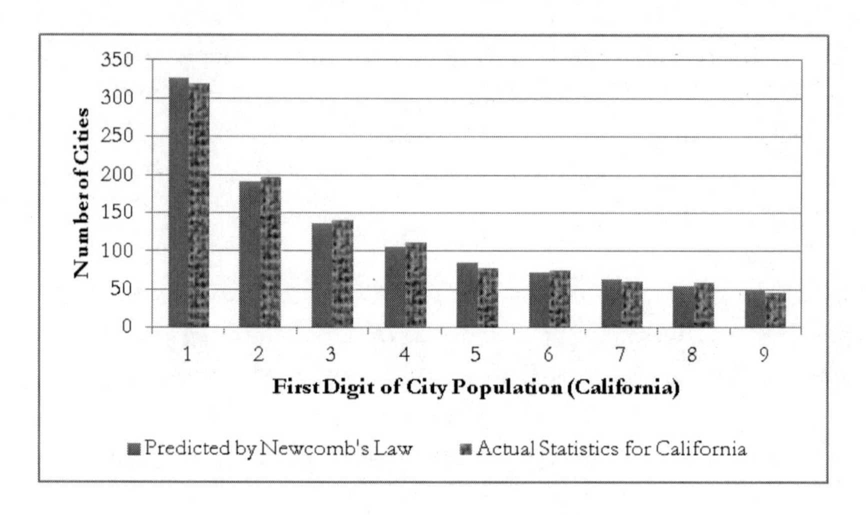


"The actual statistics match up almost identically with what's predicted by Newcomb's Law."

"What's Newcomb's Law?" I asked.

"Sorry," replied Mr. Collins. "That's the name for the magic table I just gave you: 30.1% for the first digit one, 17.6% for the first digit two, and so on."

"Who was Newcomb?"

"I'll get to that in a moment. You know, this first-digit phenomenon isn't just for population, it's also for things like house numbers. If you randomly picked one hundred streets in Cape Breton, and randomly selected one house on each of those streets, then around thirty houses will have first digit one."

"Why?"

"Let me give you a small hint. Think of various streets in Sydney, big ones like George Street and Kings Road, as well as smaller streets like the one I live on. What's the range of the house numbers? Do they all go from 1 to 99, or 1 to 999, or do some streets cut off earlier? For example, I think George Street goes from 1 to 2400, or something like that. Say you live on George Street. Is it more likely that your house number starts with one or nine?"

"I live in an apartment."

"I know," said Mr. Collins, smiling. "My cousin fixes elevators for a living. He says he comes by your building at least once a week."

I laughed, suddenly realizing that the elevator repairman did look a lot like Mr. Collins: bushy grey hair and super-thin, though he had no mustache.

"So, Bethany, let me ask the question again. Say you live in a random building on George Street, a street whose numbers go from 1 to 2400. What's the most likely first digit?"

If I lived on George Street, and the first digit of my house number was nine, my options would be limited to 9, 90 to 99, and 900 to 999. On the other hand, over a thousand houses on George Street had first digit one, specifically everything from 1000 to 1999, in addition to 1, 10 to 19, and 100 to 199.

Breanna lived on a street where the numbers went from 1 to 56. So that meant that there were eleven houses that had one as its first digit (specifically 1, 10, 11, 12, 13, 14, 15, 16, 17, 18, 19), but only one with nine as its first digit.

"I think I understand Newcomb's Law."

"Good," said Mr. Collins. "Explain your ideas to me."

After Mr. Collins agreed with my justification, a question still nagged.

"But how did they come up with the numbers in your magic table?" I asked. "You know, 30.1% of numbers have a first digit of one, 17.6% have a first digit of two, and so on?"

"Take out your calculator and do the following: for each of the numbers you see in Newcomb's table, convert it to a decimal. For example, with the percentage 30.1%, change that to 0.301. Then calculate the value of $10^{0.301}$. Do the same with all the other numbers in the table."

I followed his instructions and produced the following:

| **First Digit** | 1 | 2 | 3 | 4 | 5 | 6 | 7 | 8 | 9 |
|---|---|---|---|---|---|---|---|---|---|
| ***Proportion*** | 0.301 | 0.176 | 0.125 | 0.097 | 0.079 | 0.067 | 0.058 | 0.051 | 0.046 |
| $10^{Proportion}$ | 2.000 | 1.500 | 1.333 | 1.250 | 1.200 | 1.167 | 1.143 | 1.125 | 1.111 |

"Notice anything interesting?"

I saw it right away.

"Those numbers in the bottom row are simple fractions. It reminds me of the telescoping series that you taught me last week."

$$\frac{\mathbf{2}}{\mathbf{1}} = 2 \quad \frac{\mathbf{3}}{\mathbf{2}} = 1.5 \quad \frac{\mathbf{4}}{\mathbf{3}} = 1.333 \quad \frac{\mathbf{5}}{\mathbf{4}} = 1.25 \quad \frac{\mathbf{6}}{\mathbf{5}} = 1.2$$

$$\frac{\mathbf{7}}{\mathbf{6}} = 1.167 \quad \frac{\mathbf{8}}{\mathbf{7}} = 1.143 \quad \frac{\mathbf{9}}{\mathbf{8}} = 1.125 \quad \frac{\mathbf{10}}{\mathbf{9}} = 1.111$$

"That's correct. Way back when, before we had calculators, people had to use something called 'logarithm tables' to do calculations like 123456789 × 654321. It was a thick book with thousands of pages. Using logarithm tables was a tedious process but the calculations had to be precise – and was particularly needed by astronomers for surveying and celestial navigation.

"One day, a mathematician named Simon Newcomb looked at these logarithm tables and noticed that the earlier pages were more worn out than the later pages – in other words, there seemed to be more naturally-occurring numbers with first digits one and two than first digits eight and nine. He wondered why that was. Then he used techniques from a branch of math called 'probability theory' to show that the first digit frequencies had this beautiful property relating to powers of ten and simple fractions."

"That's really cool."

"Yes, it is. And I haven't told you the best part of the story. Guess where Simon Newcomb was from?"

"I don't know," I replied with a shrug.

"Guess."

"America? Britain?"

"He's from Nova Scotia! Simon Newcomb was born in Wallace, right next to Pugwash. It's just a four-hour drive from Wallace to Sydney."

I looked at him incredulously. "He's from Nova Scotia?"

"That's right! Our fellow Nova Scotian, Simon Newcomb, discovered the first-digit phenomenon back in 1881. Fifty years later, an American mathematician named Benford discovered the same principle, and he published it too. For some reason, the first-digit discovery is now known as Benford's Law even though Newcomb discovered it first."

Mr. Collins glanced at his watch.

"Before we end today, I want to summarize what we discussed because it's one of the most fundamental uses and applications of mathematics.

"Simon Newcomb looked at these logarithm tables and noticed the pattern that the earlier pages were more worn out than the later pages. Having found the pattern, he sought to understand and uncover its hidden structure. To do this, he had to apply his problem-solving skills to find a logical explanation for the first-digit phenomenon that 30% of the numbers start with one, about 18% start with two and so on. The mathematics was deep, and the end result was completely unexpected. That's the process by which mathematics is done in the real world. It's not people sitting around memorizing formulas, but rather employing a multitude of techniques to discover patterns to uncover truth and structure, seeing how those patterns explain what we see in society, and propose fresh ideas based on rigorous evidence to change our world.

"Newcomb couldn't have predicted this, but his discovery has led to numerous important and practical applications over a century later. One such application is fraud detection. As you know, your mother works at the Canada Revenue Agency, the federal department that administers all the country's tax laws and processes income tax reports. Well, if you look at a company's tax returns, you'll see thousands of different numbers that represent bits of financial information: assets, profits, stock prices, interest, and so on. If you look at the first digits of these numbers, what do you think happens?"

I saw where this was leading. "The first digits follow Newcomb's Law: 30% should have first digit one, about 18% should have first digit two, and so on."

"Exactly. So a legitimate tax return will have far more entries like $1,805 and $105.36 than those beginning with larger digits, such as $98.35 and $800. On the other hand, those who fudge the data and try to make the numbers appear random tend to distribute the numbers equally, with the first digits spread out evenly between 1 and 9. So using Newcomb's Law, you can identify the criminals and send them to jail."

"But if you knew Newcomb's Law, couldn't you properly fudge the numbers and get away with it?"

"No comment," said Mr. Collins.

"How do you know all this? I asked, fascinated by the story.

"Because a former student of mine has been working for the Canada Revenue Agency in Ottawa, managing a research group specializing in fraud detection. He does cutting-edge work for the federal government, using math and statistics for risk assessment, saving the Government of Canada millions of dollars each year. It's a wonderful application of mathematics, and so practical."

"Thanks, Mr. Collins. I learned a lot today."

"You're welcome. And good luck on the Gauss Contest. Are you excited?"

"Not excited. Nervous."

"You'll do great. Remember you got 96% on today's practice contest. And here's a small tip: if you think you're going to be distracted by a certain classmate and worried about what she's doing, then just sit in the front row. This way, you won't be distracted."

"Okay," I said. "And good luck with the marathon. Hope you break four hours."

"Thank you. By the way, do you know about Gauss? You know, who he was, what he was famous for?"

"Let me guess," I said. "He's from Nova Scotia too?"

"He's from Germany, but good try," said Mr. Collins. "Carl Gauss was one of the world's greatest mathematicians. He made enormous contributions to statistics, astronomy, geophysics, and created new fields of math which were centuries ahead of his time. There's a famous story about him.

"When young Carl Gauss was in Grade 4, his teacher gave all of the students a mindless and tedious addition problem. Of course, this was in the 1700s, and they had no access to calculators. So the students performed the addition term by term, and it took them all a long time. And every single student got the answer wrong, that is, every student except for Gauss. It turns out that Gauss correctly solved the problem, and he did so in mere seconds because he saw an insight that no one else did."

"What was the addition problem?"

Mr. Collins smiled. "It was to find the sum of the integers from 1 to 100. Gauss' incredible breakthrough was to create a second sum, and write the numbers backwards from 100 down to 1."

$$\begin{array}{lcl} S & = & 1 + 2 + 3 + 4 + \ldots + 97 + 98 + 99 + 100 \\ S & = & 100 + 99 + 98 + 97 + \ldots + 4 + 3 + 2 + 1 \end{array}$$

Mr. Collins added the two lines, column by column, which made each term on the right equal to 101.

$$2S = 101 + 101 + 101 + 101 + \ldots + 101 + 101 + 101 + 101$$

"The left side is 2S, twice the desired sum. The right side is $101 \times 100$, since there are 100 terms in total. Therefore, $2S = 101 \times 100$, which implies that $S = 101 \times 50 = 5050$. Gauss blurted out the answer in seconds. His classmates were amazed, and his teacher was astounded by his creativity and imagination. It was the first sign of his talent in mathematics, an ability that he nurtured and developed to become arguably the most important mathematician in the history of the world."

My mouth went dry.

"Don't worry, Bethany. It's not like I'm not comparing you to Gauss."

Mr. Collins winked.

"After all, you only had to add up the numbers from 1 to 20."

# 10

"The contest will begin in three minutes."

Our vice-principal Miss Carvery walked around the room, handing out a green sheet of paper to all of us.

I was the only person sitting in the first row. There were a dozen students writing the contest, including Gillian and Vanessa, who were sitting next to each other two rows behind me.

I arranged everything I needed so it was right in front of me: three pencils, fifteen sheets of white paper, an eraser, a ruler, a compass, and a calculator.

The green answer sheet contained twenty-five rows, one row for each multiple-choice question, with five possible answers: (A), (B), (C), (D), (E). Miss Carvery explained how each answer was to be bubbled in using pencil, and that we had exactly sixty minutes to answer the twenty-five problems: ten easy questions in Part A, ten medium questions in Part B, and five hard questions in Part C.

"Okay, it's now eleven o'clock. You can start. Good luck!"

I opened the contest booklet and took a deep breath.

*Here we go.*

**1.** When the numbers 8, 3, 5, 0, 1, are arranged from smallest to largest, the middle number is:

(A) 5 (B) 8 (C) 3 (D) 0 (E) 1

The answer was 3. I moved my pencil to the first row in my answer sheet and bubbled in (C).

**2.** The value of 0.9 + 0.99 is:

(A) 0.999 (B) 1.89 (C) 1.08 (D) 1.98 (E) 0.89

My first instinct was to bubble in the answer 0.999 but then I looked at the numbers more carefully and saw that the correct answer was actually 1.89, since ninety cents and ninety-nine cents added up to $1.89. I knew that I had avoided a potential trap, and bubbled in the correct answer, (B).

Five minutes later, I was at Question #10.

**10.** Two squares, each with an area of 25 cm$^2$ are placed side by side to form a rectangle. What is the perimeter of this rectangle?

(A) 30 cm (B) 25 cm (C) 50 cm (D) 20 cm (E) 15 cm

I drew the diagram in the contest booklet. A square with area 25 cm$^2$ has side length 5 cm. So when the squares are placed side by side to form a rectangle, the width is 5 cm and the length is 10 cm. The perimeter is 5+10, which equals 15 cm.

I bubbled in (E).

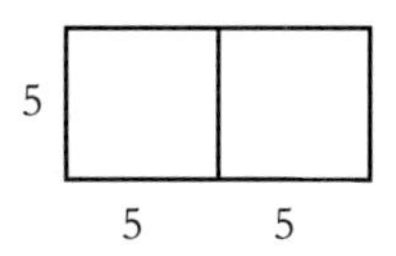


I was on a roll.

**13.** A *palindrome* is a positive integer whose digits are the same when read forwards or backwards. For example, 2002 is a palindrome. What is the smallest number which can be added to 2002 to produce a larger palindrome?

(A) 11 (B) 110 (C) 108 (D) 18 (E) 1001

I recalled Mr. Collins' advice that when it's not clear how to solve the problem directly, start from the multiple-choice answers and work backwards. This was just one of the many strategies I called "cheating" and he called "creative problem-solving". I read the question again, finding the key word.

What is the *smallest* number which can be added to 2002 to produce a larger palindrome?

I started with the smallest number among the five multiple-choice options, and worked my way up. As soon as I found a palindrome, I would have my answer.

2002 + **11** = 2013. Not a palindrome.

2002 + **18** = 2020. Close to a palindrome, but not quite.

2002 + **108** = 2110. No.

2002 + **110** = 2112. Yes!

With a smile on my face, I bubbled in (B). I glanced at the clock, amused that the time, 11:11, was also a palindrome.

**14.** The first six letters of the alphabet are assigned values $A$=1, $B$=2, $C$=3, $D$=4, $E$=5, and $F$=6. The value of a word equals the sum of the values of its letters. For example, the value of *BEEF* is 2+5+5+6 = 18. Which of the following words has the greatest value?

(A) *BEEF* (B) *FADE* (C) *FEED* (D) *FACE* (E) *DEAF*

I recalled another strategy from Mr. Collins, to take a few seconds right at the beginning of each question and look for an insight that would make the problem easier and less mechanical.

All of a sudden, I saw what to do: each of the five words had the letters *F* and *E* appearing, so these would effectively "cancel" each other when determining which of the five words had the greatest sum. I realized that finding the greatest value of these five words:

(A) *BE**EF*** (B) ***F**AD**E*** (C) ***FE**ED* (D) ***F**AC**E*** (E) *D**E**A**F***

was equivalent to finding the greatest value of these five "words":

(A) *BE* (B) *AD* (C) *ED* (D) *AC* (E) *DA*

From here, I could tell that $ED$ = 5+4 = 9 had the greatest value. The answer was (C). The questions got progressively more difficult.

**20.** The word "stop" starts in the position shown in the diagram

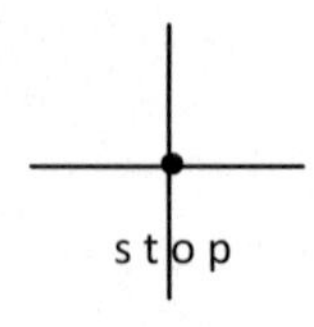

It is then rotated 180 degrees clockwise about the origin, and this result is then reflected in the $x$-axis. Which of the following represents the final image?

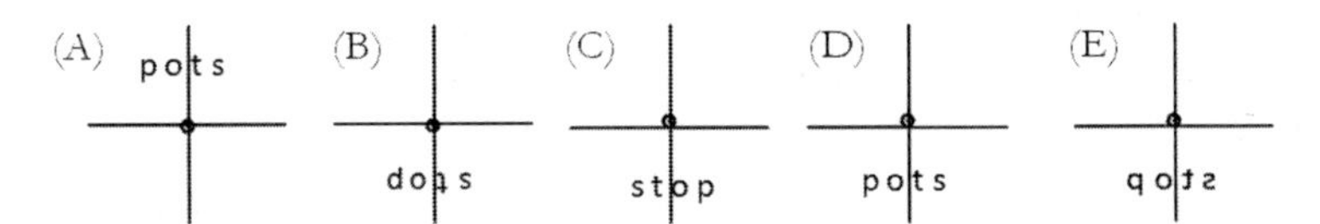


There were four letters to keep track of in my head, in addition to one rotation and a flip. It was too much to hold in my mind.

I sat and pondered for at least one minute.

An idea hit me. Remembering Mr. Collins and his sugar packets at Le Bistro, I thought of something I could do.

I took out a fresh sheet of paper, drew two large perpendicular lines that filled the entire page, signifying the horizontal $x$-axis and vertical $y$-axis, and wrote down **s t o p** in large block letters with my pencil, making sure the letters were dark enough that I could see through them when I needed to flip the page over.

I held the paper in my hand. First I rotated the sheet 180 degrees clockwise about the origin, which gave me the following picture:

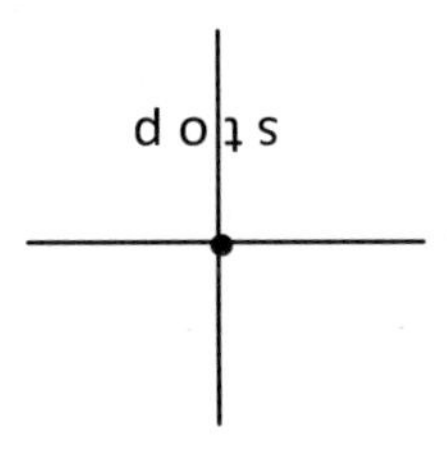

For the reflection, I pinched the left and right side of the paper with my thumbs lying on the same line as my horizontal $x$-axis, and just turned over my wrists. The paper flipped over and I saw the final image by holding my piece of paper to the bright light above me:

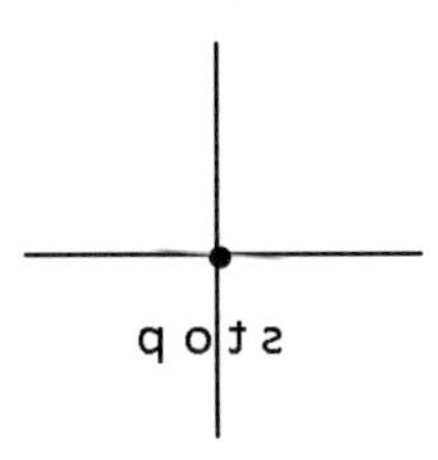

Looking at the five multiple-choice options, I saw the answer I was looking for, and bubbled in (E).

I glanced at the clock, and realized that I had finished the first twenty problems in twenty-four minutes, giving me plenty of time to work on the five hard Part C questions. I quickly checked my answer sheet and was relieved to see that I had bubbled everything in correctly.

Question #21 was a simple problem of counting handshakes; Question #22 dealt with the surface area of a rectangular box; Question #23 was a probability question involving coloured marbles; and Question #24 asked for the area of a trapezoid shaded inside a particular triangle. I was confident that I had correctly solved all four, and was super-careful to check each step.

*11:45 a.m.*

Fifteen minutes left for the final question.

**25.** Each of the integers 226 and 318 have digits whose product is 24. How many three-digit positive integers have digits whose product is 24?

(A) 4 (B) 18 (C) 24 (D) 12 (E) 21

I knew all the ways we could multiply two integers to get to 24. The combinations were 1×24, 2×12, 3×8, and 4×6. Now how could I adapt this to three integers?

I started listing the options. Taking out a fresh sheet of paper, I wrote down all the ways three digits could multiply to give 24.

2×3×4 2×2×6 1×4×6

That's all I could come up with. I ignored options such as 1×1×24 and 1×2×12, since 12 and 24 weren't digits. All I needed to do was determine how many three-digit integers could be made from each of the sets [2,3,4], [2,2,6], and [1,4,6], and I would be all done!

I knew that [2,3,4] could be made into six possible three-digit numbers, namely 234, 243, 324, 342, 423, and 432. I knew there had to be 6 = 3×2×1 possible rearrangements, since there were three choices for the first digit, two choices for the second digit, and one choice for the final digit.

Similarly, [1,4,6] could be made into six possible three-
the exact same argument: 146, 164, 416, 461, 614, and 64

Finally, [2,2,6] could only be made into three possible
numbers since the digit 2 appeared twice. The possible o
262, and 622.

The correct answer was 6+6+3 = 15, and I was all done with some time to spare! All I had to do was fill in the right bubble.

But 15 was not listed as any of the multiple-choice answers. What happened? Did the contest organizers screw up? Where was 15?

Or did I make a mistake? Did I miss a case?

*Did I forget about the O-ring?*

I started to get anxious. What did I forget? What else was there besides 2×3×4, 2×2×6, and 1×4×6?

Did I read the question wrong?

*Each of the integers 226 and 318 have digits whose product is 24. How many three-digit positive integers have digits whose product is 24?*

No, I definitely read the problem right. So what happened? My mind started to race.

*Relax, Bethany. Calm down.*

I took a deep breath and closed my eyes for twenty seconds.

When I opened my eyes, I re-read the question. *Each of the integers 226 and 318 . . .*

And then I saw it.

Right in the statement of the problem, the number 318 was written. I had forgotten the case 3×1×8, the other instance of three digits multiplying to 24.

From the missing case [1,3,8], I quickly determined the six additional three-digit numbers whose digits multiplied to 24, namely 138, 183, 318, 381, 813, and 831.

So my answer wasn't 15, but 15+6.

Did 21 appear as one of the five possible multiple-choice options? To my relief, it did.

I bubbled in the right letter and dropped my pencil on the table. I closed my eyes and wanted to scream.

*Yes, I got them all!*

I knew I had nailed everything in Part A, and so I didn't bother to re-check those ten questions. I skimmed through the ten questions in Part B, making sure I didn't make any careless computation errors.

*11:55 a.m.*

I went through the five questions in Part C, double-checking and triple-checking that I didn't do anything incorrectly. Everything was fine. I glanced up at the clock, seeing that the minute hand was now touching the hour hand.

"Thirty seconds," said Miss Carvery.

My heart was pounding. I was going to get a perfect score.

I opened the contest to the first page, where all the Part A questions were. My face froze when I saw Question #10 and the diagram that I had marked on the contest booklet.

**10.** Two squares, each with an area of 25 $cm^2$ are placed side by side to form a rectangle. What is the perimeter of this rectangle?

(A) 30 cm (B) 25 cm (C) 50 cm (D) 20 cm (E) 15 cm

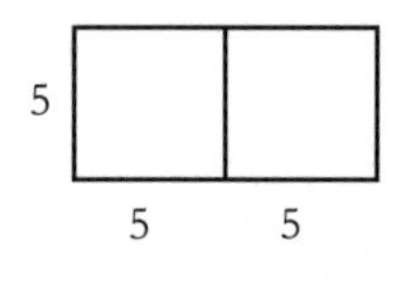


P = 10 + 5 = 15

The length was 10 and the width was 5. Since the perimeter represented the sum of the lengths of all *four* sides, the perimeter wasn't 10+5=15, but 10+5+10+5=30. Oh no!

Quickly grabbing my eraser, I found Question #10 on the answer sheet, erased my incorrect answer and bubbled in the correct letter.

"Okay, the contest is done. Stop writing!" said Miss Carvery.

I let out a deep breath, and dropped my pencil on the table.

"Please leave your answer sheet on the table," said Miss Carvery. "Congratulations to you all. The pizza will be arriving in a few minutes."

My heart continued to pound. I had trouble breathing.

I messed up Question #10 but was just barely able to fix it. What other careless mistakes did I make?

I didn't bother to check my answers to the Part A questions since I was sure I got them all, but did I? If I screwed up Question #10, what else did I get wrong?

*M-A-R-R-O-O-N.*

I felt like throwing up.

Miss Carvery walked by my desk to pick up my answer sheet.

"Bethany, are you okay?"

I shook my head.

"Do you need some fresh air?"

I nodded. I stood up and ran to the washroom. I hurried into a stall, locked the door, knelt down next to the toilet bowl, and lifted the seat.

With a great heave, I barfed up my entire breakfast. There were pieces of dried toast and peanut butter everywhere. I started to breathe a bit slower and felt my stomach calm down.

I sat down for a couple of minutes on the cold tiled floor, and leaned over to flush the toilet.

"Bethany."

It was Miss Carvery. She knocked on the stall.

"Could you open the door please?"

I turned the lock and Miss Carvery came in. The vice-principal sat down next to me and put her hand on my shoulder. She gave me a bottle of water, which I gladly accepted.

"Do you want me to stay with you?"

I nodded. Miss Carvery told me to take deep breaths, which I did, many times. I started to calm down.

We eventually made it back to the classroom. Walking next to Miss Carvery, I noticed that all the students were sitting around the tables in the back, comparing their answers while eating pizza.

Gillian saw us and pounced immediately.

"Bethany cheated, Miss Carvery."

*What?*

"That's a serious accusation," said Miss Carvery. "What evidence do you have?"

Gillian marched to my desk and held up the sheet of paper with **s t o p** written in large block letters.

"I saw her take this page and hold it up to the light, seeing what the reflection would look like. That's how she solved Question #20. That's cheating."

I was speechless.

"Gillian, take it easy. Bethany's not feeling well."

"But Miss Carvery, it's not fair! I did the questions properly. If Bethany gets the highest mark in the school, then it's because she cheated. You have to disqualify her, Miss Carvery!"

"Gillian, calm down. I'll take care of this."

Miss Carvery turned to me and motioned towards the door. "Bethany, come to my office. I want to talk with you privately."

Gillian looked smug and glared at me, before walking back to join Vanessa.

I followed Miss Carvery to her office, the one with VICE-PRINCIPAL written in big bold letters. I was scared and nervous for what was about to unfold.

Before Miss Carvery closed her door, she motioned to one of the administrative staff and gave him a ten dollar bill. She whispered something that I couldn't hear.

Miss Carvery turned to face me. I needed to set the record straight.

"I promise I didn't cheat."

"I believe you, Bethany. Please, have a seat."

"Yes, I did hold that page up to the light, but there's no rule saying I couldn't. Mr. Collins told me I could use props if I wanted, if it would help me solve the problem faster. He said it wasn't cheating – he called it 'creative problem-solving'."

Miss Carvery laughed. "That sounds like exactly something Taylor Collins would say."

"You know him, right?" I asked.

"Indeed. I'm a graduate of Sydney High School. Mr. Collins was my teacher too."

"So you don't think I cheated?"

"Of course not. Gillian is just overreacting, but let's just keep that between us. Don't worry about her."

"She's so mean."

"Gillian can say mean things, but you don't know her situation. Let me just say that Gillian has a lot of pressure to be the top student in the class. You threaten her, Bethany. You're the only person who can challenge her. You're the only person who is just as bright and clever as she is. If Gillian talks down to you, it's not because she's mean; it's because she's intimidated by you."

I heard a knock on the door. Miss Carvery opened it, and returned holding a tray with two sandwiches and two small cartons of milk. She handed one of the sandwiches to me.

"After what happened, I thought you would prefer eating this, rather than pepperoni and bacon pizza."

I smiled in appreciation, as I took off the plastic wrapping and took a bite.

"Gillian is intimidated by me?" I asked, my mouth full of tuna and celery.

"Yes. And Gillian knows that you live in a great home. Perhaps that also makes her jealous."

I was confused. "But Gillian lives in the biggest house in Cape Breton. She has a pool!"

"A house is different from a home, Bethany. You have something that she doesn't."

"What's that?"

Miss Carvery paused and took a bite of her sandwich. She hesitated.

"Your mother loves you and is supportive of everything you do."

"Thank you," I replied, realizing the implication of what Miss Carvery was saying.

Even though Mom and I argued sometimes, we were really close. We watched movies together and played Scrabble in the evening. She continued in the government job that she hated because it provided a secure and comfortable life for me. She taught figure skating to Ella Collins so that her grandfather could be my coach. Mom was awesome.

"I wish I could do something for Mom."

"You already are," said Miss Carvery. "You're living your life. And I'm sure she's secretly glad that your passion is math, not figure skating."

"Mom doesn't like to talk about figure skating," I said, tensing up. "She always changes the subject. We can talk about anything but that's the one subject that's off-limits."

Miss Carvery nodded. "Given what happened, it's natural Lucy would feel that way."

My heart stared to beat rapidly. I needed Miss Carvery to answer a question that I'd been carrying with me since the day of the Spelling Bee.

"Why were you shocked when you heard Mom was coaching Ella?"

Miss Carvery paused. She thought for a long time before answering.

"Because of what happened to your mother at the nationals. We were all devastated when she didn't make the team."

"What team?" I asked, confused.

She stared at me. Seeing my blank look, she looked away.

"It was the Olympic team, wasn't it?" I whispered.

Miss Carvery nodded.

Everything that Mom said, and didn't say, suddenly made sense.

I finally understood why Mom didn't want me to pursue the Math Olympiad, why she always changed the subject when it came to figure skating, and why she was in so much pain to talk about her past.

*Oh my God, how did I not realize this earlier?*

"Bethany, your face is all blue!" said Miss Carvery.

My mind flashed back to the conversations we had last summer, about the dangers of pursuing unrealistic goals and the risks of putting all of our eggs in one basket. I remember Mom crying as she talked about her shattered dream. I had no idea what that dream was . . . until now.

*How could I have been so blind?*

"Bethany, talk to me. What's wrong?"

I couldn't answer. My face, streaked with tears, felt the burden of Mom's pain over all these years. As she saw me getting excited about math and maybe one day becoming an Olympian like Rachel Mullen, she must have been constantly reminded of the memories of her own childhood, and all the sacrifices she was forced to make as she sought to become an Olympian herself.

I unfairly assumed that she was just a bad mother who tried to discourage my dreams, instead of a loving mother who tried to protect me from pain.

*How could I have been this selfish?*

Miss Carvery sat down beside and held my hand. She handed me a tissue, and wrapped me in a hug.

I sobbed on her shoulder and couldn't reply. She just held me, seemingly not bothered by the large wet patch forming on her suit jacket. After a few minutes, she looked gently into my eyes.

"You never knew, did you?"

## The Canadian Mathematical Olympiad, Problem #1

**Determine the value of**

$$\frac{9^{1/1000}}{9^{1/1000}+3}+\frac{9^{2/1000}}{9^{2/1000}+3}+\frac{9^{3/1000}}{9^{3/1000}+3}+\cdots+\frac{9^{998/1000}}{9^{998/1000}+3}+\frac{9^{999/1000}}{9^{999/1000}+3}$$

## Solution to Problem #1

As I close my eyes, the memories come back in a flash, one after the other.

The staircase in Mrs. Ridley's class, the visit to the hospital that brought Mr. Collins into my life, learning how to cross-train my mind, finding patterns by simplifying hard problems into easier components, the euphoria of achieving a perfect score on my very first math contest, and learning Mom's secret.

All of this happened years ago, before I even entered high school. But the experiences were so memorable that I can still remember every detail.

Years later, I find myself in this boardroom, writing the Canadian Mathematical Olympiad for a spot on Canada's IMO team. And I'm stunned that the key idea for the first problem is *The Staircase*. I re-read Problem #1.

**Determine the value of:**

$$\frac{9^{1/1000}}{9^{1/1000}+3}+\frac{9^{2/1000}}{9^{2/1000}+3}+\frac{9^{3/1000}}{9^{3/1000}+3}+\cdots+\frac{9^{998/1000}}{9^{998/1000}+3}+\frac{9^{999/1000}}{9^{999/1000}+3}$$

In my simplified problem with three terms instead of 999, I want to determine the value of $\frac{9^{1/4}}{9^{1/4}+3}+\frac{9^{2/4}}{9^{2/4}+3}+\frac{9^{3/4}}{9^{3/4}+3}$. I've already determined that middle term is equal to one-half.

$$\frac{9^{2/4}}{9^{2/4}+3}=\frac{9^{1/2}}{9^{1/2}+3}=\frac{\sqrt{9}}{\sqrt{9}+3}=\frac{3}{3+3}=\frac{3}{6}=\frac{1}{2}$$

I look down at my notes for calculating the other two terms, applying various properties of exponents.

The first term is $\frac{9^{1/4}}{9^{1/4}+3}=\frac{(3^2)^{1/4}}{(3^2)^{1/4}+3}=\frac{3^{1/2}}{3^{1/2}+3}=\frac{\sqrt{3}}{\sqrt{3}+3}=\frac{\sqrt{3}\times\sqrt{3}}{\sqrt{3}\times(\sqrt{3}+3)}=\frac{3}{3+3\sqrt{3}}$.

The last term is $\frac{9^{3/4}}{9^{3/4}+3}=\frac{(3^2)^{3/4}}{(3^2)^{3/4}+3}=\frac{3^{3/2}}{3^{3/2}+3}=\frac{3\sqrt{3}}{3\sqrt{3}+3}=\frac{3\sqrt{3}}{3+3\sqrt{3}}$.

Instead of determining the value of each term separately before computing its sum, I realize I can just add the two expressions directly, producing a fraction with the same numerator and denominator.

$$\frac{9^{1/4}}{9^{1/4}+3}+\frac{9^{3/4}}{9^{3/4}+3}=\frac{3}{3+3\sqrt{3}}+\frac{3\sqrt{3}}{3+3\sqrt{3}}=\frac{3+3\sqrt{3}}{3+3\sqrt{3}}=\mathbf{1}$$

The first and last terms, when paired, add up to one. When I see that this total is just one, I have a hunch that the same property is true in the much-harder Olympiad problem.

Sure enough, it is. I'm able to show that when we pair the terms from the two endpoints, its sum is always one. For example,

$$\frac{9^{1/1000}}{9^{1/1000}+3}+\frac{9^{999/1000}}{9^{999/1000}+3}=\mathbf{1},\quad \frac{9^{2/1000}}{9^{2/1000}+3}+\frac{9^{998/1000}}{9^{998/1000}+3}=\mathbf{1}$$

Armed with this insight, there is a simple way to determine the sum, using the staircase picture. Just to make sure I have all the details correct, I first apply it to my simplified problem with three terms. I draw three rectangles, each with height 1. I make the width in the first rectangle $\frac{9^{1/4}}{9^{1/4}+3}$ the second rectangle $\frac{9^{2/4}}{9^{2/4}+3}$ and the third rectangle $\frac{9^{3/4}}{9^{3/4}+3}$.

I attach the three rectangles together to form a staircase. Since each rectangle has height 1, the staircase has total area $\frac{9^{1/4}}{9^{1/4}+3}+\frac{9^{2/4}}{9^{2/4}+3}+\frac{9^{3/4}}{9^{3/4}+3}$.

This step has area $1\times\frac{9^{1/4}}{9^{1/4}+3}=\frac{9^{1/4}}{9^{1/4}+3}$

This step has area $1\times\frac{9^{2/4}}{9^{2/4}+3}=\frac{9^{2/4}}{9^{2/4}+3}$

This step has area $1\times\frac{9^{3/4}}{9^{3/4}+3}=\frac{9^{3/4}}{9^{3/4}+3}$

Like I did back in Grade 5, I'm able to represent the desired sum visually. But this time, instead of counting the number of squares in the staircase, I want to find its combined area. To do this, I flip the staircase diagram, and paste it to the original figure.

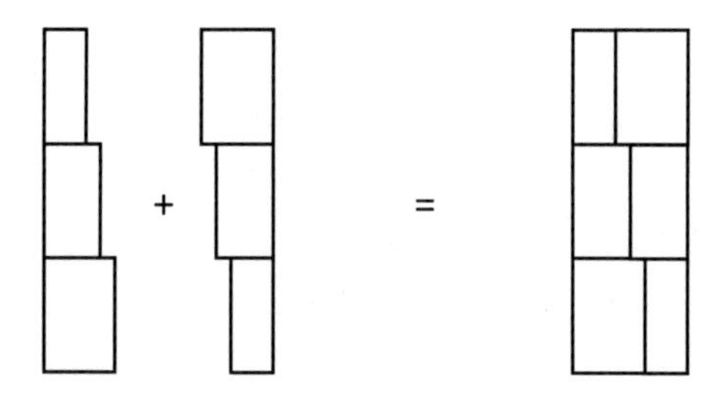


So the two staircases become one simple rectangle, since the width of each "block" is guaranteed to be 1 by how I've done my pairing. Therefore, the area of the two-staircase figure is just 3 times 1, the height of the rectangle multiplied by its width. And so the area of a single staircase is 3/2, or half of 3.

Without doing any calculations, I know that the answer to the Olympiad Problem is 999/2, by the exact same argument. I start writing my solution. To ensure I finish as quickly as possible, I decide to use "function notation", $f(x)$, and "sigma notation", $\Sigma$, a beautiful way to use mathematical symbols without losing any meaning or rigour. I complete the solution and quickly re-read what I've written.

It's perfect.

One done, four to go.

**Problem Number:** 1
**Contestant Name:** Bethany MacDonald

Let $S = \frac{9^{1/1000}}{9^{1/1000}+3} + \frac{9^{2/1000}}{9^{2/1000}+3} + \cdots + \frac{9^{999/1000}}{9^{999/1000}+3}$. We claim that $S = \frac{999}{2}$.

Let $f(x) = \frac{9^x}{9^x+3}$. We first show that $f(x) + f(1-x) = 1$ for any real value of $x$. This is true because $f(x) + f(1-x)$ equals

$$\frac{9^x}{9^x+3} + \frac{9^{1-x}}{9^{1-x}+3} = \frac{9^x(9^{1-x}+3)+9^{1-x}(9^x+3)}{(9^x+3)(9^{1-x}+3)} = \frac{9+3\cdot 9^x+9+3\cdot 9^{1-x}}{9+3\cdot 9^x+9+3\cdot 9^{1-x}} = 1$$

Now consider a "staircase" diagram with 999 steps, each of height 1, where the $k^{\text{th}}$ step has length $f\left(\frac{k}{1000}\right)$. Then the area of this staircase is $\sum_{k=1}^{999} f\left(\frac{k}{1000}\right)$, which is equal to $S$.

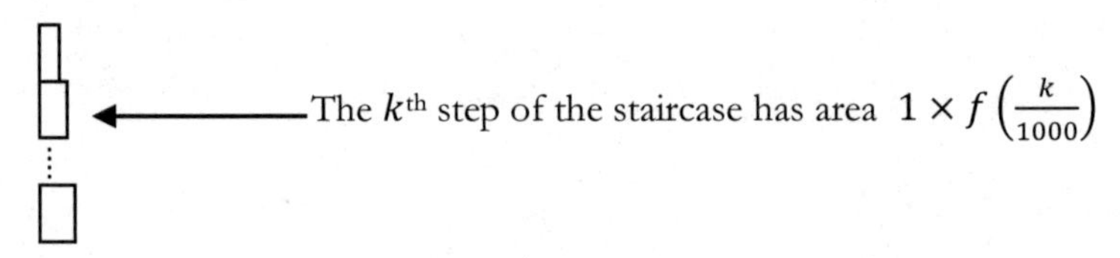


Now invert a copy of the staircase and attach it to the original so that the $k^{\text{th}}$ step in the first staircase is joined to the $(1000-k)^{\text{th}}$ step in the second staircase, as shown in the diagram below.

Each of these 999 "steps" become 1×1 squares

The area of this new figure is $2S$, since it is just the piecing together of two staircases with area $S$. By this construction, the $k^{\text{th}}$ step in the new figure has length $f\left(\frac{k}{1000}\right) + f\left(\frac{1000-k}{1000}\right) = f\left(\frac{k}{1000}\right) + f\left(1 - \frac{k}{1000}\right) = 1$.

It follows that each of the 999 "steps" in the new figure has area 1, since it is just a square with height 1 and length 1. Therefore, we have $2S = \sum_{k=1}^{999} f\left(\frac{k}{1000}\right) + f\left(\frac{1000-k}{1000}\right) = 999 \times 1 = 999$, implying that $S = \frac{999}{2}$.

Our proof is complete.

## The Canadian Mathematical Olympiad, Problem #2

**Find all real solutions to the following system of equations.**

$$\begin{cases} \dfrac{4x^2}{1+4x^2} = y \\ \\ \dfrac{4y^2}{1+4y^2} = z \\ \\ \dfrac{4z^2}{1+4z^2} = x \end{cases}$$

## Problem #2: Three Equations

I look up at the clock.

*9:41 a.m.*

Surprised that so much time has already elapsed, I try to re-focus my attention on Problem #2, knowing that I can't afford to spend over forty minutes on this question.

I've seen many problems involving a *system of equations*, a collection of equations using the same set of variables, such as:

$$\begin{cases} 2x - 1 = y \\ 2y - 7 = x \end{cases}$$

The solution to the system above is $(x, y) = (3,5)$, since substituting $x = 3$ and $y = 5$ satisfies both equations. The solution to any system of equations is an assignment of values or numbers to the unknown variables $x$ and $y$ so that all of the equations are satisfied simultaneously.

Mr. Collins first taught me how to solve systems of equations. To do this, he explained that the theory of solving equations is a fundamental part of Linear Algebra, a branch of math foundational to engineering, chemistry, and physics. Mr. Collins also told me the story of John Forbes Nash, who applied systems of equations to formulate his Nobel Prize winning Nash Equilibrium that revolutionized modern economics, leading to a richer understanding of auctions, labour negotiations, nuclear arms races, and evolutionary biology.

From those lessons from many years ago, I understand why systems of equations are important, and I know how to solve systems of equations involving two or more variables. There are several standard techniques, involving the elimination or substitution of a variable.

I have no problems with a two-variable *linear* system such as the one above, or even a three-variable linear system such as:

$$\begin{cases} 2x + y + 5z = 35 \\ 3x + 2y - z = 12 \\ 5x - 3y + 2z = 13 \end{cases}$$

But I have no familiarity with *non-linear* systems such as the one posed in Problem #2, where the three equations are $\frac{4x^2}{1+4x^2} = y$, $\frac{4y^2}{1+4y^2} = z$, and $\frac{4z^2}{1+4z^2} = x$.

I try the first thing that comes to mind, substituting the second equation into the third equation to eliminate the variable $z$, to reduce my system of three equations and three unknowns to a system of two equations and two unknowns.

But this produces a horrible expression.

$$\begin{cases} \dfrac{4x^2}{1+4x^2} = y \\[2ex] \dfrac{4y^2}{1+4y^2} = z \\[2ex] \dfrac{4z^2}{1+4z^2} = x \end{cases} \rightarrow \begin{cases} \dfrac{4x^2}{1+4x^2} = y \\[2ex] \dfrac{4\left(\dfrac{4y^2}{1+4y^2}\right)^2}{1+4\left(\dfrac{4y^2}{1+4y^2}\right)^2} = x \end{cases}$$

Even after simplifying, the second equation is far too messy. No, this can't be it.

After a couple of minutes, I have another idea, realizing that I might be able to find a few solutions by just guessing!

I consider the case $x = 1$. Substituting $x = 1$ into the first equation, I see that $y = \frac{4}{5}$. And then substituting $y = \frac{4}{5}$ into the second equation, I conclude that $z = \frac{64}{89}$. But then I realize that if I substitute $z = \frac{64}{89}$ into the third equation that $x$ would be a horribly complex fraction, instead of the simple value of $x = 1$ that I need. So that doesn't work.

I try $x = 0$. Substituting this value of $x$ into the first equation, I see that $y = 0$. And then substituting this value of $y$ into the second equation, I get $z = 0$. Finally, substituting this value of $z$ into the third equation, I come back to $x = 0$.

*Oh good! This works!*

The simple solution $(x, y, z) = (0,0,0)$ satisfies all three equations, and is therefore a solution to this system. I've got one solution.

I try another case, $x = -1$. Substituting this value of $x$ into the first equation, I see that $y = \frac{4}{5}$. Substituting this value of $y$ into the second equation, I get $z = \frac{64}{89}$.

But wait, isn't this what I had before? This won't work, and I know that for sure. If I substitute this value of $z$ into the third equation, $x$ would take on some positive value, and so it can't end up as $x = -1$.

*Wait, does this mean* $x$ *has to be positive?*

No, not necessarily, since I just showed that $(x, y, z) = (0,0,0)$ is a solution.

Or maybe this means $x$ just has to be at least zero? Yes, I think so.

I draw a table of values, to see if I can detect any interesting patterns between the three variables. Following up on my hunch that $x$ has to be at least zero, I decide to only try non-negative values of $x$, with the hope of finding a pattern that unlocks the problem.

In addition to the values $x = 0$ and $x = 1$ that I just tried, I list out some other values for which $y$ is easy to calculate.

| $x$ | $y$ | $z$ | **Solution?** |
|---|---|---|---|
| 0 | 0 | 0 | YES |
| 1 | $^4/_5$ | $^{64}/_{89}$ | NO |
| 2 | $^{16}/_{17}$ | | |
| 3 | $^{36}/_{37}$ | | |
| $^1/_2$ | $^1/_2$ | | |
| $^1/_3$ | $^4/_{13}$ | | |
| $^1/_4$ | $^1/_5$ | | |
| $^3/_2$ | $^9/_{10}$ | | |

As I start filling out the numbers in the third column, I notice something strange. The numbers in the first column are always larger than the numbers in the corresponding second column, except for two instances where they are equal: $x = y = 0$ in the first row, and $x = y = \frac{1}{2}$ in the fifth row.

Is that a coincidence? I scribble an idea.

$$\textit{Does } \frac{4x^2}{1+4x^2} = y \textit{ force } x \geq y?$$

I perform a simple algebraic calculation to show this is indeed true, that as long as $x$ is non-negative, the inequality $x \geq y$ holds whenever $\frac{4x^2}{1+4x^2} = y$.

In other words, I've shown that in the system of three equations and three unknowns, the first of my three equations forces $x$ to be at least as big as $y$. What can I do with this information?

All of a sudden, I see it.

*It's symmetric.*

Just as I did with the Staircase problem, I remember something from my past that triggers an idea. This time, it's an insight I learned during that life-changing week in Ottawa, right after I finished Grade 9.

At the Canada Math Camp.

# 11

The sharp descent of the plane popped my ears.

I looked out the window, and saw the wheels approaching the paved runway. The plane landed with a thud and crawled to a slow pace towards the terminal. We had arrived in the nation's capital.

I opened my clipboard and re-read my invitation letter to Ottawa for what must have been the hundredth time.

> Dear Bethany MacDonald,
>
> On behalf of the Canadian Mathematical Society, it is my pleasure to invite you to this summer's Canada Math Camp, to be held at the University of Ottawa between June 23 and June 29.
>
> This camp will bring together twenty outstanding mathematics students from coast to coast, who have the potential to represent Canada at the International Mathematical Olympiad in two to three years' time. Based on your results in the recent Canadian Open Math Challenge, we are delighted to offer you an invitation to attend this year's camp.
>
> There is considerable cost to organizing a week-long event such as this. Thanks to the generous sponsorship of the Royal Oil Foundation, we are able to offer this camp at a cost of just $300 per individual. Please note this $300 registration fee includes all meals and accommodation but does not cover the costs of travel and transportation to and from Ottawa. From past experience, we have found that most students have had their travel and camp fee paid for by their school board.
>
> Attached is a registration form. Please send the completed form to Marlene Thomas, the Director of this year's Canada Math Camp, at the address shown. I look forward to meeting you this June in Ottawa.
>
> Sincerely,<br>
> J. William Graham,<br>
> Executive Director<br>
> Canadian Mathematical Society

In the three years that I had trained with Mr. Collins, my confidence grew week after week. We spent months preparing for the Canadian Open Math Challenge, a contest written by ten thousand high school students from all across the country. I had no chance against the Math Olympians, two of whom solved every problem and got a perfect score of 80 out of 80. But among the Grade 9s and 10s, I finished nineteenth in Canada, which was good enough to secure me an invitation to Ottawa.

As the last passenger to exit the plane, I followed the long line of people to the baggage claim. The escalator took us down to the first floor of Macdonald-Cartier International Airport. Within ten minutes, Mom's faded green suitcase came into sight.

Following the instructions in my invitation letter, I hopped on Bus #97, and soon arrived at the University of Ottawa. Dragging my suitcase, I followed the directions and within a few minutes I had arrived at the main residence. On the grassy field I saw teenage boys running around playing Frisbee, and standing close by was an older lady grilling hamburgers and hot dogs on a large barbeque.

There was a large sign, right by the entrance to the residence: *Canada Math Camp*.

After two months of anticipating this moment, I was finally here.

Walking into the residence, I was met by a young man who was wearing the same dark blue T-shirt as the lady by the barbeque. He gave me a bookbag containing my room key and various other items.

I went up to Room 216, and opened the door. The dorm room was small and cozy, just like my bedroom back home, with a single bed on one side, a big table on the other, and a small dresser. I unpacked my suitcase and dumped the contents of the bookbag onto my bed: a red binder, a map of the university, paper, pens, and a name tag. I put on my name tag and walked outside into the hallway.

"And you must be Bethany."

I turned to face an auburn-haired girl in her early twenties, wearing the same dark blue T-shirt. She looked up at me with a big smile on her dimpled face.

I recognized her instantly.

"I'm Rachel," she said, greeting me with a warm handshake. "I'll be one of your teachers this week."

I stared at her in shock.

"You're Rachel Mullen," I whispered. "The IMO gold medallist."

"Yes," she replied, a bit embarrassed. "I'm sure we'll be chatting a lot over the next few days."

"Thank you," I replied, unsure of what else to say.

"Well, do you want to come downstairs? Dinner has just started."

I wordlessly followed her down the stairs, and stood behind her in the line for the barbeque. I couldn't believe that Rachel, my inspiration for starting this journey three years ago, was going to be my teacher at this camp.

"Are you excited to be here?" asked Rachel.

"Definitely," I said. "I'm so happy to meet you . . . to meet all of you."

When we arrived at the front of the line, Rachel introduced me to the lady by the barbeque.

"Bethany, this is Marlene Thomas, the director of the Canada Math Camp."

"Welcome, Bethany. I hear you're from Cape Breton. That's my favourite place in Canada."

I smiled and shook hands with Marlene, and put out my plate to receive two hot dogs.

Standing by the condiments table, I wanted to soak up every minute with Rachel. She told me that she was an undergraduate at the University of Waterloo, majoring in pure math. She said that three of her teammates from the IMO were at Waterloo with her, in the same program.

"What was the IMO like?" I asked.

"Amazing. It was my first time travelling outside of Canada, and I loved every minute. I made friends from all over the world and still keep in touch with them."

"And you got a gold medal!"

"Thanks," said Rachel, shrugging. "On the last problem of the contest, I found the key idea right at the end, and finished writing my solution with two seconds left. It was luck."

"It wasn't luck," I said.

"Maybe," she said with a grin. "So, how about you, Bethany? How did you get interested in math, and how did you hear about the IMO?"

Rachel wanted to know my story. I knew exactly where to begin.

"Well, a few years ago, I was watching TV with Mom, and CBC had this documentary . . ."

"Yes, there's another girl here!"

We were interrupted by an Asian girl with long black hair and round glasses, who ran up towards us.

"Bethany," said Rachel. "This is Grace Wong. She's from Vancouver."

"You are so tall," said Grace, who put her hand on her head and saw that it was at the same level as my shoulder. Rachel snickered.

"I'm going to let you two introduce yourselves," said Rachel. "Bethany, it was great chatting with you. I'm really looking forward to working with you this week."

"Me too," I replied, trying not to show my disappointment.

I watched Rachel move to a crowd of four boys and introduce herself. I looked around where all of us were standing around, eating dinner. Most of the campers were in groups of three or four, except for one boy who was sitting by himself on a bench reading a book, and another who was walking silently around the back of the field, oblivious to the rest of us.

"Hey, Bethany, can you believe we're the only two girls at this camp?"

"Really?"

She was right. Other than Grace, every single camper was a boy. I stared at their faces, seeing a lot more diversity than what I was used to at Pinecrest Junior High.

Grace tapped me on the shoulder. I turned towards her and looked down.

"How tall are you, anyway?" she asked.

"Five-ten, I think. And still growing. Unfortunately."

"Lucky you," she said with a big smile that showed off braces. "I'm going to be four-eleven for the rest of my life. So, where are you from?"

"Sydney."

"Australia?" asked Grace, confused.

I laughed out loud. "No, not Australia. Cape Breton."

"Where's Cape Breton?"

"You don't know?" I asked, surprised. "Cape Breton's in Nova Scotia."

Grace looked embarrassed. "I should know my Canadian geography better. I've lived in Canada my whole life but I've never been east of Ontario. Sad, eh?"

"Not at all," I responded. "Today's the first time I've been outside of Nova Scotia."

She looked relieved. "Well, you should definitely come and visit Vancouver some time. It's an amazing city. We can bike around the seawall in Stanley Park. You can even stay at my place."

"Sounds awesome," I said, surprised to hear that from someone I just met. I immediately felt comfortable around Grace.

Chatting with Grace for a few minutes, I could tell that we were polar opposites: Grace was loud, outgoing, and super-extroverted. And I could tell that she easily made friends.

Marlene called all of us, and asked us to finish our dinner and gather together.

We stood in a large circle on the grassy field. I quickly did a head count and saw there were twenty-two people, including the two adults, Marlene and Rachel.

"Good evening. My name is Marlene Thomas, and I'm the director of this year's Canada Math Camp. I'm delighted to team up with Rachel Mullen. We'll be your coaches this week. Welcome to Ottawa."

Marlene asked us to introduce ourselves, and share something about us that was unique and interesting. She began by telling us that she was a high school math teacher, working part-time on a Masters degree in Education. She told us that as a 55-year-old, she was by far the oldest person in her program, and was shocked to discover her thesis supervisor was the same age as her youngest daughter.

The three boys to Marlene's left introduced themselves as Ryan, David, and Ric, three Grade 10s who went to different high schools in Toronto but were best friends since kindergarten. Ric said that the half-Jamaican half-Chinese Ryan was nicknamed "Bruce LeRoy" and referred to David Hartman as "David Fartman". Marlene cringed and shook her head, while I noticed Rachel trying hard to conceal a smile.

"Is Ric a short form for Richard?" asked Marlene. "Is that your real name?"

With that, Ryan and David started laughing out loud, and it was Ric's turn to be embarrassed. While Ric turned red, Ryan began air-strumming an imaginary guitar and was joined by David as they started singing the lyrics of a Johnny Cash song I had heard before.

Ric's face flushed. After a short pause, with the other two still smirking, Ric spoke up: "My Grade 1 teacher named me Ric, short for Ricochet. She thought I always bounced around from place to place. She said I never stopped moving."

"Or stopped talking," added Ryan.

David spoke up. "His Korean name is Soo, spelled S-O-O. Since he hates that name, we call him Ric. But when Ric gets on our nerves, we just remind him that he's a *Boy Named Soo*."

Marlene laughed. She moved on to the next person in line.

I tensed up.

He was tall and slim, with long brown hair, big brown eyes, light freckles on his pale face, and a cute dimple.

"I'm Cooper Robertson. I'm in Grade 10 at Lakewood Academy in Hamilton, Ontario. My favourite sport is basketball, and I played on the varsity team for Lakewood this year."

"That's amazing," said Marlene. "You're in Grade 10, and you're already on the varsity team?"

"Yeah," replied Cooper with a slight shrug.

I wasn't paying attention to the next half-dozen people who introduced themselves, lost in my thoughts. I caught a few words but kept glancing over at Cooper. Someone mentioned a love for jazz piano, another for photography. Two people were from Quebec, one was from somewhere in the Prairies, and the rest were from Ontario or Alberta. Half were in Grade 9, the other half were in Grade 10. Most went to public schools, but a few went to private schools and one was homeschooled.

The next person to speak was Grace . . . then me.

I didn't know what to say about myself that was "unique and interesting". I went to school, did Math Club twice a week, and learned from Mr. Collins on Saturdays. I hung out with Bonnie and Breanna and would often go to the mall with them. But most of the time, I just came home from school and ate dinner with Mom, and then I'd go to my room and read. Once in a while, Mom and I would watch a movie on Netflix after I finished my homework. My life was neither unique nor interesting.

My mind was a complete blank.

"I'm Grace Wong," said Grace in her loud perky voice. "I'm a Grade 9 student at the Vancouver Independent School. I love singing, and I lead worship at my church every Sunday."

"Grace, that's wonderful. We have a talent show on Thursday night. I hope you'll sing something for us."

"Yeah, totally."

"Thank you, Grace," said Marlene. "Okay, moving on to Bethany."

I nodded. I noticed Cooper looking at me. Without realizing it, I was staring back at him. I opened my mouth to speak but no words came out. People started fidgeting. I turned red in embarrassment.

"Bethany?" prompted Marlene.

"Sorry," I said, composing myself. I looked away from Cooper, my mouth dry all of a sudden. "My name is Bethany MacDonald."

"Where do you go to school?"

"Pinecrest Junior High."

"Thank you, Bethany. It's hard to hear you, so please speak up. Bethany, where is Pinecrest Junior High?"

"In Sydney," I said, a bit louder, suddenly self-conscious of my deep voice. But unlike at school, no one was laughing at me. "That's in Cape Breton, not in Australia."

I could hear Grace chuckle.

"Tell us something interesting about yourself," said Marlene.

"I'm not sure what to say."

"How about your hobbies? Are you an avid photographer like Dominic? Or a world traveler like Raju?"

Marlene gave me an idea of what to share. It was certainly something unique to this group. I looked towards Marlene and Rachel.

"Today was my first time on a plane. Today was the first time I've ever left Nova Scotia."

"Thank you for sharing, Bethany," said Marlene, smiling and moving to the next person in the circle.

As people introduced themselves, I felt more and more out of place, as a small-town girl who had never travelled, living in an apartment with a single mother who hadn't been on a plane since I was born. I felt a bit jealous hearing the backgrounds of some of the campers, who had lived in different countries, exploring the world with their families.

I discreetly glanced at Cooper, and was surprised to find him staring at me.

I quickly looked away, and was relieved when Rachel began to speak.

"To reiterate what Marlene said earlier, I'm thrilled that all of you could be here this week. The twenty of you come from all over Canada, and nearly every province is represented. Some of you have recently moved to Canada, while some of you may be ninth- or tenth-generation Canadians. Though you may look different, and have different experiences and backgrounds, all of you have a passion for math. What Marlene and I want to do is take that passion and make it shine even brighter.

"The purpose of this camp is more than teaching you new mathematics that you can use to succeed in contests. As far as we're concerned, the real value of this camp is bringing you together, so that you can learn from one another and form a community of young scholars. Many of you are isolated in your schools, being the best math student in your city or province, without anyone to push you or challenge you. Bringing all of you together in a setting like this gives you a chance to meet other like-minded people, and become life-long friends with those who share common interests and passions.

"I can say this from personal experience. Six years ago, just after I finished Grade 9, I got invited to the Canada Math Camp. I remember how scared I was, as a small-town girl flying to Ottawa."

Rachel turned towards me and smiled.

"In fact, that was the first day I'd ever left Manitoba."

I smiled back.

"I could say a lot more but I know many of you travelled long distances to get here. So you have free time for the rest of the evening. Please be at breakfast by eight tomorrow. Thanks, everyone."

I followed the crowd as they walked into the residence. Many people went into the lounge on the first floor, to play Ping-Pong or cards, or watch a hockey game playing on the big screen. I was feeling tired and decided to go upstairs.

"Bethany, wait for me," said Grace. "I'm coming up too."

Grace's dorm room was right next to mine. She looked around and noticed that several boys were close-by, heading into their rooms. She looked at me coyly, and dropped her voice to a whisper.

"I noticed someone staring in our direction during the introductions. He was looking at you the whole time. A couple of times, I'd stare back and he'd move his eyes away. But it was pretty obvious; he couldn't help himself. You know what that means?"

"No," I whispered back.

"It means he likes you!"

No boy had ever shown interest in me. I could feel a lump in my throat.

"Who was it?" I asked, trying to act as calm as possible.

Grace looked around. No one was within earshot. She looked at me and grinned, flashing her braces.

"The really cute tall guy. Cooper."

# 12

We walked into the seminar room, stuffed from breakfast.

The classroom was U-shaped, so that we could all see the speaker at the front while also being able to see one another.

Grace and I took two seats on the side of the room away from the door. Moments later, several people sat down beside us, with Ryan and Ric taking up the two seats on my immediate left.

Grace flashed her braces and gave me a knowing wink.

I quickly turned my head to the left and saw Cooper glancing in my direction. We locked eyes for a brief moment. He looked away and pretended to stare at his binder.

I tried to play it cool but my stomach began to churn.

*Calm down, Bethany.*

Because Cooper was on the left, and the chalkboard was on the right, I could turn my chair and not be distracted.

By nine o'clock, all of us had filed in and taken our seats. I noticed Marlene was sitting right between Jean-Philippe (J.P.) and Dominic, the two francophone students. Others were chatting or whispering amongst themselves, while several remained still and silently took out their paper and pencils in anticipation of the morning lecture.

One student in the back, Albert Suzuki, had an intense look in his eyes and looked especially intimidating. I casually said hello to him lining up at breakfast but he just ignored me. I saw him eating by himself at the far end of the cafeteria reading a thick textbook. Many of the boys were pointing to him and whispering.

And I thought some of the boys at Pinecrest were strange.

Rachel finished setting up her notes at the front of the class. She grabbed a stack of paper and asked each person to take a copy.

"Okay, everyone. Here is the first problem of the Canada Math Camp."

> Weighing the baby at the clinic was a problem.
> The baby would not keep still and caused the scales to wobble.
> So I held the baby and stood on the scales while the nurse read off 140 pounds.

> Then the nurse held the baby while I read off 120 pounds.
> Finally I held the nurse while the baby read off 240 pounds.
> What is the combined weight of the nurse, the baby, and me?

Several students started giggling, including Grace and me. Most word problems from math class were bland. But this one was cute.

"Yes, the wording is silly," said Rachel. "Let's translate this word problem into the language of mathematics, a language you're all familiar with. Let $b$ be the weight of the baby, $n$ the weight of the nurse, and $m$ the weight of me. Then we have three equations."

She picked up a piece of chalk and wrote three equations on the board. The question asked us to find the combined weight of the three people.

$$m + b = 140$$
$$n + b = 120$$
$$m + n = 240$$
Find the value of $m + b + n$

"Take some time now and solve this question on your own. When you are done, sit quietly and wait for the others to finish. We'll discuss the solution when everyone in the room has gotten the answer."

I picked up my pen and starting writing down the three equations. Before I had even finished writing the second equation, I heard a loud sound to my left. I looked up.

"I'm done," said Albert. "It's trivial."

*Done already?*

Rachel glared at him. "Albert, I asked that you sit quietly when you are done. It's not a race."

I looked down at my sheet of paper, and finished writing down the three equations. Flustered that Albert managed to solve the problem so quickly, I reminded myself of Rachel's words. This was not a race. I was going to proceed methodically, making sure I didn't make a mistake.

This was a standard problem I had seen many times before. We had a system of three equations in terms of three unknown variables, $m$, $n$, and $b$.

The first step was to eliminate one of the variables to reduce it to a system of two equations and two unknowns.

Subtracting the second equation ($n + b = 120$) from the first equation ($m + b = 140$), I was able to eliminate the variable $b$. This gave me the new equation $m - n = 140 - 120$, which was equivalent to $m - n = 20$.

I could now combine this equation $m - n = 20$ with the third equation on the board, $m + n = 240$, reducing my problem to two equations and two unknowns.

I knew this could be answered using a method known as *solving by substitution*, where we solve an equation for one of the variables by expressing $m$ in terms of $n$, and then substitute this expression into the other equation to solve for $n$.

The equation $m - n = 20$ could be written as $m = n + 20$. Substituting this expression into the other equation, I got $m + n = (n + 20) + n = 240$, which was equivalent to $2n + 20 = 240$. From there, I determined that $2n = 220$.

Dividing both sides of the equation by two, I saw that $n = 110$. So the nurse weighed 110 pounds.

I could feel Grace looking over my shoulder, and I pretended not to notice her.

Now that I had calculated the nurse's weight, how could I determine Rachel's weight and the baby's weight? I saw that I had written down $m = n + 20$ just a few lines above. That meant $m = 130$, implying that Rachel weighed 130 pounds.

I had solved two of the three variables. All I needed to do was determine the baby's weight. From the first equation I had written down, $n + b = 120$, I knew the nurse's weight added to the baby's weight was 120 pounds. Since the nurse weighed 110 pounds, the baby had to weigh 10 pounds.

Therefore, I could conclude that $b = 10$. Just to double-check my answer, I substituted my three values back into the original equations to make sure they were all correct.

$$m + b = 130 + 10 = 140$$
$$n + b = 110 + 10 = 120$$
$$m + n = 130 + 110 = 240$$

Since the problem asked for the value of $m + n + b$, the answer was simply the sum of the three weights, which was 130 + 110 + 10 = 250. I was done.

I smiled and looked up.

To my horror, most of the students were staring at me. Their pens were on the table, and I could tell they were waiting for me to finish.

"She's finally done!" exclaimed Albert.

"That's enough, Albert!" snapped Marlene, from the other end of the room.

My face flushed. Was I the last one to finish?

I couldn't bring myself to look to my left, in case Cooper was staring at me.

Rachel looked horrified. She took a deep breath and glanced in my direction, and then towards Marlene.

"I'm sorry everyone; that was my fault. My intention was only to illustrate how this problem can be solved extremely quickly in one way, while standard approaches take much longer. Raju, I noticed how you answered this question. Please come down and share your solution with us."

Raju Gupta, a Grade 9 student from Ottawa, came down to the board and took Rachel's place at the front of the classroom. I stared blankly at the four lines that Rachel had written earlier.

$$m + b = 140$$
$$n + b = 120$$
$$m + n = 240$$

Find the value of $m + b + n$

Raju grabbed a piece of chalk and turned towards us.

"Just add the three equations, then divide by two. The answer pops right out."

On the chalkboard, Raju added three lines.

$$(m + b) + (n + b) + (m + n) = 140 + 120 + 240$$
$$2m + 2b + 2n = 500$$
$$m + b + n = 250$$

As Raju sat back down, a few people clapped. My jaw dropped, amazed at how he made the answer "pop right out".

"Very good," said Rachel. "Look at what Raju did. He found an elegant and simple solution by adding the three equations, instead of solving each variable one by one. Raju, why did you tackle the problem this way?"

"Well," said Raju, "the question didn't ask for the weight of each person, but the total sum. I realized you could get double this total, $2m + 2b + 2n$, by just adding the three equations. So that's what I did."

"Right. But why does your solution work?"

"What do you mean?"

"Adding the three equations. How did you know this would give you $2m + 2b + 2n$, twice the total weight?"

"I'm sorry, I don't understand the question," said Raju.

"Because the system of equations is symmetric," said Albert.

"That's right," said Rachel.

I was completely lost.

I glanced at the other students in the classroom. Some appeared riveted by the discussion, while others were staring off into space. It wasn't clear to me whether they were bored or confused, uninterested or overwhelmed.

"What Raju and Albert did was exploit the *symmetry* in the three equations, as they realized that each pair of weights was represented exactly once. Instead of solving each variable separately, they realized that by adding up the three equations, they could get the answer directly. Often we can exploit the symmetry within a mathematical problem to derive a clean and elegant solution. Just out of curiosity, how many of you solved the problem this way?"

I looked around the room. Six people raised their hands, including Cooper.

I felt somewhat relieved that only six of the twenty students saw the clever insight, and was thankful it wasn't everybody. I would have never thought of adding up the three equations like that.

My ability to find time-saving shortcuts enabled me to be successful on math contests, but I was sure that the nineteen other students were out of my league, and that all of them were much smarter.

I suddenly felt an uncomfortable mixture of dread and anxiety, two feelings that I hadn't encountered in a math classroom for years. How could I be so sure of myself at Pinecrest and feel such confidence during my weekly sessions with Mr. Collins, but feel none of that here at this camp?

Grace was glancing over at me.

"Here are three new problems," said Rachel. "Please work on these problems for the next ten to fifteen minutes, and I'll have volunteers present solutions. If you'd like to work in pairs, you're most welcome to do so."

Grace turned to me. "Bethany, can we work together?"

"You bet," I said instantly.

"Great," she said, looking happy. "I prefer working in pairs. It's far less stressful."

"Tell me about it."

# 13

"We can do this," said Grace.

I nodded silently, trying to convince myself that I indeed could. We stared at the three problems on the sheet.

> 1. Take seven coins and toss them into the air. Count the number of coins that turn up heads. What is the probability that the number of heads is even?
>
> 2. The Toronto Blue Jays and the Houston Astros are playing in the World Series, a best-of-seven baseball tournament where the winner is the first team to win four games. Assuming the teams are evenly matched, is it more likely for the World Series to end after six games, or end after seven games?
>
> 3. Solve the following system of equations: $\begin{cases} x^2 = 2y - 1 \\ y^2 = 2x - 1 \end{cases}$

We took a look at the information given in Problem #1: seven coins tossed into the air, with each coin having an equal chance of landing heads or tails. We wanted to find the probability that the total number of heads is even.

"Any ideas?" asked Grace.

I took a deep breath. "Simplify the problem by considering a smaller odd case? Maybe we can pretend there are three coins or five coins, and see if we can find a pattern?"

"Sounds great," said Grace. "For three coins, there are eight cases, since $2 \times 2 \times 2 = 8$."

"I agree," I said, grabbing a fresh sheet of paper from my notepad.

I began making a table, listing out the eight possible cases and marking the three columns with the headings "1st Coin", "2nd Coin", and "3rd Coin".

| 1st Coin | 2nd Coin | 3rd Coin |
|---|---|---|
| H | H | H |
| H | H | T |
| H | T | H |
| H | T | T |
| T | H | H |
| T | H | T |
| T | T | H |
| T | T | T |

"Can I add something?" asked Grace. I nodded and slid the paper towards her.

Grace added a fourth column, "Total Heads", and wrote down the number of times heads came up in each of the eight cases. We looked at our sheet together.

| 1st Coin | 2nd Coin | 3rd Coin | Total Heads |
|---|---|---|---|
| H | H | H | 3 |
| **H** | **H** | **T** | **2** |
| **H** | **T** | **H** | **2** |
| H | T | T | 1 |
| **T** | **H** | **H** | **2** |
| T | H | T | 1 |
| T | T | H | 1 |
| **T** | **T** | **T** | **0** |

"One, two, three, four," I said, counting the rows with an even number of heads. "For three coins, the probability is 4/8, or one-half. That's the answer for three coins."

"Awesome. Here, let's do another case. Let's do four coins," she said.

"Can we try five coins?" I asked. "Just in case there's something to do with seven being odd, I think it makes sense to try three coins and five coins, to see the pattern."

"Yeah, I agree."

Grace took out a fresh sheet of paper and started writing down the possible cases. We both knew that there would be thirty-two cases, since $2\times2\times2\times2\times2 = 32$. All thirty-two cases could occur with equal probability.

I marked the cases that resulted in an even number of heads.

| HHHHH | **HHHHT** | **HHHTH** | HHHTT | **HHTHH** | HHTHT | HHTTH | **HHTTT** |
|---|---|---|---|---|---|---|---|
| **HTHHH** | HTHHT | HTHTH | **HTHTT** | HTTHH | **HTTHT** | **HTTTH** | HTTTT |
| **THHHH** | THHHT | THHTH | **THHTT** | THTHH | **THTHT** | **THTTH** | THTTT |
| TTHHH | **TTHHT** | **TTHTH** | TTHTT | **TTTHH** | TTTHT | TTTTH | **TTTTT** |

"I got sixteen," I said. "So the answer is 16/32, or one-half. It's one-half again."

"Hey, look!" said Grace pointing to the sheet. "Look at each of the columns. Of the four rows in any column, exactly two have an even number of heads."

"Oh yeah, I see it too! And also, it's always the first and the last, or the second and the third. But why?"

We stared at the paper, trying to explain the pattern.

"Hey, I got it!" said Grace. "Look at each column. The last three flips are always the same in each column. Each entry in the first column ends with HHH, each entry in the second column ends with HHT, and so on. For example, look at the second column: HH+HHT, HT+HHT, TH+HHT, and TT+HHT."

"Yeah," I said slowly, not really catching on.

"Well, the last three flips are always the same in each column, but what's different are the first two flips. If the last three flips have an even number of heads, then sticking HH or TT at the beginning keeps the total number of heads even since HH adds two and TT adds zero."

"You lost me."

"Say the last three flips are HHT," said Grace. "That's two heads right there, an even number. So if we stick HH or TT at the beginning, we're adding an even number of heads, so that keeps the total number of heads even. Check it out. HH+HHT is four heads, TT+HHT is two heads. Both are even."

| First Two | Last Three | # of Heads |
|---|---|---|
| **HH** | **HHT** | **Even** |
| HT | HHT | Odd |
| TH | HHT | Odd |
| **TT** | **HHT** | **Even** |

"I get it," I said. "Anytime the last three flips have an even number of heads, then sticking HH or TT at the beginning keeps this total even, while sticking HT or TH at the beginning makes this total odd."

"Exactly," said Grace. "And if the last three flips have an odd number of heads, then sticking HT or TH at the beginning makes it even, since HT and TH both add one head."

She pointed to the first column, which contained the entries HH+HHH, HT+HHH, TH+HHH, and TT+HHH. The number of heads were 5, 4, 4, and 3, respectively.

| First Two | Last Three | # of Heads |
|---|---|---|
| HH | HHH | Odd |
| **HT** | **HHH** | **Even** |
| **TH** | **HHH** | **Even** |
| TT | HHH | Odd |

"I get it!" I said. "So that proves each column must have two rows with an even number of heads. Either we have HH and TT, the first and fourth rows, or we have HT and TH, the second and third rows."

| | | | | | | | |
|---|---|---|---|---|---|---|---|
| HHHHH | **HHHHT** | **HHHTH** | HHHTT | **HHTHH** | HHTHT | HHTTH | **HHTTT** |
| **HTHHH** | HTHHT | HTHTH | **HTHTT** | HTTHH | **HTTHT** | **HTTTH** | HTTTT |
| **THHHH** | THHHT | THHTH | **THHTT** | THTHH | **THTHT** | **THTTH** | THTTT |
| TTHHH | **TTHHT** | **TTHTH** | TTHTT | **TTTHH** | TTTHT | TTTTH | **TTTTT** |

"Awesome!" said Grace. "We've shown the answer to thc five coin case is one-half, since exactly half the rows in each column have an even number of heads. What about seven coins?"

"I have an idea," I said, reminded of a technique that Mr. Collins taught me last year.

I took out a fresh sheet of paper, and tried to recall Mr. Collins' explanation of "mathematical induction", analyzing a scenario by building upon the results of previous scenarios. In this problem, I was using my knowledge of the result for $n = k$ (five coins) to determine the result for $n = k + 2$ (seven coins).

The calculations for seven coins were a bit tedious, since there were $2^7$ = 128 cases to consider in total. But I was able to show Grace why there had to be 64 cases with an even number of heads, and 64 cases with an odd number of heads. Once again, the correct probability was one-half.

We gave each other a high-five. "Yes!"

A few people turned to look at us. Albert glared back.

I felt a surge of confidence and was eager to tackle the second problem with Grace.

> 2. The Toronto Blue Jays and the Houston Astros are playing in the World Series, a best-of-seven baseball tournament where the winner is the first team to win four games. Assuming the teams are evenly matched, is it more likely for the World Series to end after six games, or end after seven games?

"It's like the first problem, isn't it?" I asked. "You know, evenly matched teams? Isn't that a coin flip?"

"For sure," said Grace, pausing slightly. All of a sudden she saw something. "Ah, Rachel is so clever. Yes, they are coin flips. Check out the team names. Houston and Toronto. H and T. *Heads* and *Tails*!"

"Brilliant," I said. "So the problem can be modeled by flipping coins one at a time. We stop as soon as we've flipped four heads, or flipped four tails."

Grace held up the sheet of paper we had created together, for the five coin case.

| HHHHH | HHHHT | HHHTH | HHHTT | HHTHH | HHTHT | HHTTH | HHTTT |
|---|---|---|---|---|---|---|---|
| HTHHH | HTHHT | HTHTH | HTHTT | HTTHH | HTTHT | HTTTH | HTTTT |
| THHHH | THHHT | THHTH | THHTT | THTHH | THTHT | THTTH | THTTT |
| TTHHH | TTHHT | TTHTH | TTHTT | TTTHH | TTTHT | TTTTH | TTTTT |

"Let's use this! We have 32 scenarios here – these are all the possibilities after five games have been played. These are all the possible sequences of coin flips. I know the World Series is best-of-seven, not best-of-five, but this gives us a good start."

Grace pointed to an arbitrary entry, HTHHH. "In this case, after five games, the series is over, and Houston has won four games to one."

I pointed to the first entry, HHHHH. "In this case, after four games, Houston's already won four times, so there's no need to play the fifth game."

Grace pointed to another entry, THTHT. "In this case, after five games, the series is not over, and Toronto leads three games to two."

Eventually we came up with the following table:

| World Series Scenario | # of Cases | Example |
|---|---|---|
| Houston Wins in Four | 2 | HHHHT |
| Houston Wins in Five | 4 | HHHTH |
| Toronto Wins in Four | 2 | TTTTT |
| Toronto Wins in Five | 4 | THTTT |
| Houston Leads 3-2 | 10 | HHHTT |
| Toronto Leads 3-2 | 10 | THTHT |

We double-checked that there were 32 cases in total. Yes, 2+4+2+4+10+10=32.

There was a probability of 10/32 that Houston would lead after five games, and a probability of 10/32 that Toronto would lead after five games. Also, there was a probability of (2+4+2+4)/32 = 12/32 that the World Series would be over after five games.

Rachel got our attention. "Three more minutes everyone. Three more minutes."

I looked at Grace. "We can do this."

"What's the question again?" asked Grace.

"Is it more likely for the World Series to end after six games, or end after seven games?"

"I have an idea," said Grace.

Grace explained that after five games, we have one of three scenarios: either the World Series is already over, or Houston leads 3-2, or Toronto leads 3-2.

"Say Houston is leading 3-2. If Houston wins that game, then it's 4-2, and so the World Series ends after six games. But if Toronto wins that game, then it's tied 3-3, and so the World Series must end after seven games. There's a 1/2 chance that either team will win that sixth game."

"I get it," I replied. "It doesn't matter who wins that seventh game. All that matters is whether the Series ends after six games or seven. The same is true if Toronto is leading 3-2. If Toronto wins that game, then it's over after six. Otherwise, Houston wins, and it's over after seven."

I grabbed a pen and made a few simple calculations:

| **Length** | **Probability** |
|---|---|
| 4 or 5 Games | $12/32 \times 1 = \mathbf{12/32}$ |
| 6 Games | $10/32 \times 1/2 + 10/32 \times 1/2 = \mathbf{10/32}$ |
| 7 Games | $10/32 \times 1/2 + 10/32 \times 1/2 = \mathbf{10/32}$ |

"We're done!" I said. "It's equally likely for the World Series to end in six games or in seven games."

"Yup," said Grace, giving me another high-five. "The probability is 10/32 in both cases. We got it!"

"Okay, time's up," said Rachel. "Let's look at the three questions."

Grace and I smiled at each other.

"J.P.," said Rachel, pointing towards one of the students from Quebec. "I noticed you had a nice solution to Question #1. Please share it with us."

J.P. stood up and walked down to the front. I looked over at Rachel who turned her head towards me. She looked into my eyes and silently mouthed three words.

"Way to go."

# 14

"Let's see how you all solved Question #1," said Rachel. "There are at least five different solutions."

I looked up at Rachel in surprise, and reminded myself of the problem.

> 1. Take seven coins and toss them into the air. Count the number of coins that turn up heads. What is the probability that the number of heads is even?

J.P. picked up a piece of chalk and scribbled two rows of letters right in the centre of the blackboard.

1st : HTHTTHT
2nd : THTHHTH

"If you write down all possible sequences of seven coin flips, you can always pair them up like this."

"Pair them up like how?" asked Rachel.

"Letter by letter," responded J.P., pointing to the first row of letters, then down to the second row. "Each time there's an H, write the letter T just below. And whenever there's a T, write down H."

"Good," replied Rachel, looking at all of us to make sure we understood. "So for any sequence of flips, what you did was find another sequence of flips whose letters are different in each position. Then you matched up those two sequences. Right?"

"Yes," he replied.

"So if the first sequence is HHHTTTT, then what's its pair?"

"TTTHHHH."

"And if the first sequence is TTTHHHH, then what's its pair?"

"HHHTTTT."

"Exactly. Now that we all understand the argument, please continue."

J.P. pointed back to the board.

1st : HTHTTHT
2nd: THTHHTH

"Each pair of seven-letter sequences contains fourteen letters in total. Based on how we did our pairing, exactly seven letters have to be H and seven letters have to be T. So if the first row has an even number of heads, then the second row must have an odd number of heads, since the total number of heads is seven. And also, if the first row has an odd number of heads, then the second row must have an even number of heads."

"Very nice," said Rachel. "So what does that mean?"

"There are $2^7 = 128$ possible sequences, and so there are 64 pairs. In every pair, exactly one of the two sequences has an even number of heads. Since there are 64 sequences with an even number of heads, the answer is 64/128, which is one-half."

"Excellent," said Rachel. "How many of you understood that?"

All hands went up, including mine. I liked how J.P.'s solution involved pairing up sequences, which reminded me of the Staircase.

"Okay, that's one solution. Any others?"

Grace raised her hand. "Bethany and I came up with a different proof."

Rachel nodded. "Excellent. Which of you wants to present?"

Grace turned to me. "It was your idea. Go ahead."

I shook my head. "Can you do it?"

"You sure?"

"Yeah, definitely."

"Okay," said Grace, shrugging and walking up to the board.

Grace clearly explained our inductive solution, showing how to get the answer from three coins to five coins to seven coins. Everyone clapped when she was done.

"Amazing, Grace," said Rachel.

She then chuckled to herself. I got the joke and started laughing too.

David raised his hand. "I've got another solution."

"Great. Please come on down."

"No need," said David, remaining in his seat. "I can do it from right here."

"Go ahead."

"I figured it out by actually flipping seven coins from my wallet," began David, pointing to a line of seven coins he had placed in front of him. "The evenness and oddness of the number of heads depends only on the *last* flip."

"How so?" asked Rachel.

"Well, the first six flips are irrelevant; they don't matter at all. It all comes down to the last coin. Whether that last coin lands heads or tails, exactly one of those two possibilities will lead to the final head count being even. The final answer must be one-half."

"Excellent," said Rachel. "Do you all see how David solved that?"

Most of us nodded, me included.

"Is his solution identical to mine?" asked J.P..

"Not quite, but it's close. He paired the sequences slightly differently, but it's essentially the same idea. In your solution, HHHTTTT was paired with TTTHHHH, by flipping *each* letter. In David's solution, only the *last* letter was flipped. So HHHTTTT was paired with HHHTTTH. In both his solution and yours, exactly one sequence from each pair has an even number of heads, so that's why the final answer has to be one-half."

As J.P. nodded in agreement, Rachel looked around. "Anybody else?"

Cooper raised his hand. He stood up and walked down to the front. I stared at Cooper, but in a way I hoped wasn't too obvious. I glanced down at his pale legs below his basketball shorts, staring at his calf muscles.

When I looked up, I noticed that Cooper had written two equations on the board.

P(# of Heads is even) = P(# of Tails is even)
P(# of Heads is even) = P(# of Tails is odd)

Cooper pointed to the first equation. "By symmetry, the probability of getting an even number of heads must be the same as the probability of getting an even number of tails. It's a fair coin, so this must be the case."

He pointed to the second equation. "But there are seven coins in total, which is an odd number. If there is an $x$% chance of getting an even number of heads, then by definition, there has to be an $x$% chance of getting an odd number of tails."

I nodded, completely following Cooper's logic. Since there were seven coins, whenever there were an even number of heads, there were an odd number of tails.

Cooper continued: "You just add up the two equations. Check it out."

$$2 \times P(\# \text{ of Heads is even}) = P(\# \text{ of Tails is even}) + P(\# \text{ of Tails is odd})$$
$$2 \times P(\# \text{ of Heads is even}) = 100\%$$
$$P(\# \text{ of Heads is even}) = 50\%$$

"The right side has to be one hundred percent, since in any sequence the number of tails is either even or odd. The left side is double the answer. Therefore, the probability the number of heads is even must be exactly fifty percent."

All of us applauded. Ric shook his head in amazement.

"Dude, that's so dope."

Rachel ignored the comment. "Outstanding solution, Cooper. Now suppose instead of seven coins you had 999 coins. What's your answer then?"

"It's still fifty percent, since 999 is an odd number. It's the exact same solution."

"Correct."

Cooper walked back towards his seat. Ric leaned over and gave him a fist bump.

"That was sick, bro."

"Shut up, Ric," interjected Ryan, shaking his head in disgust. "You always say that, like you're some brotha' from the 'hood. Give us a break. You're a rich Asian kid from Rosedale."

I let out a loud cackle.

Cooper turned towards me and smiled. Our eyes met.

I felt a tingle down my spine.

"Well, we've got four solutions," said Rachel. "I really like Cooper's solution since it generalizes, and works for any odd number of coins, not just for seven coins. Okay, any other solutions?"

Albert raised his hand. "I've got two more."

Rachel forced a smile. "Okay, Albert, please show us how you did it. Just give us one of your solutions."

"Let $n$ be the total number of flipped coins. I claim the answer is one-half, regardless of whether $n$ is even or odd."

With that, he proceeded to write down four lines on the board. Surprisingly, his handwriting was neat.

Let $e_n$ be the probability that if you flip $n$ coins, the # of heads is even.
Let $o_n$ be the probability that if you flip $n$ coins, the # of heads is odd.
By definition, $e_1 = o_1 = \frac{1}{2}$ and $e_{n-1} + o_{n-1} = 1$.
Then $e_n = e_{n-1}e_1 + o_{n-1}o_1 = \frac{e_{n-1}}{2} + \frac{o_{n-1}}{2} = \frac{e_{n-1}+o_{n-1}}{2} = \frac{1}{2}$.

Albert then stepped back to let the audience admire his handiwork.

I had no idea what he had just done. I could tell from the other faces that many, if not most, were equally lost.

"Can you take us through this?" asked Rachel.

"It's self-explanatory."

"It's not self-explanatory. Walk us through your solution."

"What part do you not understand?" said Albert, lifting his chin up high. "Do I need to explain the words *even* and *odd*?"

"No," said Rachel, clearly irritated. "I understand your solution and I think it's brilliant, but this camp isn't for me – it's for the twenty of you. Not everyone sees things as quickly as you do, so for the benefit of everyone, let's go through this line by line. Could you explain it for us?"

"There's nothing to explain."

"Fine," she said, tight-lipped. "Have a seat, Albert. I'll do it."

Rachel pointed to Albert's solution.

"I think we're all okay with the first two lines, since Albert is just defining $e_n$ and $o_n$, and we want to solve for $e_n$. His third line is also clear. If you flip one coin, the probability that it lands heads is one-half, which is the same as the probability that it lands tails, so $e_1 = o_1 = \frac{1}{2}$. And by Cooper's argument from before, if you flip any number of coins, such as $n - 1$ coins, the total number of heads must either be even or odd, but not both. That's why $e_{n-1} + o_{n-1} = 1$. Does that make sense?"

We all nodded.

Rachel explained Albert's insight and explained why it was a more general version of David's solution. The key was to look at the last coin. After $n - 1$ flips, if the number of heads is even (which has probability $e_{n-1}$), then this number stays even provided the last flip is a tail (which has probability $e_1$). And similarly, if the number of heads is odd (which has probability $o_{n-1}$), then this number becomes even provided the last flip is a head (which has probability $o_1$).

I stared at Rachel, completely lost.

"That's why $e_n = e_{n-1}e_1 + o_{n-1}o_1$," continued Rachel. "Albert set up a *recursion*, more formally known as a recurrence relation, from which the variable $e_n$ could be expressed as a function of the variables $e_{n-1}$ and $o_{n-1}$. From there, he got the answer."

A couple of people politely clapped.

"Nice solution," said Raju, clapping respectfully.

Albert ignored him.

Just then, I noticed one of the undergraduate student helpers come in, rolling a cart stacked with trays of snacks and juice.

Rachel turned to face us. "Okay folks, let's take a five-minute break. Come on up and grab some cookies and drinks, and we'll go over the next solution in a bit."

I walked over to the cart and picked up a cookie and a small box of orange juice, and brought it back to my desk.

My head hurt. But in a good way.

# 15

I tapped Grace on the shoulder. “Just going to the washroom. Be right back.”

“Wait for me, I’ll come too!”

Grace and I walked into the washroom, and entered adjacent stalls.

“Wow, what a morning, eh?” said Grace, her perky voice carrying from her stall into mine.

“Crazy.”

I couldn’t believe how clever those boys were. Grace and I came up with one solution to that seven-coin problem, but I had no idea that the same problem could be solved in four other ways, all of them using different mathematical techniques.

Flushing the toilet, I walked to the sink. Grace had just begun washing her hands. She looked over.

“Cooper’s really cute, isn’t he?”

I tensed up. “I guess so.”

“Do you want me to find out if he’s seeing someone?”

“No!” I shouted. “Please don’t do that.”

“No problem,” said Grace, laughing.

I followed Grace back to the classroom, consumed by a hundred thoughts swirling around my head. I purposely chose not to look in Cooper’s direction, and finished my cookie and juice in silence.

“Okay,” said Rachel loudly. “Let’s present Question #2. Who’s got it?”

To remind myself, I looked down at the second problem. Oh yeah, I definitely remembered this one.

> 2. The Toronto Blue Jays and the Houston Astros are playing in the World Series, a best-of-seven baseball tournament where the winner is the first team to win four games. Assuming the teams are evenly matched, is it more likely for the World Series to end after six games, or end after seven games?

Ryan raised his hand. “I got it.”

Ryan came down and presented his solution. His explanation was virtually identical to ours, and he showed that the six-game and seven-game scenarios were equally likely.

"Nicely done, Ryan. Anyone else solve it that way?"

Grace and I raised our hands, as did a half-dozen others.

"Okay, any other solutions?" asked Rachel.

Raju spoke up. "I've got a short one."

He stood up from his seat, and turned to face us. "After five games, if the World Series is not over, some team must be leading three games to two."

Rachel interrupted him. "Do all agree with what he just said? Raise your hand if you agree with Raju."

A few people raised their hands immediately. I thought about it for a second. If the World Series is not over after five games, then neither team has won four games. That means no team can be leading five to zero, or four to one; it has to be three to two. I nodded, and raised my hand too.

"Thank you, Raju. Please continue."

"We agree that some team must be leading three games to two. By *symmetry*, it doesn't matter which team, since it's equally likely for either team to win any particular game. Let's assume Toronto is leading 3-2. If Toronto wins the sixth game, which occurs with probability one-half, then the World Series ends after six games. If Houston wins the sixth game, which also occurs with probability one-half, then the series is tied 3-3, and the World Series must then have a seventh game. This proves that both scenarios are equally likely."

"Very nice," I said quietly.

I was amazed that the problem could be solved that elegantly, using a simple symmetry argument without any calculations or computations.

Rachel then moved us along to Question #3. I looked down at my piece of paper, to read the other question from the morning activity.

3. Solve the following system of equations: $\begin{cases} x^2 = 2y - 1 \\ y^2 = 2x - 1 \end{cases}$

Ric came down and proceeded with a complicated proof that required two blackboards of writing equations. He solved the first equation for $y$, substituted that identity into the second equation to derive a messy identity in terms of $x$, eventually producing the equation $x^4 + 2x^2 - 8x + 5 = 0$ after a lot of effort. He then showed that the left side could be factored as $(x^2 + 2x + 5)(x - 1)^2$, and justified that this expression could only equal 0 when $x = 1$.

After ten minutes, he was still writing.

"Okay, I'm nearly done. Thank God. I've explained why $x = 1$. From the first equation, $x^2 = 2y - 1$, so that means $1^2 = 2y - 1$, which simplifies to $2 = 2y$, or $y = 1$. So the only solution is $x = 1$ and $y = 1$."

A bunch of us basked Ric with loud applause. He took a bow.

Of course, we all knew there was a better way to solve the problem.

Dominic raised his hand. "I have a different solution."

Rachel nodded. "Great, come show us."

Dominic pointed to the two equations in the problem. He had the clever insight of subtracting one equation from the other, deriving the identity $x^2 - y^2 = 2y - 2x$, which he factored to show that $(x - y)(x + y + 2) = 0$. He explained that whenever two terms multiplied to 0, at least one of those two terms had to be 0. This forced either $(x - y) = 0$ or $(x + y + 2) = 0$. From there, Dominic was able to solve the problem quite easily.

"Nice solution, Dominic," said Ric.

Cooper raised his hand. "I've got another one."

Cooper came down and grabbed a piece of chalk.

He began writing some equations on the board. After he concluded his explanation, I studied his solution and figured out what he did, step by step.

The two equations are $x^2 = 2y - 1$ and $y^2 = 2x - 1$.

If you add the two equations, we get $x^2 + y^2 = (2y - 1) + (2x - 1)$.

This is the same as $x^2 - (2x - 1) + y^2 - (2y - 1) = 0$.

And this becomes $(x^2 - 2x + 1) + (y^2 - 2y + 1) = 0$.

This is equivalent to $(x - 1)^2 + (y - 1)^2 = 0$.

Therefore, we have two perfect squares that add up to zero.

Since every square is at least zero, we must have $(x - 1)^2 = 0$ and $(y - 1)^2 = 0$, which implies that $x = 1$ and $y = 1$.

"Brilliant," said Rachel, starting the applause.

Cooper shrugged modestly and strolled back to his seat.

Rachel looked at the clock. I was surprised to discover that we were nearly done.

"Great job, everyone. All of you did exceptional work today, coming up with a diverse set of solutions to the three problems. Some solutions were long and complicated, while others were remarkably short and elegant. As all of you noticed, by recognizing and exploiting symmetry, you can come up with concise solutions to difficult problems, as Albert did on Question #1, Raju on Question #2, and Cooper on Question #3. Keep this in mind as you work on the Problem Set this afternoon. All three of these solutions are the best possible; you can't find more elegant solutions than this."

"That's not true."

We all turned to stare at Albert.

"What's not true?" asked Rachel.

"That Cooper's solution to Question #3 is the best possible."

"You have a better solution?" she challenged.

"I do."

"Could you show it to us?"

"No," he responded.

Rachel shook her head, and I could tell she was angry. There was an uncomfortable silence.

"Albert," said Marlene in a voice that left no room for compromise. "We've got a few minutes before lunch. Show us how you did it."

Albert shrugged, and walked down to the front, picking up a piece of chalk. As he did before, he just wrote down four lines.

$x^2 = 2y - 1$ implies $2y = x^2 + 1 = (x-1)^2 + 2x \geq 0 + 2x$, so $y \geq x$.
$y^2 = 2x - 1$ implies $2x = y^2 + 1 = (y-1)^2 + 2y \geq 0 + 2y$, so $x \geq y$.
Since $x \leq y$ and $x \geq y$, we must have $x = y$.
Therefore $x^2 = 2x - 1$ and $y^2 = 2y - 1$, and so $x = 1$ and $y = 1$.

We stared at the board in disbelief.

"That's really nice, Albert," admitted Rachel. "I like how you exploited the symmetry to show $x \leq y$ and $x \geq y$, which forces $x = y$. I never thought about it that way."

With his chin held high, Albert walked back to his seat.

I admired Albert's intellect, even if he was an arrogant jerk. I was stunned by the insight that explained why $x \leq y$ and $x \geq y$. I understood how that forced $x = y$, with the simple argument that if I'm at least as tall as you and you're at least as tall as me, then we must be the exact same height.

Though I completely followed his solution, there was no way I could have come up with an insight like that myself. How could anyone, let alone a teenager, come up with that?

Every year, only six students are chosen to represent Canada at the International Math Olympiad.

Just three hours into the opening day of the Canada Math Camp, I had reached two conclusions: Albert Suzuki would become a Math Olympian, and Bethany MacDonald would not.

I could never ever be that good.

No way.

# 16

"Where do you want to sit?" Grace asked.

I looked around. All of the tables could seat four or six, and most of the students had already sat down and begun eating. To my right, I saw that Cooper was sitting at a six-person table with Raju, J.P., and Dominic, with two seats empty. To my left, I saw Rachel and Marlene occupying a four-person table.

Grace pointed to Cooper's table. "Want to sit there?"

My face turned red.

"No, let's sit with Rachel and Marlene."

"Okay," said Grace, giggling. "No problem."

Marlene and Rachel saw us approaching and invited us to sit with them. At least with these two, I could relax and be myself. I chose the seat facing the wall so I wouldn't be distracted by Cooper.

"How was the morning?" asked Marlene.

"Tiring," I said, biting into my sandwich. "I'm feeling drained but good."

"Yeah, me too," added Grace. "I knew the camp would be intense, but I didn't expect this. I've never concentrated so hard in my life. And we still have five days to go!"

"It was definitely intense," responded Marlene with a nod. "By the way, both Rachel and I were so impressed by the work you two did this morning."

"Thanks," replied Grace. "I really liked the questions we worked on. You know, today was not at all like my math class at school."

"How so?" asked Marlene. "Take me through a typical math class at your school in Vancouver. What happens?"

"Well, I have this amazing math teacher named Mr. Cheung. We start the class with a fun riddle or logic puzzle. Then he checks the homework and reviews the material we covered in the previous class, to make sure we all understand. He gives a short lesson on some topic, and works through some examples on the board, and then we get a worksheet with a bunch of questions, which we're always allowed to do in groups. We spend the rest of the period doing the worksheet, and Mr. Cheung moves around the class to see how we're doing, helping anyone who's stuck. Just before the bell rings,

Mr. Cheung assigns homework, tells us a silly math joke that makes us all groan, and then we go off to the next class."

Marlene smiled. "Mr. Cheung sounds like a fabulous teacher. If you compare Mr. Cheung's class to what you experienced this morning, what was different?"

"We were the ones going up to the board and presenting the solutions."

"That's one difference," said Marlene. "What else?"

I spoke up. "We solved each problem in multiple ways. For the seven-coin problem, I think we had five different solutions."

"Great. What else?"

Grace chuckled. "The problems were a lot harder."

"That's true," said Marlene. "In most high school math classes, students don't work on challenging multi-step problems. Teachers often get students to memorize procedures and formulas, and apply them to solve routine exercises. For homework questions, you know exactly what technique to use because it's just a reinforcement of the concept you learned that day. That's how we've been teaching for decades. As a result, students become quite strong at procedures and techniques and calculations, but remain weak at open-ended problem-solving."

"Is Mr. Cheung a bad teacher?" asked Grace.

"Not at all. He sounds like a great teacher. He explains concepts in an engaging way and cares about his students. That's wonderful. But in his class, do you ever struggle with a problem?"

"What do you mean?"

"In a math class, have you ever spent an entire period struggling through a difficult problem, where you try dozens of different approaches but nothing seems to work?"

Grace shrugged. "Not really."

"And how about you, Bethany?"

I finished the last of my sandwich, and drained my chocolate milk. I looked up at Marlene.

"Not at school. But every Saturday, I meet with Mr. Collins, my math coach. We usually spend an entire hour working through a single problem."

"It's great you have such a resource, Bethany," added Marlene. "Mr. Cheung teaches many others in addition to Grace, and he's expected to cover the entire curriculum. Because the curriculum is so packed, it would be extremely challenging for him to spend an entire class period working through the mathematics in a single problem."

"What would you change?" asked Grace.

"I've always believed that the point of teaching isn't to cover the material, but to *uncover* the material, to have students discover how the different areas of mathematics are connected. This math camp is great because we're not bounded by having to teach a set number of topics this week, so we can focus on problem-solving. We have a lot more flexibility here than I do at my high school in Ottawa, where I teach hundreds of students. I believe that by teaching less, the students learn more."

"What do you mean?" I asked, bringing my chair in closer.

"Well, take a look at this morning," said Marlene. "Rachel gave out three questions that could be solved elegantly by exploiting symmetry: the seven-coin problem, the World Series problem, and the system of two equations. There was no prescriptive method to solve any of the three questions. She probably spoke for five minutes all morning. And look at the diversity of solutions that were presented.

"For the seven-coin problem, the two of you solved it *inductively* by figuring out the pattern for the five coin case and then building it up to seven. David solved the problem *kinesthetically*, flipping coins from his wallet and realizing that the even/odd outcome only depended on the last flip. J.P. solved the problem *visually*, where he first wrote down all the possible outcomes on a piece of paper, which then allowed him to discover the key idea that they could be paired up. Cooper solved the problem *logically*, exploiting the fact that seven was odd to show the answer had to equal half of one hundred percent. And Albert solved it *symbolically*, representing the desired answer as a function of $n$, and using the technique of recursion to come up with a four-line solution. Didn't you learn a lot by teaching each other?"

"Definitely," said Grace. "I had no idea that you could solve that problem in more than one way. And also I learned a lot by working with Bethany, trying these difficult problems together."

"I agree completely," said Marlene. "The fact that you two struggled with difficult problems that had no obvious approach, and figured them out anyway, stretched you this morning. If a problem takes you two minutes to solve, then you won't learn much. But if you spend two hours working on a problem, you'll have learned a lot, regardless of whether you actually find a solution."

"Marlene's right," said Rachel. "There's so much value in struggling with a difficult problem and staring at it for a long time. It improves your focus, perseverance, and concentration. In a typical high school math class, all you see are routine exercises, which you might spend ten seconds thinking about. But with the types of problems you'll see at this camp, as well as on math contests, you'll be thinking about them for hours. My experience at this camp six years ago prepared me so well."

"Prepared you for what?" asked Grace.

I looked at Grace. "She's talking about the IMO."

"Actually, I'm not talking about the IMO," said Rachel. "This camp prepared me for what I'm doing now."

"Sorry," I said, embarrassed. "What are you doing now?"

"I'm involved in a summer research project with one of my math professors at Waterloo. We're working on an unsolved problem, and we've spent the past six weeks looking at it. We've come up with some ideas but we've gotten nothing concrete. Even worse, we're not sure if our problem can even be solved given the techniques currently known by the leading experts in the field. With math contests, at least you know that every problem has a short elegant solution. With research mathematics, that's certainly not the case, and we could be spending months, if not years, trying to prove a theorem that's not true."

I stared at Rachel in admiration.

Rachel continued. "In mathematical research, the key is not to get the right answer, but to ask the right question. That's often ninety-nine percent of the work."

"What do you mean?" asked Grace.

"Often in mathematics, a problem is solved by asking the right question, which opens up a new perspective no one had thought of before. And I'd argue this is true, not just of mathematics, but of life itself. We're shaped not by the answers we receive, but by the questions we ask. That's why we need to constantly challenge traditional ways of thinking and question authority – to replace incorrect ideas based on misguided assumptions."

I nodded.

"What's your area of math?" asked Grace.

"It's called Group Theory."

"Group Theory?"

"It's a strange name, but the concept of groups is key to abstract algebra. It's about figuring out the symmetries in various algebraic structures, and how these representations enable you to solve complex problems in other branches of math, like algebraic number theory and topology."

"Sorry, you lost me," said Grace. "What's a group?"

Rachel looked at us. "Have either of you ever solved a Rubik's Cube?"

Grace and I shook our heads. I'd played with the multi-coloured 3×3×3 cube as a kid, but Mom and I had no idea how to solve it.

"The different moves you can make on a Rubik's Cube correspond to a structure known as permutation groups. Different rotations, twists, flips, and so on. You can use group theory to classify all possible moves you can make, and figure out what happens when you combine a sequence of consecutive moves. In fact, when I took my first Group Theory class, the professor explained the theory using the Rubik's Cube. I got hooked right away."

I looked at Rachel. "You can solve the Rubik's Cube?"

Rachel smiled. "Yes, I most certainly can. And what's more, I can explain it mathematically. There are certain move combinations that flip two side cubes, shuffle three corner cubes, and so on. Group theory describes precisely how to do it. Once you know all the moves, anyone can learn how to solve a jumbled cube in just a few minutes. Do you want me to teach you?"

"Yes, please!"

"Great. Well, find me after dinner. I've got my cube upstairs. I'd be happy to show you."

"What's the problem you're working on? asked Grace.

Rachel paused. "It's one special sub-case of group theory's most famous unsolved question, known as Burnside's Problem. My question deals with something known as finite groups."

I looked at Rachel. "If you solve this, does it lead to something useful?"

"What do you mean?"

"You know, like, some real-world application?"

Rachel shook her head. "Not really. But if I can solve the question I'm working on, maybe the techniques can be used to solve even harder cases of Burnside's Problem. I guess we're chipping away at it, one block at a time."

"So even if this problem has no practical application, it still excites you?"

*Oops. Maybe I'd worded that too strongly.*

I nervously looked at Rachel.

"Yes, it definitely excites me!" said Rachel, not feeling the least bit offended by my question. "Why does math need an application for it to be worthwhile?"

I paused. I'd never thought of that.

All the math I had learned from Mr. Collins always had an application to the real world. In fact, we never spent a Saturday session without discussing the relevance of math to everyday society.

"I'm not sure," I said. "I always thought math was important because it was useful."

"Do you like music or art?"

"Both," I replied.

"Well, what is the point of music and art? You can adore Beethoven's Fifth Symphony or stare in wonder at an M.C. Escher masterpiece: not because it has any practical application, but because it is beautiful. I believe mathematics works the same way. Just like music and art, we're mesmerized by the elegance and beauty of math, not because of its utility."

"That makes sense," I said, nodding.

"I think abstract ideas are so beautiful," continued Rachel. "For me, it doesn't matter if the ideas just exist in the mind. Of course, an abstract idea sometimes exists in reality, such as when mathematicians conceived of the notion of 'black holes' in the eighteenth century, and then in the twentieth century, a group of astronomers actually found them. Math enables us to explain what's happening in the world, and provides beautiful methods to model and solve real-life problems. But that's not what math is."

Marlene jumped in.

"I know where you're coming from, Bethany. Just as Mr. Collins does for you, I spend a lot of time talking about real-life applications in my classes, as a way to get students excited about the subject. But I agree with what Rachel says. The subject is worth doing for its own sake. I know you and Grace already know that. After all, there's nothing practical about math contests."

"Except getting invited to camps like this," interrupted Grace. "And meeting boys that are smart and good-looking. Or, in Bethany's case, meeting a smart good-looking boy that likes you back."

Grace tapped me on the shoulder. "Right?"

My face flushed, and I saw Rachel and Marlene exchange a look.

I glared at Grace, and she responded with a grin.

"Don't worry," said Marlene, touching my hand with her finger. "Your secret is safe with us."

"What secret?" I said.

"I have no idea," said Marlene, as Grace giggled.

I turned to Rachel, desperate to change the subject. "Rachel, you were telling us what math is."

"Yes, I was," said Rachel, smiling. "You want to talk about that, or if you want, we could . . ."

"I want you to talk about that," I replied, way more forcefully than I should have.

Rachel chuckled, and paused before speaking.

"Well, for some mathematicians, having a practical application is what motivates them. If it isn't directly related to some current real-life problem, for them it has no value. For others, they're happy to know their theories are paving the way for future technology, like some of my grad school friends at Waterloo who specialize in quantum cryptography. But for an aspiring pure mathematician like myself, my goal is to advance knowledge. The joy of discovery is my motivation."

"That's awesome," I replied.

*The joy of discovery.*

"So, Rachel," said Grace. "You've been working all summer on a complicated problem that might not be solvable. What if you don't get it?"

"Then I'll learn some more math and try it again next summer," said Rachel.

"But is it worth it? I mean, what if you try for the next five years and still don't get it? What if it's impossible? Would the effort still be worth it?"

"Absolutely," said Rachel.

"How do you know?" I asked.

"Because I spent nearly as many years as a teenager trying to do something that some people told me was impossible. When I was your age, just six years ago, I was invited to this camp in Ottawa. That's when I learned about the IMO. It was here I decided I wanted to make the Canadian IMO team someday. Of course, it was wonderful to have made the team in my last year of high school, but even if I hadn't, everything I learned, not just about math but about myself, was an experience I wouldn't trade for anything."

"But what if you hadn't made the team? Wouldn't you have been disappointed?"

"Of course. But I would have moved on, and the experience would have made me stronger the next time I pursued something really hard. So there would have been no regrets either way."

My mind flashed back to a conversation I had with Mom a few years ago.

I looked at Rachel. "Did your parents ever try to discourage you from trying to make the IMO team? Did they ever say it was unrealistic, or too risky?"

Rachel smiled. "No, it was actually the opposite. When I told them about the IMO after I came back from this camp, they encouraged me to go for it. My Mom told me this great story that I'll never forget, the story of two people at an amusement park, one on a roller-coaster and the other on a merry-go-round. The person on the roller-coaster has a crazy unpredictable ride filled with many twists and turns, ups and downs. It's thrilling, scary, and sometimes awful, but at the end of the day, the person on the roller-coaster has an amazing experience and a great story to tell. On the other hand, the person on the merry-go-round just turns slowly around a circle, leisurely returning to the same point again and again. There's no risk involved. It's safe and comfortable, and it's not at all exciting."

"I'm confused," interrupted Grace. "What does this have to do with the IMO?"

"Everything. Because this story isn't about two people at an amusement park, it's about two people going through life. As my Mom said, many people choose the merry-go-round, taking the path of least resistance and living an uninspired life with minimal challenge that revolves around security and stability. But a small percentage choose the roller-coaster, daring to live out their dreams, and seeking a life of courage rather than a life of comfort. Striving for a crazy dream like the IMO is an incredible risk, because so much sacrifice is required. But for me, a life without risks isn't a life worth living."

I was hanging on Rachel's every word. My body chilled with excitement.

"Do you think Bethany and I can make the IMO team someday?" asked Grace.

I looked at Grace with a raised eyebrow.

*Who was she kidding?*

"Yes, I most certainly think you two can make the IMO team," replied Rachel without hesitation.

"I agree," added Marlene. "You two are creative, and have great imagination and insight. You've just finished Grade 9, and have plenty of time to grow and develop. We've never had two girls on the team together. I would love to see you both make it."

Marlene thought Grace and I had the potential to make the IMO team. More importantly, Rachel did too.

I looked up at Rachel. Knowing that she believed in my potential gave me a confidence I hadn't experienced before.

Sure, Albert was a prodigy, and Raju and Cooper were not too far behind. But the IMO team had six people. It would be so cool if Grace and Cooper and I could make the team together.

"Bethany, I meant to ask you something," said Rachel, snapping me out of my daydream. "We were chatting at the barbeque last night, and you said you saw something on TV that inspired you to make the IMO team someday. What was the program about?"

I spoke slowly, choosing my words carefully.

"Well, it was a TV clip that profiled someone who loved math. I was moved by how this mathematician developed self-confidence and joy from doing math. At that moment, on my twelfth birthday, I knew I wanted to follow in this person's footsteps."

I smiled at Rachel.

"And that's why I'm here at this camp."

"That's awesome," said Rachel. "Just out of curiosity, who was the mathematician in the TV clip? Newton? Gauss? Einstein?"

"It was you."

# 17

"Break time!"

Grace and I sighed in relief. It was now 4:00 p.m. and we had worked for three hours straight.

"Great job everyone," said Rachel. "Make sure you give me your solutions on the way out. See you at dinner."

We handed our three solutions in to Rachel, and walked out of the classroom. Of the ten questions on the Problem Set, Grace and I had managed to solve the first four all on our own. We individually wrote up solutions to Questions #2, #3, and #4.

"I am so tired," said Grace.

"Yeah, me too," I replied. "I need a nap."

As we walked up the stairs, we could hear Raju and several others behind us.

"How many did you get?" asked Raju.

I could hear Dominic tell Raju that he solved the first six, and J.P. followed up by saying that he solved the first seven.

"How about you, Raju?"

"All but the last one."

"I bet Albert got #10."

"Here comes Albert. Hey, did you solve all ten?"

"Of course I did," he replied.

"Damn it," muttered Raju.

Grace tapped me on the shoulder. "Let's get out of here."

We walked a bit faster, and reached the second floor. Grace took out her key and motioned for me to come into her room.

"Wow, what a day," she said, kicking off her shoes. I stared at her tiny feet, which were smaller than the size of my hands.

"Grace, I feel so out of place here. Everyone's so smart."

"But remember what Rachel told us at lunch. There's a reason we were invited to this camp."

"Yeah, but we only solved four problems – some of the guys, they solved seven, eight, nine, ten."

"We don't have to compare ourselves to them."

"Yeah," I said, standing up and walking towards the door. "I better get going before I crash on your bed. Can you wake me up in an hour?"

"No prob. And Bethany?"

I turned around.

Grace smiled. "I'm so glad we're friends."

"Thanks. Me too."

I returned to my room and lay down on the bed. A hundred thoughts were running through my head: about Albert, Raju, Rachel, Grace, and Cooper.

Eventually, the thoughts subsided, and I drifted off.

I heard a knock at my door.

"Bethany?"

"Yeah?" I said, groggily.

"It's Grace. You asked me to wake you up."

I opened my eyes. The alarm clock said it was already 5:00 p.m. The lights were still on, and my shoes were still on my feet. Wow, I really had nodded off.

"Just a sec."

A few seconds turned into a few minutes, but eventually I was ready to take a stroll around campus. We walked down the stairs together and noticed a small group watching TV, while others were playing pool. Albert was by himself reading a textbook.

Near the door, we saw Raju, David, J.P., and Dominic huddled over a small table, with some cards and poker chips. They looked serious. I wondered if they were playing for real money, like some of the boys at Pinecrest.

We walked outside and headed towards the edge of campus. Within a few minutes, we found an outdoor basketball court, and saw some familiar faces. Cooper, Ric, and Ryan were playing against three guys who looked much older, probably students at the university.

As we approached the court, one of the guys on the other team dribbled around Ric and took a shot. The ball went through the hoop, hit the concrete, took a bounce, and sent the ball rolling towards us.

Cooper ran to get the ball. He looked startled to see Grace and me standing there.

I picked up the ball and gave it to Cooper, wordlessly. Our eyes met.

His reaction was subtle, but I saw it. A faint smile.

Cooper ran back to the court, and handed the ball to a player on the opposing team.

"Six-six," said someone on the other team. "Next point wins."

The university student passed the ball to one of his teammates, but Cooper rushed in and stole the ball. Cooper then passed the ball to Ryan, who was positioned near half-court, far away from the basket.

"Ric, get out of there," shouted Cooper, motioning for Ric to move away from his position underneath the basket. As Ric moved towards the far edge of the court on the side opposite Cooper, he was followed by the defender covering him.

I stared at Cooper, noticing how he was looking at Ryan to get his attention. Cooper lightly nodded, and Ryan nodded back.

At that instant, Cooper sprinted towards the basket, going around his defender who reacted a split second too late. Ryan lobbed a high pass over the outstretched arms of the player covering him, and in one motion, Cooper leaped high to catch the ball and dunk it through the hoop.

"Yeah!" shouted Ryan and Ric, running towards Cooper. They gave each other high-fives, and then jogged towards the three defeated university students to shake hands.

"Wow," I whispered.

Ric walked towards us, followed by Ryan and Cooper. They grabbed their water bottles, and sat down next to us on the grass. Cooper, drenched in sweat, had a gleam in his eye.

"You guys played awesome," said Grace.

"Thanks," said Ryan. "Well, it's easy to do well when you're playing with Cooper. The guy's a machine."

Cooper shrugged modestly. "We're a good team."

Ryan looked at us and pointed towards Cooper in admiration. "Six foot two white guy, fifteen years old, and he can dunk."

Ric looked at us. "You two came at a great time. You saw that perfect alley-oop? That was sick."

"Alley-oop?" I asked.

Cooper looked up at me. "It's that last play we did – catching the pass in the air and slamming it in."

"I see," I replied.

Another trio of players had assembled, ready to challenge the winners. Grace and I stuck around to watch the boys play another game. This one was over in minutes, with Cooper's team winning 7-1. The following game was a bit closer, but the result was still the same, the good guys victorious again.

"You want to continue walking?" said Grace.

"Yeah," I said. "Let's go."

I looked back and saw the three boys turn towards the court, to face off against another team. I noticed Cooper glance back in our direction . . . in my direction. I felt a lump in my throat.

Grace noticed our exchange, and gave me a tap on the shoulder.

"See? He definitely likes you."

"But why me?"

"What do you mean? Why not you?"

I turned to face her. "Boys don't like me. All the popular girls wear nice clothes. They're outgoing. They're petite and skinny."

"Who cares about that?" replied Grace. "Look at you. You're intelligent, you're beautiful . . ."

"No, I'm not."

"I disagree," said Grace. "You're great just as you are, and Cooper sees that. That's why he's attracted to you. Besides, why should you be outgoing and extroverted? It's obvious that Cooper isn't."

"But why would Cooper like me? I'm sure there are lots of pretty girls at his school."

"I don't think so," said Grace, laughing. "He goes to Lakewood Academy, an all-boys boarding school."

Grace pointed towards a bench on the edge of campus, overlooking at the Rideau Canal. We sat down next to each other and talked . . . and talked . . . and talked.

To take my mind off Cooper, I asked Grace to tell me about her life in Vancouver. She was happy to oblige, with stories of her school, her church, how she got into math, her passion for piano and singing, and her desire to become a professional musician someday.

I told Grace about my life in Sydney, about The Staircase, the TV documentary about Rachel, the weekly sessions with Mr. Collins, and the perfect score on the Gauss Contest last year. I shared the story of how Mom offered to coach Ella Collins so that her grandfather could tutor me, and how Mom coached Ella until just a few weeks ago, when Ella and her family moved to Newfoundland.

"You have an amazing mother," said Grace. "Do you have a dad?"

I shook my head and told her what I knew, that my dad had left Mom just before I was born, and moved to Alberta to be with someone else. Last year, through a friend of a friend, Mom found out my dad had died in a car accident. While the news left her bitter, I didn't feel any emotion since I had never once met him. Not one phone call. Not one letter.

"Do you have a picture of him?" asked Grace.

"No," I said. "But apparently he was six-four and was a star hockey player. He would have made the NHL if he didn't get injured."

"That explains why you're so tall. Did you ever play hockey?"

"Mom wouldn't let me take skating lessons," I said. "I don't even know how to skate."

Grace told me that her Mom and younger brother also died in a car accident. The accident happened when Grace was only three, and since then, lived alone with her dad, a church pastor. Even though she didn't know her Mom and brother, she missed them a lot.

I had known Grace for less than twenty-four hours, yet we had this instant connection. It didn't matter that we were complete opposites in personality and appearance. We were sharing our hurts and hopes. We were sharing our dreams. I had never talked this way with anyone before, not even with Mom.

"I got lucky," I said. "If I wasn't doodling in class that day, the Staircase thing wouldn't have happened. If the TV wasn't playing in the background on my birthday, I wouldn't have learned about Rachel and the IMO. If I didn't go to the hospital that day to visit Grandpa, I wouldn't have met Mr. Collins. It's all one big fluke – a bunch of lucky coincidences."

"Do you really believe that?" asked Grace. "What if all that was meant to happen?"

"What do you mean? Like fate?"

"No, not fate. Faith. I don't think those events were coincidences. What if all those things were supposed to happen exactly the way they did?"

"Yeah, right," I said, sarcastically. "Like there's someone in the sky planning everything we do."

"Exactly," responded Grace. "That's what God does. What do you think, Bethany? Do you believe in God?"

"Not really," I replied. "Mom and I don't go to church."

We were having such a nice conversation. I didn't want to talk about religion.

"You know, it's getting really dark," I said. "We should probably head back for dinner. What time is it, anyway?"

"I have no idea," replied Grace, shrugging.

Grace and I headed back and walked into the cafeteria, where we saw everyone already eating dinner. Nearly everyone had an empty plate in front of them.

Marlene saw us enter, and looked disappointed. "We said dinner was at six o'clock. Where were you?"

I looked up at the clock, surprised that it was already 6:45 p.m. "I'm sorry. We just lost track of time."

"That's fine," said Marlene. "Make sure that doesn't happen again. Quickly grab whatever's left, since Rachel will be making an announcement in a couple of minutes."

While we munched on a plate of rubbery chicken and stale French fries, Rachel stood up.

"Okay everyone, I think all of you noticed we posted the model solutions to today's Problem Set on the bulletin board in the main hall. If you haven't checked it out, do so sometime this evening, and make sure you read the solutions to the problems you didn't get. Congratulations on your effort today. For those of you wanting to see a movie, we'll be showing *A Beautiful Mind* in ten minutes, in the classroom. If you're not interested, relax in the residence lounge and we'll serve you a snack at nine-thirty, right when the movie's done. Okay, that's it."

Everyone stood up and cleared their plates, while Grace and I were still finishing our dinner. The food was cold and gross, so I took my plate and headed towards the garbage bin. When I noticed who was standing in front of me, I paused.

"Hi, Cooper."

"Oh, hey Bethany," replied Cooper. He looked nervous.

That gave me confidence. It was my turn to speak.

"Are you watching the movie?"

"Nah, I've seen it a bunch of times before. Just going to chill out in my room."

"Cool. See you later then."

I calmly headed back to my seat and winked at Grace, who was watching our little exchange. She gave me a thumbs-up. I waited for Grace to finish dinner, trying hard to conceal my grin.

When Grace had cleared her plate, we walked towards the main hall in the residence, and found the bulletin board.

We noticed that Rachel had selected one correct solution for each of the ten problems. The solutions to Questions #8, #9, and #10 were written by Cooper, Raju, and Albert, respectively. I skimmed their write-ups, amazed by what they had written.

Then I noticed that Rachel had selected my solution for Question #4! Excited, I stared at the bulletin board, and re-read what I wrote.

**Question #4**

Eve and Oddie play a game on a 3 by 3 checkerboard, with black checkers and white checkers. The rules are as follows:

i) They play alternately, with Eve moving first.

ii) A turn consists of placing one checker on an unoccupied square of the board. A player may select either a white checker or a black checker on her turn.

iii) When the board is full, Eve obtains one point for every row, column, or diagonal that has an *even* number of black checkers, and Oddie obtains one point for every row, column, or diagonal that has an *odd* number of black checkers.

iv) The player obtaining at least 5 of the 8 points wins.

Describe a winning strategy for Eve (i.e., show that she can always win regardless of how well Oddie plays).

**Solution by Bethany MacDonald**

Here's the strategy. Eve plays in the centre square first, with a black checker. Now whatever Oddie does next, Eve plays in the *diametrically opposite* corner with the other colour. For example, if Oddie places a white checker in the top left corner, then Eve places a black checker in the bottom right corner. Eve repeats this "copycat strategy" until the game ends.

Just one row and one column pass through the centre square. There are two diagonals passing through the centre. Each of these four "lines" (see diagram) must contain two black checkers and one white checker, because the checker in the centre is black, and because Eve's copycat strategy ensures that exactly one of the other two checkers is black.

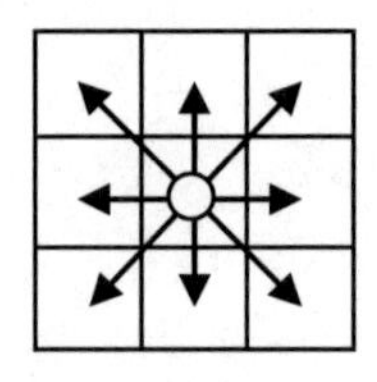

Eve is guaranteed four points from these four lines. To complete the proof, we just need to prove that Eve gets one more point. The top and bottom rows contain six checkers. By the copycat strategy, there must be three checkers of each colour. If the top row has $x$ black checkers, then the bottom row must have $3 - x$ black checkers.

In any scenario, Oddie wins one row and Eve wins the other, since one row has an odd number of black checkers and the other row has an even number of black checkers. This is because the total number of black checkers in these two rows is three.

Similarly for the left and right columns, Oddie and Eve must split the two points. This copycat strategy guarantees that Eve wins any game, six points to two. Therefore, we conclude that this is a winning strategy.

Our proof is complete.

I was touched by what Grace had said earlier, about me being intelligent and beautiful. I didn't have much confidence in myself, and it meant a lot that someone I barely knew thought so highly of me.

Come to think of it, others thought that way of me: Marlene and Rachel. And Cooper.

Maybe self-confidence isn't something you have, but something you choose to have.

Maybe Grace had figured that out a long time ago. Maybe that was a lesson I was finally beginning to learn. That it was okay to be myself and not think of myself as weird or different. That my self-worth wasn't defined by Gillian Lowell and her friends.

I tapped Grace on the shoulder. "I'm going to talk to someone. See you in a bit."

Grace smiled. "Good luck!"

I knew exactly what I wanted to do. I walked up to the second floor and found the room I was looking for.

After a few seconds to compose myself, I knocked twice.

"Yeah."

I took a deep breath and opened the door. The person in the room was smiling back at me.

"Hey, Bethany."

"Can I come in?" I asked.

"Of course you can."

I smiled back.

"Rachel, can you teach me how to solve the Rubik's Cube?"

# 18

Marlene walked into the classroom, rolling a large cart filled with snacks and juice boxes.

I breathed a sigh of relief.

Our instructor for Wednesday morning was Vlad, a graduate student at the University of Ottawa. He wrote a lot on the board, but I couldn't read his handwriting. Because of his thick accent, I couldn't understand most of what he said.

After a couple of minutes he noticed Marlene standing by the door.

"Okay, I'll stop here," said Vlad. "Take a snack break. In a few minutes I'll give out the Problem Set."

Vlad's "teaching style" was the polar opposite of Rachel's. While Rachel had spoken for all of five minutes the past two mornings, Vlad had lectured non-stop until 10:33 a.m., mercifully taking a break only when Marlene came into the room.

The topic was Number Theory. While I had learned factorization and rules of divisibility from my time with Mr. Collins, I was thoroughly unprepared for the ninety-three-minute barrage that Vlad unleashed on us this morning. I looked at the headings in my notes.

*Modular Arithmetic, Euclidean Algorithm, Chinese Remainder Theorem, Quadratic Residues, Fermat's Little Theorem, Wilson's Theorem.*

So many new topics, all covered so quickly, taught by a guy with a strong accent and messy handwriting.

My notebook filled seven pages, with numerous symbols I had never seen before. I had copied everything I could, as accurately as I could and as quickly as possible before Vlad erased it and moved on to the next topic. I was utterly lost.

Just after Vlad announced snack time, Ric leaned over.

"What the heck was that? I didn't catch any of it."

"Me neither," I whispered.

"When did you stop following him?" asked Ric.

"Maybe two minutes into the lecture. Right after he defined the *mod* symbol. How about you?"

He shrugged in defeat. "Right after he said 'Good morning'."

I laughed and looked over at Grace, sitting beside me.

"How much of that did you catch?"

She shrugged and shook her head. "Twenty percent at most. That was insane."

We walked over to the snack tray and grabbed a couple of cookies. To my relief, the boys looked just as shell-shocked as we did . . . well, most of them.

"Wow, that was so awesome," exclaimed Raju, to no one in particular.

Grace stared at him. "You got all that?"

"Well, I already knew the stuff from the first half of the lecture, but the second half was all new. I really liked how Vlad explained it so rigorously."

Grace moved closer to him and dropped her voice to a whisper.

"But he was so confusing!"

"Yeah, but I think he did that on purpose so that we'd have all that content. Then, in the Problem Set we can figure out how to apply it."

"But it was impossible to keep up with him!"

"Yeah, it was really fast," nodded Raju in agreement. There were a couple of steps that I didn't understand but I'll look at my notes and try to figure it out for myself."

"Well there were a couple hundred steps I didn't understand," said Grace.

Raju gently chuckled. "Well, we're getting the Problem Set now. Do you want me to explain the lecture to the two of you?"

"Definitely," I said.

Grace nodded. "Thanks, Raju. We'll go bring over our stuff."

I went to my seat, pleasantly surprised. Raju was really competitive, and almost as smart as Albert, but at least he was a nice guy. We lifted our desks and positioned them next to Raju's.

Raju glanced towards the snack table.

"Cooper, do you want to join us?"

I gulped.

Cooper turned and saw Grace and me sitting with Raju. He paused.

"Sure," he said.

He brought over his table and chair. The four of us formed a square.

Grace turned to me and grinned. I responded by kicking her under the table.

I tried to act casual and nonchalant, but found it hard. I knew he liked me, but I didn't want him to know that I liked him. Or maybe I did.

I wondered whether he knew that I knew that he liked me.

*Why does this have to be so complicated?*

"Raju, can you help us with these questions?" I asked, staring directly at Raju.

"No prob."

Raju first simplified the symbolic mess that Vlad used, explaining why the difficult concept of "modular arithmetic" could be viewed as the simple process of "clock arithmetic", where numbers wrap around after reaching a particular value. He used his watch to illustrate the concept of "modulo 12", explaining how six hours after 11:00 a.m. is not read as 17:00 a.m., but 5:00 p.m.

Similarly, since the clock wraps around every twelve hours, Raju explained why forty hours after 11:00 isn't 51:00, but 3:00, and how this can be expressed mathematically using the *congruence equation* $11+40 \equiv 3 \pmod{12}$.

That made the concept much clearer. There were many ideas in Number Theory that were far too complicated, but Grace and I understood modular arithmetic. Working with Raju's guidance, we were able to solve the first three questions on the Problem Set. Cooper wasn't saying much, lost in his own thoughts, scribbling away on a different problem. I stole a glance at his sheet of paper, and saw him working on Question #5.

Grace gave Raju a thumbs-up. "Thanks for helping us. I think we're good on our own now."

Grace and I worked on Question #4 together. By the time we had solved that question, it was almost noon.

Rachel walked up to the front.

"All right everyone, let's take a break from the Problem Set. You've worked really hard this morning, and we want to exercise a different part of your mind before we break for lunch."

Rachel waited until all of us had put our pens away. Grace and I turned our desks to face the front.

Rachel spoke up. "Marlene, the floor's all yours."

Marlene came up and turned to face all of us.

"Okay, I need a volunteer."

A bunch of people raised their hands.

"Ric, your hand was up first. Come on down."

Ric got up from his seat, and walked to the front of the classroom.

Marlene turned to face us. "I'm now going to give you a logic puzzle, and we'll work on it as one large group. As soon as you've solved the puzzle, you can break for lunch. The faster you solve this puzzle, the faster you eat. Sound good?"

A bunch of us nodded.

"What we need are two different activities that Ric would find humiliating and embarrassing, and would cause him great discomfort if he were forced to do either of them. Ryan, start us off. You've known Ric for years. Say you had to think of the cruellest and meanest way to punish him. What would you get him to do?"

Ryan snickered. After a short pause, his face lit up. "I'd make him stand inside a shopping mall wearing a ballerina dress."

Raju chuckled. "Yeah, send him to the Rideau Centre. And get him to pirouette around in a circle in his ballerina dress, singing *I'm a Little Teapot.*"

"Yes, that would be cruel," said Marlene. "Okay, that's the first punishment for Ric. Now we need another form of punishment. Any ideas?"

J.P. raised his hand. "You could lock him up in his room, and force him to do something he really hates."

"Yeah, lock him up, and get him to listen to the same song over and over again," said David with a chuckle. He looked at Ric. "Hey, any ideas for a good song, Soo?"

Ric scowled at David. I snickered, remembering *A Boy Named Soo.*

"Hey, you can get him to listen to Miley Cyrus," said someone. "Over and over."

"I have an idea," said Marlene. "Miley Cyrus has a famous dad, who wrote *Achy Breaky Heart.* We could make him listen to that again and again."

I laughed because I knew that Billy Ray Cyrus song. It was so annoying.

"Here's the puzzle. You are allowed to make a *single statement.* This statement must either be true or false. If the statement is TRUE, then Ric has to spend this afternoon in a busy area of the Rideau Centre pirouetting around in a ballerina dress singing *I'm a Little Teapot.* If the statement is

FALSE, then Ric has to spend this afternoon locked up in his room upstairs, with a loud stereo playing *Achy Breaky Heart* over and over again.

"Okay, here's the question: is there some statement you can make that would enable Ric to escape either form of punishment?"

"Is this a trick question?" asked Dominic.

"No, it's not," replied Marlene. "If the statement is true, he gets Punishment A, and if the statement is false, he gets Punishment B. No trick at all."

David spoke up. "Let's try it. My statement is 'One plus one equals three'. What happens?"

"Well, your statement is false, so Ric gets Punishment B. He gets locked up in his room and has to listen to *Achy Breaky Heart* a billion times."

"Oh yeah."

One student raised his hand. "How about '*Napoleon Dynamite* is the greatest movie of all time'."

Marlene shook her head. "That's subjective. It's a true statement for some but a false statement for others. You can't ask that, since there is no definite answer."

J.P. raised his hand. "What about 'Black holes exist'. No one knows the answer for sure, but there's a definite answer. What happens then?"

Marlene smiled. "Excellent point. I should have mentioned, your statement must either be demonstrably true or false. Of course, the answer is either true or false, but no one knows for sure. So you can't ask that."

Ryan raised his hand. "The Maple Leafs will win the Stanley Cup next year."

Marlene laughed. "Being a die-hard Ottawa Senators fan, I can tell you with absolute certainty that your statement is false, like it's been false every year since 1967."

"This is stupid," said Albert, shaking his head and crossing his arms.

"Why?" asked Marlene.

"Because the problem has no solution. The statement you give has to be true or false. That's what you just said. If it's true, you get A. If it's false, you get B. No matter what you say, you have to get either A or B. There's no statement you can say that avoids either possibility."

Marlene smiled. "Not true."

While Albert looked angry, I noticed Cooper fidgeting in his seat. He raised his hand, tentatively.

"Yes, Cooper?"

"I've seen something like this before, but can't remember exactly. It's some statement that involves the actual punishment."

"What do you mean? This is excellent, Cooper. Elaborate."

"I don't remember."

Raju spoke up. "I remember something like this too. The answer involves the actual punishment itself. I think the right statement is something like 'Ric will get Punishment A'."

"Good idea," said Marlene, turning toward the board. "Let's investigate."

Marlene wrote down Raju's sentence: *Ric will get Punishment A*.

She turned to us. "Remember the conditions of the logic puzzle. If this statement is true, Ric will get Punishment A. If this statement is false, Ric will get Punishment B.

"Consider the first case where the statement is true," Marlene said, pointing to the board. "*If* Ric will get Punishment A, *then* Ric will get Punishment A. It's logically consistent. So the statement is TRUE, and Ric doesn't escape the cruel punishment since he's stuck pirouetting around the Rideau Centre for the entire afternoon. Sorry Raju, your answer isn't correct. But you're really close."

I nearly jumped out from my seat.

"Ric will get Punishment B," I blurted out.

"No, that doesn't work," said Raju from behind me. "That's equivalent to my statement."

"Oh," I said, leaning back in my chair.

"Bethany," said Marlene. "Repeat what you just said."

"It doesn't matter," I replied softly. "My idea's wrong."

"I want to hear your thought, regardless of whether you think it's right or wrong. What was your statement?"

"Ric will get Punishment B," I mumbled.

"Please speak up, Bethany. Say it again."

"Ric will get Punishment B," I said, a bit louder.

"You just said your idea was wrong. Why is your idea wrong?"

"Because Raju says it's equivalent to his statement, that Ric will get Punishment A."

"That's a bad answer, Bethany. Your idea isn't wrong just because somebody else said it was wrong. Let's try that again. Why is your idea wrong?"

I felt goose-bumps on my arms. I racked my brain for a better response. Suddenly, I remembered the problem-solving topic from Monday.

"It's by symmetry, right?"

"What do you mean?" she pressed.

"Raju's statement is equivalent to mine. So if his answer doesn't work, by symmetry, my answer can't work either. Isn't that right?"

"Let's see," said Marlene with a smile. She turned to everyone else. "Let me write down Bethany's statement on the board."

*Ric will get Punishment B.*

Marlene was no longer looking at me, and I breathed a sigh of relief.

She looked around the room. "Okay, let's analyze this statement. Any volunteers?"

J.P. raised his hand. "If the statement is TRUE, that Ric will get Punishment B, then Ric will get Punishment A. If the statement is FALSE, that Ric will not get Punishment B, then Ric will get Punishment B. Wait. That makes no sense."

I tensed up, knowing with absolute certainty that my statement was the solution to Marlene's logic puzzle.

Before I could say anything, Cooper jumped in.

"That works! Bethany's statement can't be true, since Ric can't be at two different places doing two different things, both at the same time. So the statement must be false. That means Ric does *not* get Punishment B. But, if the rules of punishment are forced to apply, then you get a contradiction – since Ric can't get something and not get that thing. Therefore, the only logical outcome is that Bethany's statement is false and the rules of punishment are dropped. Ric gets to walk away, a free man."

"You got it," said Marlene.

Ric breathed a sigh of relief, and headed back to his seat. "Yes, I escape!"

Raju leaned over, and gave Cooper a fist-bump. "Good job, man."

Cooper leaned over to me and brought his hand close to my shoulder. "Hey, it was your idea."

I turned around to complete the high-five.

But instead of slapping my hand to his, my fingers gently found his fingers, and our palms touched. My fingers moved down his hand to feel the smoothness of his palm. I drew my hand away.

"Congratulations, everyone," said Marlene. "You're free to go to lunch."

With a loud cheer, we stood up and proceeded out of the classroom, a few minutes before noon.

Marlene tapped me on the shoulder. "Let's chat."

I moved off to the side while everyone else, including Cooper, left the room.

"Here, let's have a seat," said Marlene, motioning towards two seats right by the front door.

"What did I do wrong?" I asked, sitting down.

"Nothing, Bethany. I just wanted to ask you something. You saw the movie we put on last night. What did you think about it?"

"I loved it," I replied. "It was inspiring."

We saw *Stand and Deliver*, the true story of a high school math teacher in East Los Angeles who inspired his underachieving students to pass the Advanced Placement Calculus Exam, even though nearly every student lived below the poverty line, and virtually no one had a parent with a university degree.

"I found it inspiring too. Why do you think those students succeeded?"

"It's because they had an amazing teacher. He believed in them."

"But other teachers believe in their students too. How come only one high school in the entire United States passed more students in the AP Calculus Exam than this gang-ridden school in East Los Angeles?"

"Well, the students all worked really hard. Summer school, coming in on Saturdays, and so on."

"That's right. What else?"

I racked my mind, but couldn't think of anything. I looked up at Marlene and shrugged.

Marlene spoke up. "You said the teacher believed in the students. What else? Who else believed in the students?"

"The students themselves."

"That's right. The students believed in themselves. They believed they could master university-level mathematics, even though most of them couldn't do basic arithmetic when they began at that school. The students believed that excelling in mathematics would allow them to win prestigious college scholarships and break the cycle of poverty for them and for their families. In fact, a bunch of these students ended up getting Ph.D. degrees and are working at places like NASA right now. True story."

Marlene leaned over and held my hands.

"Bethany, these students reached their potential because *they believed in themselves.*"

I saw what Marlene was getting at.

"I believe in myself," I whispered.

"Do you really mean that?" she asked. "Of course, you know you're good at math. You're probably the top math student at your school, and maybe soon you'll be the top math student in Cape Breton, or maybe in all of Nova Scotia. But do you believe you have the potential to be just as strong as Albert or Raju?"

"No," I admitted.

"Why not?"

"I'm not super-smart like Albert or Raju. I'm not a natural genius like those two."

"I disagree. I've known Raju for years, because he also lives here in Ottawa. Raju practices old contests and reads problem-solving books that teach sophisticated techniques for solving Olympiad-level questions, and he studies on his own for a couple of hours every night. At first he couldn't solve any of the Olympiad problems, but now he can get a few of them completely on his own. I'm sure having Albert at this camp has spurred Raju to push even harder.

"And Albert's father is one of Canada's most accomplished mathematicians. I'm sure Albert's dad has been coaching him ever since he was young – but Albert's gotten to the level he's at because he's put in the effort himself. You see him at this camp, reading textbooks while everyone else is playing sports and watching TV. Raju and Albert aren't natural geniuses – in fact, I don't believe in the concept of 'genius'.

school's program would have already
itself with his third-choice school and

Even if students were qualified to get
preferences strategically. This create
choice because that way they would a
students had an incentive to tell the tr
school best suited for them, based bo
The school systems he helped create
theoretical work.

The system works by tentatively acc
gone through all the other applicatic
given school (because of higher tes
those students happened to rank th

"The idea is to level the playing fie
have to spend the time learning the
players are not hurt by the fact that th

This same sort of system is used to match new medical school graduates to medical residency programs, which was once a messy process that led to a lot of unhappy candidates. Now all residency assignments are posted simultaneously. In the mid-1990s Mr. Roth redesigned the system to help match married couples who were jointly looking for jobs at hospitals.

Mr. Roth has also helped build a system that assigns kidney donation swaps. For example, a man needs a kidney, and his wife is willing to donate one of hers but she is not a match. Across the country there is a couple in the same position, and it turns out that the wives are a match for the husbands in the opposite couple. In this simple case, the two couples essentially barter their kidneys: Wife A gives her kidney to Husband B, and Wife B gives her kidney to Husband A. It is rare that two couples will serendipitously match each other's kidney donation needs this way, and there are often more pairs of donor-recipients involved. Mr. Roth's system helps find the most efficient exchange of organs so that the most patients can be saved with the fewest number of pairs involved in a given trade.

Mr. Shapley was born in Cambridge, Mass. He received his bachelor's degree from Harvard and his a Ph.D. in mathematics at Princeton, where he studied alongside John Nash, a fellow Nobel laureate. He is married and has two sons.

Mr. Roth received his bachelor's from Columbia and his master's and doctorate from Stanford, all in operations research. He is married with two children.

orial Prize in Economic Science on
eople and companies find and select one

constraints on prices. The laureates'
matches when prices are not available

es new doctors to hospitals and more
on, Chicago and Denver use an
he recently accepted a new position at

"Al has spent the last 30 years trying to make economics more like an engineering discipline," said Parag Pathak, an economics professor at M.I.T. who has worked on school-matching systems with Mr. Roth. "The idea is to try to diagnose why resource allocation systems are not working, and how they can be engineered to produce something better."

Mr. Shapley, 89, a mathematician long associated with game theory, is a professor emeritus at the University of California, Los Angeles. He made some of the earliest theoretical contributions to research on market design and matching, in the 1950s and 1960s.

In a paper with David Gale in 1962, Mr. Shapley explained how individuals could be paired together in a stable match even when they disagreed about what qualities made the right match. The paper focused on designing an ideal, perfectly stable marriage market: having mates find one another in a fair way, so that no one who is already married would want (and be able) to break off and pair up with someone else who is already married. In the 1980s, Mr. Roth applied this work to matches for medical residency programs and eventually school choice. He was interested in how to keep matches fair and how to keep more sophisticated players from manipulating the system to their advantage.

In older matching systems, a student would apply to his first-choice school, which was often popular. If the student did not get in, then the application would be sent on to the student's second choice. But if that was also a popular choice, then that

The New York Times

October 15, 2012

## 2 From U.S. Win Nobel in Economics

By CATHERINE RAMPELL

Two Americans, Alvin E. Roth and Lloyd S. Shapley, were awarded the Nobel Men
Monday for their work on market design and matching theory, which relate to how p
another in everything from marriage to school choice to jobs to organ donations.

Their work primarily applies to markets that do not have prices, or at least have strict
breakthroughs involve figuring out how to properly assign people and things to stable
to help buyers and sellers pair up.

Mr. Roth, 60, has put these theories to practical use, in his work on a program that match
recently for a project matching kidney donors. Public school systems in New York, Bost
algorithm based on his work to help assign students to schools. A professor at Harvard,

filled up by the time his application was even considered, and the process would repeat so on.

into one of their top schools, they could be shut out because they did not rank their an incentive to try to game the system by listing a less popular school as their first t least have a chance of getting in somewhere. Mr. Roth designed a system in which uth about where they wanted to go. A centralized office could then assign them to a th on their own preferences and the preferences of the schools they were applying to. use a "deferred acceptance algorithm," which was developed by Mr. Shapley's

epting students to their top-choice school. It holds off on the final assignment until it has ns to make sure there are not other students who have a higher claim to a spot at that t scores, a sibling at the school or whatever other criteria the school prioritizes), even if e school lower on their list of preferences.

ld," Mr. Pathak said. "You want to make sure that not only do sophisticated players not strategies and different heuristics that will get them ahead, but also that unsophisticated ey are not aware

"Raju and Albert are that talented because a couple of years ago they decided they were capable of getting to the Math Olympiad, they desired to pursue the goal despite the amount of sacrifice required, they chose to believe in their abilities, and they committed themselves to putting in the work to turn their present potential into a future reality."

"So it's all about how hard you work?"

"No," replied Marlene. "It's about believing in your potential. That has to come first. Once you've got that, it's easier to put in the work, because every day you know you're walking one step closer to your goal. The hours you put in, the sacrifice, the effort . . . become liberating rather than draining. I want to tell you this because I see your potential, and I want you to see your potential too."

"Thank you."

"Bethany, you write beautiful solutions. You express yourself so clearly, and you have creative ideas that demonstrate insight and mathematical maturity. Sure, you don't know as much Olympiad-level content as Albert or Raju right now, but they've been doing this much longer. Who's to say that won't change in a couple of years?"

"Maybe," I replied. "But anything I can do, those two can do better."

"Not true. In the logic puzzle we just did, was it Albert who solved it?"

"No," I replied.

"That's right. Albert dismissed the problem, insisting there was no solution. Was it Raju who solved it?"

"No."

"That's right. Raju heard an idea from another student and immediately rejected it, incorrectly jumping to the conclusion that her solution couldn't be correct. Who solved it?"

"Me," I mumbled.

"That's right. It was you, Bethany. You can be just as good as them, but only if you have the confidence to believe it."

"Can I ask you a question, Marlene?"

"Yes, most certainly."

"Remember that symmetry problem from the first day? You know, the one where Albert found this brilliant solution, showing $x \leq y$ and $x \geq y$ to force $x = y$? Well, even Rachel didn't see that, and she's an IMO gold medallist. Do you think I'd ever be able to come up with an insight like that?"

Marlene paused.

"You've now learned that technique, and it's now a part of your problem-solving repertoire. Let me turn the question back to you. Let's say, at some point in the future, you're writing a tough Olympiad-level math contest, and there's a problem on it that can be solved using Albert's method of finding and exploiting a symmetric inequality. Would you be able to solve it?"

I thought carefully about Marlene's question, and considered the wide gap that existed between Albert and me, and the contrasting words of support I had received from Marlene, Rachel, and Grace.

After a long time of thinking and reflecting, I knew the answer.

"Yes," I said, looking directly into Marlene's eyes. "I'd solve it."

# 19

I sat on the bed, and began crying.

Today was the last full day of the camp; all of us were leaving tomorrow. Most of the students, including Grace and Cooper, were catching early-morning flights and trains, while I was departing a bit later on Saturday, taking a mid-afternoon flight back to Halifax.

The past five days had gone by too quickly. Tonight's banquet dinner was going to be bittersweet.

I was going to miss this place so much, and I wasn't ready to say goodbye to all the people I would remember for the rest of my life.

Grace, Rachel, Marlene, Raju. And Cooper.

"Hey, Bethany," said Grace, knocking on my door.

I wiped my eyes with a tissue and tried to compose myself.

"Come in."

Grace walked in. "Aren't you getting ready for the banquet?"

"Yeah," I whispered.

"Your boyfriend will be so impressed."

"He's not my boyfriend."

"Well, he should be. It's obvious how you two feel about each other."

Grace was right. Over the past few days, we had been eating every meal together with Cooper and Raju, and sometimes with Ric, David, and Ryan as well. I knew how Cooper felt about me, since he told Raju he liked me but made Raju promise to keep it a secret.

Thanks to Grace, that secret found its way to me. During our meals, I tried to act cool about it.

"What are you wearing?" I asked Grace, changing the subject.

"Here, come see," she said, inviting me into her room. There was an ironing board in the centre, along with a light blue summer dress. Grace would look amazing in it.

Grace wanted to see my outfit, so we popped back into my room. I showed her what I was planning to wear, a simple dress in my favourite colour, dark purple.

Mom had convinced me to go dress shopping just before I flew to Ottawa. I had noticed the tall purple dress as soon as I walked into the store, and it fit me perfectly. At the time, I didn't want Mom to buy me such an expensive dress, but now I was grateful to have it.

The closing event for the Math Camp was a celebration banquet, and Marlene informed us that there would be important guests in attendance. The instructions in the invitation letter had clearly stated "jacket and tie for boys" and "dress or skirt for girls".

This was my first formal dress-up occasion. I knew Cooper would look amazing, and I couldn't wait to see him in a suit.

I struggled to control my emotions as I changed into my dress. I knew I'd be back in Nova Scotia within twenty-four hours, and I didn't want to think about that.

Grace took forever to put on her makeup. When she finally declared herself "ready", we walked down to the first floor and stepped outside.

The group had just finished lining up for the picture with Rachel and Marlene on the left and right, both wearing tops and skirts. The boys were all in the middle, wearing suits and ties of different colours. I noticed Cooper staring at me as I slowly walked towards him. He smiled nervously.

I liked how I had that effect on him.

"Ah, finally our ladies have arrived," said Marlene, to loud sarcastic applause from Ric and a few others. "We're ready to take the picture. The two of you get right in the middle there. Grace, first row. Bethany, back row. Hurry."

We walked to the middle, surrounded by all the boys. I quickly smiled at Cooper, who was standing next to me on my left.

I giggled when I saw his tie. Dark purple.

"Hey," I whispered.

We posed for a dozen pictures, both serious ones and a few funny-face shots at the end. The boys were relieved; apparently some of them had arrived early and were standing in that position for a while, impatiently waiting for us girls to make their entrance.

Cooper inched closer to me and our wrists made contact. His fingers slowly wrapped around mine.

"Is this okay?" asked Cooper.

"Yeah," I said, smiling up at him.

"Okay, you two!" exclaimed Grace. "There are other people here you know."

I quickly let go of Cooper's hand and felt my face flush. Grace started giggling.

Marlene came over towards us. "Rachel and I have seated our guest of honour at your table, directly across the two of you. Just be yourself. Nothing to be nervous about."

Marlene paused, noticing who was standing next to me.

"Cooper, we've put you at that table too, seated right next to Bethany."

She laughed out loud. "I assume you don't mind?"

Cooper fidgeted and looked down at his shoes. "No, I don't mind at all."

"And Bethany doesn't mind either," added Grace, winking at Marlene.

The three of us walked to the banquet hall together, with Marlene and Rachel trailing behind us. I wanted Cooper to reach over and hold my hand, but I knew this wasn't the right time.

We arrived at the banquet hall and saw a fancy set-up with three forks and two spoons at each seat. To our left, we saw an older blond-haired man sitting at a table, engaged in conversation with J.P. and David. He looked familiar, but I couldn't recall where I had seen his face before.

To our right, we saw a table for six holding our name cards. Ric and Raju were already seated.

"Ric, you look amazing," said Grace. He had his hair slicked back, and wore a dark pinstripe suit that matched his crisp white shirt and black tie. He even had cufflinks on.

"Thanks," replied Ric. "Marlene put me right next to the guest of honour. She warned me to behave myself."

As we laughed, Cooper looked over at Raju, and frowned.

"Raju, what the hell? You're wearing running shoes and jeans?"

"Uh, yeah," he replied, a bit sheepishly. "The invitation letter just said jacket and tie for boys. It didn't say anything about dress pants and dress shoes."

I snickered and shook my head.

"Oh, Raju, Raju," said Cooper. "So smart, yet so dumb."

"Who's the guest of honour?" asked Grace.

Ric turned the name card over. "It says 'Aviva Goldberg'. Who is he?"

Cooper shrugged. "Maybe some math prof at the university?"

"What kind of a name is Aviva, anyway?"

"It's a palindrome," said Raju. "You know, his name reads the same whether you read it from the left or from the right. Maybe his childhood nickname was 'The Palindrome'."

Grace rolled her eyes.

"What a funny name," said Ric. "As a kid, he probably got beaten up a lot for having such a weird name."

"Yeah," nodded Raju. "But it could've been a lot worse. His parents could have named him Soo."

While Ric scowled, everyone else at our table began giggling, and soon I had my hand on my stomach, trying to control my laughter. One by one we'd calm down and then someone else at the table would start laughing, and that had a contagious effect.

Everyone turned around. It was impossible not to notice us.

Just then, I noticed a heavy-set woman march through the entrance door, dressed in a dark black suit. She was much taller than me, probably even taller than Cooper. She scanned the seats and we locked eyes. The lady headed towards me.

She stopped just behind the empty seat. All of us stopped laughing right away. We stared at this large lady in astonishment and fear. Was she going to kick us out of the banquet hall for being too noisy?

"Hi, everyone. Sorry I'm late. I'm Aviva Goldberg."

After a few paralyzing seconds, Grace broke the silence.

"Hi, Ms. Goldberg, my name is Grace Wong," she said slowly, standing up and shaking the lady's hand.

The rest of us complied, standing up and introducing ourselves, trying to act as natural as possible.

"Well, I'm really happy I can join all of you tonight," she replied. She took her seat, and invited all of us to do the same.

"Please, tell me about yourselves."

One by one, we shared our favourite memories from the past five days. Ric was happy to comply, and told some funny stories that lightened the mood.

Raju looked at Ms. Goldberg. "Do you know that your first name is a palindrome?"

I rolled my eyes.

"Yes, I do," Mrs. Goldberg replied. "Last year, I attended this banquet dinner, and you know, there was a student at my table who made the exact same observation. It's interesting that only at math camps do people notice these types of things. But the young gentleman I met last year thought Aviva was a man's name. Isn't that ridiculous?"

I coughed into my napkin.

Ms. Goldberg mercifully changed the subject, and introduced herself, explaining how she had been employed by Royal Oil for nearly twenty-five years, rising from a junior engineer to the President of the company's charitable foundation. She explained how she oversaw $12 million in contributions each year to hundreds of projects across Canada, from aboriginal development to sponsoring amateur hockey players to supporting initiatives in math education.

Ric looked at Ms. Goldberg. "Can I ask you a question?"

"Absolutely."

"I know why your company sponsors hockey, since everyone watches hockey. But you're a gas company. No offense, but why sponsor a math camp?"

Just as Ms. Goldberg was about to respond, Marlene clicked her glass with a spoon to get our attention. Ms. Goldberg turned to Ric. "Great question. I'll answer it in my speech."

Marlene introduced the Dean of the Faculty of Science at the University of Ottawa, and thanked him for hosting tonight's banquet dinner. After a few short words of welcome in both official languages, the Dean introduced J. William Graham, the Executive Director of the Canadian Mathematical Society.

Ah, so it was Dr. Graham speaking to J.P. and David earlier. I remembered the blond-haired man from Rachel's clip on CBC.

Dr. Graham explained how mathematics was the driver of numerous technological breakthroughs in recent decades, and explained the mission of the Canadian Math Society in promoting mathematics education among the country's youth through contests, camps, scholarships, web-based resources, outreach programs, and the selection and training of Canada's IMO team. He concluded his speech by introducing Ms. Goldberg.

"Thank you so much for your kind words," said Ms. Goldberg. "As Dr. Graham said, my name is Aviva Goldberg, and I'm the President of the Royal Oil Charitable Foundation. We've proudly supported the CMS Canada Math Camp since 1999. It's an honour for us to be involved yet again. You know, just before the speeches started, I was asked by a young man at my table why we sponsor this camp. What's in it for us? Why would a gas company sponsor a math camp?

"I want to answer Ric's question by sharing my story. Throughout my twenty-five-year career, I've used math in various ways, such as developing resource recovery technologies to reduce our environmental impact, to managing a large operational team and allocating limited resources in the best possible way to maximize productivity. I've discovered that math is critical to the success of my work; more importantly, I've discovered that math is critical to the success of our country.

"But today only a small percentage choose to pursue careers in math, science, and engineering – fewer than twenty percent of university degrees in Canada are awarded in those three areas. When I became the President of our corporate charitable foundation, I had the opportunity to pay it forward, and ensure that our time and money was invested in Canada's most important asset – our young men and women.

"In the 1970s, only five percent of medical students were female. I remember a time when men said that women couldn't be good doctors, that they lacked the skills and intelligence of their male counterparts. Of course, that myth has been completely shattered, and now nearly sixty percent of medical students in Canada are female. Who's to say math can't end up that way too? Right now, thirty percent of Ph.D. degrees in mathematics are given to women, and this percentage will surely increase in the years ahead.

"Over the past decade, we've contributed more than $20 million to support math and science education in Canada, and have done particularly well promoting math among young girls, to increase the participation and involvement of females in fields such as engineering and computer science. We're confident this investment will reap huge dividends, so the country that invented the telephone, Canadarm, and Blackberry can continue to lead innovation in the twenty-first century. That, in turn, helps companies like ours, as the young people we've invested in choose to stay and work in Canada.

"To answer your question, Ric, that's what's in it for our gas company. It's so we can contribute towards creating the Canada we want. But you don't have to be a big company to do that. Individuals can do that, no matter how old you are. I encourage you to do the same thing, to give your time and your talents to the causes you believe in, and be the change you wish to see in this world. Thank you."

We clapped as Ms. Goldberg returned to her seat. After we finished our meal, she shook hands with each of us. Ms. Goldberg said she was especially happy to meet Grace and me.

We said goodbye and walked back towards the residence. After a couple of minutes, Grace turned to me and Cooper.

"Can you two help me? I think I dropped my watch this afternoon. I think I know where it is. Can you help me look for it?"

Cooper shrugged. "It's dark. We'll never find it."

I was confused. "But, you don't wear a watch."

"Yeah, I do," replied Grace. "I mean, I was wearing one this afternoon. Here, just come with me."

Cooper and I exchanged puzzled glances, and followed Grace towards some trees on the edge of campus.

"Where are we going?" I asked.

"Just trust me. We'll be there in a minute."

The three of us arrived in a quiet area, in a grassy field surrounded by large trees. There were a couple of lights on so we could make out the path ahead. I glanced around, and saw no one.

Satisfied at reaching her destination, Grace smiled at me and Cooper.

"I found what I was looking for. I'm heading back to the residence. See you later!"

"Wait!" I shouted, as Grace ran past us.

I realized what Grace had done, purposely leaving me alone with Cooper for the first time.

Me and Cooper. Alone.

I stared nervously at Cooper, and felt a lump in my throat.

After an awkward silence, Cooper moved his right hand towards my left. He touched one finger, and then another.

"Can I do this?"

"Yeah," I said, moving closer and putting my hand in his.

We stared into each other's eyes, and he leaned over to hold my other hand.

"I really like you, Bethany."

"I know," I said, smiling up at him. "And I really like you too."

"Will you be my girlfriend?"

My heart skipped a beat. I couldn't believe this was happening.

"Of course," I whispered.

Our faces were inches apart now. He tilted his head to the right, and moved closer. I stood there, unable to move any part of my body.

Our lips touched. I closed my eyes and fell into his arms.

*Wow.*

## The Canadian Mathematical Olympiad, Problem #2

**Find all real solutions to the following system of equations.**

$$\begin{cases} \dfrac{4x^2}{1+4x^2} = y \\[2ex] \dfrac{4y^2}{1+4y^2} = z \\[2ex] \dfrac{4z^2}{1+4z^2} = x \end{cases}$$

## Solution to Problem #2

The stories of that magical week come flashing back.

I see faces of people I met three summers ago: Grace, Rachel, Marlene, Raju, Albert . . . and Cooper. The memories are still so fresh in my mind.

I know I'm taking precious seconds reminiscing about one particular memory that week.

My first kiss. My only kiss.

I then remember what happened a few months after the camp.

*Focus, Bethany! Focus!*

I take a deep breath. I know how to solve the problem; all I need to do is calm down, concentrate on the task in front of me, and get back into my zone. I know I can do it.

*9:52 a.m.*

Putting my pen on the table, I open my eyes and stare at my notes to Problem #2. Yes, I definitely know how to solve this.

The first key insight is realizing the equation $\frac{4x^2}{1+4x^2} = y$ forces the inequality $x \geq y$. In any solution $(x, y, z)$ to the system of equations, the value of $x$ has to be at least as large as the value of $y$.

The second key insight is seeing and exploiting the symmetry in the system of equations. Because of this symmetry, the same argument can then be repeated on the other two equations.

Specifically $\frac{4y^2}{1+4y^2} = z$ forces the inequality $y \geq z$ and $\frac{4z^2}{1+4z^2} = x$ forces the inequality $z \geq x$.

I recall Albert's brilliant solution at the Canada Math Camp, where he showed that $x \geq y$ and $y \geq x$ to conclude that $x = y$. There is only one additional step required in this problem, since there are three variables instead of two. But the underlying idea is the same, setting up a *chain of inequalities* to show that all the variables must be equal.

$$x \geq y, \;\; y \geq z, \;\; z \geq x \;\text{ imply }\; x \geq y \geq z \geq x$$

If Xavier is at least as tall as Yvette, Yvette is at least as tall as Zachary, and Zachary is at least as tall as Xavier, then all three people must have the exact same height. These three symmetric inequalities necessitate that $x = y = z$.

Therefore, in any solution $(x, y, z)$ to the system of equations, we must have $x = y = z$. So a difficult three-variable problem becomes a simple one-variable problem. In other words,

$$\begin{cases} \dfrac{4x^2}{1+4x^2} = y \\ \\ \dfrac{4y^2}{1+4y^2} = z \\ \\ \dfrac{4z^2}{1+4z^2} = x \end{cases} \quad \to \quad \begin{cases} \dfrac{4x^2}{1+4x^2} = x \\ \\ \dfrac{4y^2}{1+4y^2} = y \\ \\ \dfrac{4z^2}{1+4z^2} = z \end{cases}$$

From here, I know exactly how to complete the solution.

All I have to do is prove the first key insight, that the equation $\frac{4x^2}{1+4x^2} = y$ implies the inequality $x \geq y$. As I work out some algebraic calculations to establish this, I run into a problem.

The inequality is true, but only if $x$ is non-negative. If $x$ is negative, the inequality $x \geq y$ no longer holds. For example, if $x = -1$, then $y = \frac{1}{5}$.

I feel beads of sweat on my neck.

Can fix this? Is there a simple way to show that $x$ can't be negative?

Staring at my notes, my eyes move towards the third equation $\frac{4z^2}{1+4z^2} = x$. I realize that no matter what $z$ is, $x$ can't take on a negative value.

If $(x, y, z)$ is a solution, then we must have $x \geq 0$. And by symmetry, we must have $y \geq 0$ and $z \geq 0$.

As long as I explain why the three variables must all be at least zero, the rest of my proof works. I can show the inequalities $x \geq y,\ y \geq z,\ z \geq x$ hold, enabling me to complete the problem. Yes, this works for sure!

I start writing my solution, and smile when I've finished writing the last word.

*10:07 a.m.*

Beautiful.

**Problem Number:** 2
**Contestant Name:** Bethany MacDonald

We claim the only solutions are $(x, y, z) = (0,0,0)$ and $(x, y, z) = \left(\frac{1}{2}, \frac{1}{2}, \frac{1}{2}\right)$.

Consider the first equation $\frac{4x^2}{1+4x^2} = y$. Since $x$ is a real number, both $4x^2$ and $1 + 4x^2$ are non-negative, implying that $y \geq 0$. By symmetry, the equations $\frac{4y^2}{1+4y^2} = z$ and $\frac{4z^2}{1+4z^2} = x$ imply $z \geq 0$ and $x \geq 0$ respectively.

Suppose $(x, y, z)$ is a solution to this system of equations. Then from above, we must have $x, y, z \geq 0$. Now we prove that $x \geq y$, $y \geq z$, and $z \geq x$, enabling us to deduce that $x = y = z$. To prove the first inequality, note that $0 \leq x(2x - 1)^2$ since $x \geq 0$ from above. Expanding, we have $0 \leq x(2x - 1)^2 = x(4x^2 - 4x + 1) = 4x^3 - 4x^2 + x$.

This implies that $4x^2 \leq 4x^3 + x$, which is equivalent to $x(1 + 4x^2) \geq 4x^2$. Dividing both sides by the positive term $1 + 4x^2$, we have $x \geq \frac{4x^2}{1+4x^2}$. By symmetry, we also have $y \geq \frac{4y^2}{1+4y^2}$ and $z \geq \frac{4z^2}{1+4z^2}$.

We are given that $\frac{4x^2}{1+4x^2} = y$, $\frac{4y^2}{1+4y^2} = z$, and $\frac{4z^2}{1+4z^2} = x$. From the above paragraph, this implies the three inequalities $x \geq y$, $y \geq z$, and $z \geq x$. Clearly, this chain of symmetric inequalities forces $x = y = z$.

Therefore, all solutions to this system of equations must be of the form $(x, y, z) = (k, k, k)$, for some real number $k$. We now prove that either $k = 0$ or $k = \frac{1}{2}$.

Substituting $x = k$ and $y = k$ into the first equation, we have $\frac{4k^2}{1+4k^2} = k$. Therefore, $4k^2 = k(1 + 4k^2) = k + 4k^3$ implying that

$$0 = 4k^3 - 4k^2 + k = k(4k^2 - 4k + 1) = k(2k - 1)^2$$

If $k$ is a real number satisfying $0 = k(2k - 1)^2$ then either $k = 0$ or $(2k - 1)^2 = 0$. In other words, $k = 0$ or $k = \frac{1}{2}$.

Therefore, we have proven the only solutions are $(x, y, z) = (0,0,0)$ and $(x, y, z) = \left(\frac{1}{2}, \frac{1}{2}, \frac{1}{2}\right)$. By inspection, we see both triplets are indeed valid solutions to the given system of equations.

Our proof is complete.

## The Canadian Mathematical Olympiad, Problem #3

**Twenty-five men sit around a circular table. Every hour there is a vote, and each must respond *yes* or *no*.**

**Each man behaves as follows: on the $n^{th}$ vote, if his response is the same as the response of at least one of the two people he sits between, then he will respond the same way on the $(n + 1)^{th}$ vote as on the $n^{th}$ vote; but if his response is different from that of both his neighbours on the $n^{th}$ vote, then his response on the $(n + 1)^{th}$ vote will be different from his response on the $n^{th}$ vote.**

**Prove that, however everybody responded on the first vote, there will be a time after which nobody's response will ever change.**

## Problem #3: Circle Voting

I roll my eyes. Twenty-five *men*.

I read Problem #3 a couple more times. To make sure I understand what's being asked, I consider a simpler example and try to find a pattern. I draw a circle and put five female names around it, randomly assigning a Yes or No to each woman.

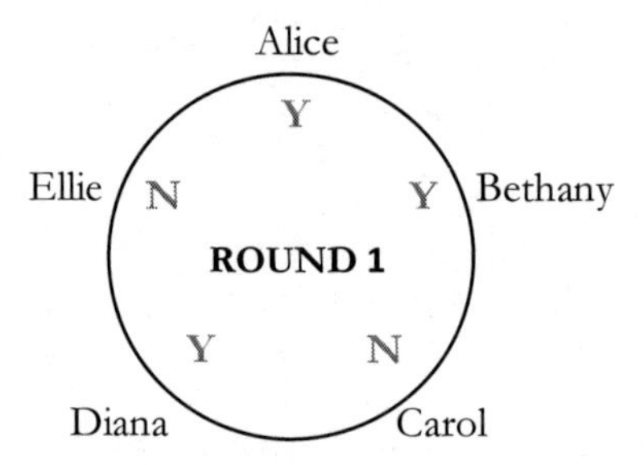


On the second vote, each woman behaves as follows: If her first vote agrees with at least one of her neighbours, then she will vote the same way. If her first vote disagrees with both of her neighbours, then she will change her vote and conform to the opinions of her neighbours.

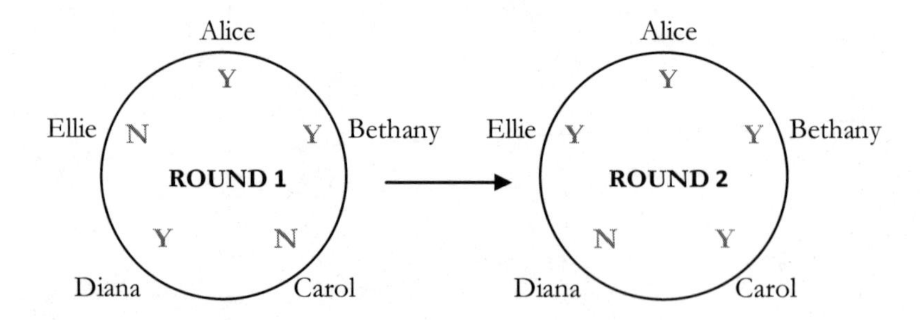


Alice stays as Yes since her vote agrees with Bethany.
Bethany stays as Yes since her vote agrees with Alice.
Carol changes to Yes since her vote disagrees with Bethany and Diana.
Diana changes to No since her vote disagrees with Carol and Ellie.
Ellie changes to Yes since her vote disagrees with Diana and Alice.

From the results of Round 2, I see only Diana changes her vote in Round 3. The other four women keep voting Yes since at least one neighbour also voted Yes in Round 2.

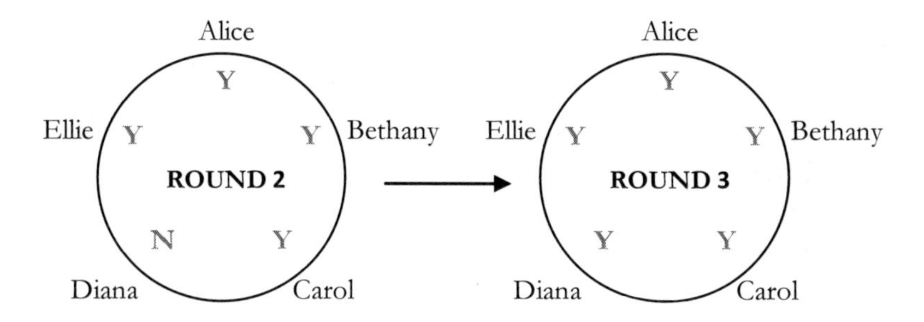


After three rounds of voting, there is a unanimous Yes. Every subsequent vote will stay the same, and nobody's response changes after Round 3.

I know what I want to prove. I want to show that no matter how the women vote in Round 1, at some point everyone will reach consensus: that all the women will eventually vote Yes, or all the women will eventually vote No. If I can prove that, I'm done.

I try another example, where the first three women vote Yes and the last two women vote No, and make a diagram of what happens from Round 1 to Round 2.

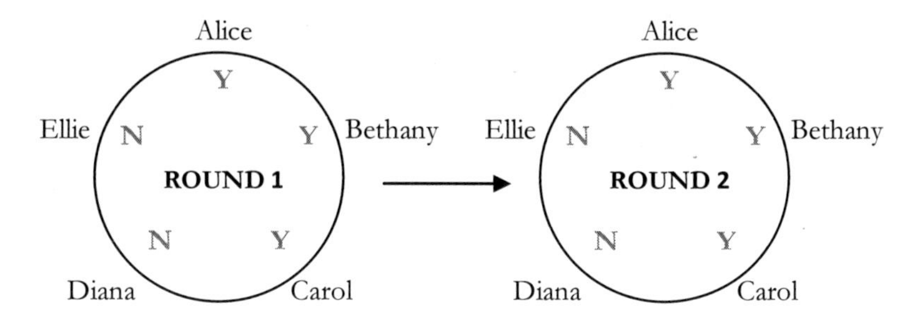


I pause and tap my pen on the table. Something's wrong.

The results of Round 1 are identical to Round 2. If Round 1 is identical to Round 2, then Round 2 must be identical to Round 3, and so on. The five women will *never* reach consensus.

I hesitate, wondering what's going on. In this case, consensus won't ever be reached. The first three women (Alice, Bethany, Carol) form a "Yes-coalition" where they stick together and vote Yes forever, and the last two women (Diana, Ellie) form a "No-coalition" where they stick together and vote No forever.

Is the question wrong? I re-read the problem statement: "Prove that, however everybody responded on the first vote, there will be a time after which nobody's vote will ever change."

I feel relieved. The problem doesn't require the women to reach consensus. The goal is to show that after some vote, no woman will change her mind. In my first example, no woman changes her mind after Round 3, while in my second example, no woman changes her mind after Round 1.

This process of writing letters around a circle reminds me of a problem I've solved before. But I can't remember the specifics.

I suspend that thought and return to my last example. I see that once a woman is in a Yes-coalition, she will vote Yes forever; similarly, once a woman is part of a No-coalition, she will vote No forever.

Unlike the TV show *Survivor*, I realize that a coalition only needs two people, and once a coalition is formed, no one will ever backstab another, and the coalition will never break.

Just like Grace and me.

If someone is not part of a coalition, that means both her neighbours vote the other way, and she will conform to their way of thinking on the next vote.

I sketch out another scenario with more people, but decide to arrange the votes in a straight line to make it easier to visualize.

Round 1: N N N N N **Y N Y N Y N** Y Y Y Y Y

I see that the five women on the left and the five women on the right belong to a coalition. These women will continue voting the same way in all subsequent rounds.

However, each of the six women in the middle switch their votes in the next round.

Round 1: N N N N N **Y N Y N Y N** Y Y Y Y Y
↓ ↓ ↓ ↓ ↓ ↓
Round 2: N N N N N N **Y N Y N** Y Y Y Y Y Y

.

Looking at the sequence, I see that two "non-coalition" women get sucked into a coalition, leaving only four people not belonging to a coalition. I see the process continues until everyone is drawn into the No-coalition on the left or the Yes-coalition on the right.

Round 1: N N N N N **Y N Y N Y N** Y Y Y Y Y
↓ ↓ ↓ ↓ ↓ ↓
Round 2: N N N N N N **Y N Y N** Y Y Y Y Y Y
↓ ↓ ↓ ↓
Round 3: N N N N N N N **Y N** Y Y Y Y Y Y Y
↓ ↓
Round 4: N N N N N N N N Y Y Y Y Y Y Y Y

I understand the heart of the problem. At the beginning, some of the women belong to a coalition, depending on how her neighbours vote. Everyone who does not belong to a coalition will eventually be drawn into one.

But how on earth am I supposed to explain this?

From my examples, it makes intuitive sense that every woman will eventually get sucked into a No-coalition or a Yes-coalition, but it's not clear why. I don't know how to justify this rigorously.

The Canadian Mathematical Olympiad demands formal proofs. Even a small gap in logical reasoning will be detected by the examiners, leading to major point deductions. I can't afford that.

I look up at the clock.

*10:20 a.m.*

The contest is nearly halfway done. All I've got to show for it are solutions to the two easiest problems.

I take a deep breath, trying to regain my focus. Suddenly, I have a strange thought.

*What if no one belongs to a coalition at the beginning?*

I wonder whether my theory would still hold. To check, I create a simple four-woman scenario, and see what happens from Round 1 to Round 2.

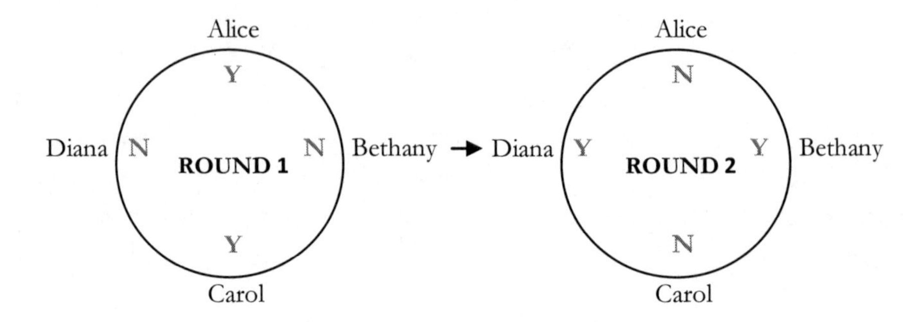


*Oh no.*

My coalition idea doesn't work. In Round 1, the votes are Y, N, Y, N around the circle. No one belongs to a coalition. In Round 2, everyone switches her vote, so that the votes are N, Y, N, Y around the circle. Since no one belongs to a coalition after Round 2, that means that Round 3 has to produce the same pattern as Round 1, and this pattern must repeat forever.

Each woman changes her vote in every round. No one ever belongs to a coalition.

In confusion, I read the problem again:

> Prove that, however everybody responded on the first vote, there will be a time after which nobody's vote will ever change.

What's going on? The statement in the Olympiad Problem isn't true!

I've found a simple counterexample with four women where everyone changes her vote each round, whether it's in Round 1 or Round 1000.

This isn't making any sense.

I close my eyes and try to calm down. After thirty seconds, I'm still feeling anxious and jittery.

I look up and get the attention of the person supervising my exam. It's no longer the superintendent in the designer suit; it's a blonde-haired woman reading a report and making some changes with a red pen.

"Excuse me?" I say, speaking out loud for the first time in ninety minutes.

The young blonde looks up from her papers, and appears startled by the interruption.

"Yes, Bethany?"

"Can I use the washroom?"

"Sure. It's down the hall and to your right."

As I enter the stall and sit on the toilet, I press my fingers to my temples and put my head down.

"It's too hard," I say out loud, to no one in particular.

*My dream of becoming a Math Olympian . . . is over.*

I kick the stall door in frustration.

"I can't do this."

Tears form in my eyes.

"It's impossible."

I bolt up in shock, remembering a time when I heard someone say those words.

*It's impossible.*

It was Grade 10. On my first day of high school.

In my very first class.

# 20

"Hey guys, what's up?"

I walked up to Bonnie and Breanna, in front of our new lockers at Sydney High School.

Breanna looked away.

"Oh, hi," replied Bonnie, not bothering to look in my direction.

What was wrong? We were supposed to hang out together all summer, but I only saw Bonnie and Breanna a few times after coming back from the Canada Math Camp.

Last time we saw each other, I talked a lot about Cooper. By the end of the day, I could tell they were annoyed. They were jealous. But I knew they'd get over it, especially after they got boyfriends of their own.

We headed towards Room 204 for Grade 10 Math, our first class at Sydney High School. As we walked into the room, Bonnie turned towards us.

"What the heck is this?"

Instead of a typical classroom where the chairs and desks faced the front, the teacher had arranged the seats in groups of four. The chairs in each group formed a square. Half of the students were already there, and had already formed groups.

I saw Gillian chatting and laughing with Vanessa, while the twins Alice and Amy sat across from them and were whispering among themselves. I winced, realizing Gillian and her clique were all in this class.

"Let's sit there," I said, walking towards the only group of unoccupied seats. Bonnie and Breanna followed me.

Within a few minutes, every seat in the classroom was taken, except the one right next to me.

The bell rung to signal the beginning of high school.

Our teacher entered, carrying a large grocery bag. He adjusted his tie, and stared at us with eyes that radiated focus and intensity. The talking in the room ceased, and we all turned to face our new teacher: he was in his mid-forties, had long jet black hair slicked back into a ponytail, and his skin had a distinct olive hue that was unlike anybody's I had seen before.

"Good morning, class. My name is Mr. Marshall, and I'll be your teacher for Mathematics 10 this semester. Welcome to Sydney High School."

Somebody walked into the classroom.

Mr. Marshall glared at the boy. "Why are you late?"

"I was just chatting with . . ."

"You were chatting with someone instead of coming to class on time?"

"But the bell rang just a few seconds ago . . ."

"No excuses, Patrick!" replied Mr. Marshall, in a voice that gave me goose bumps. "Will you ever be late for class again?"

"No," he mumbled.

"Speak up, young man, I can't hear you. Will you ever be late for class again?"

"No, sir."

Mr. Marshall continued to stare right through Patrick. As I watched the exchange between the two, I made a mental note never to be late for Mr. Marshall's class.

"There's an empty seat right there," said Mr. Marshall, motioning to the chair next to me.

Patrick shuffled to his seat, head bowed. It was the first minute of the first class of the first day of high school, and he was already in trouble.

Mr. Marshall briskly walked around the classroom holding his grocery bag, handing a metal can to each group. When he came around to us, Mr. Marshall handed Bonnie a small can of Chunk Light Tuna.

"If you received a can, hold it high up in the air so that everyone else can see."

We looked around and saw cylindrical steel cans of various sizes, ranging from our small tuna can to a huge can of French pea soup that could feed an entire family.

"Here's what I'd like you to do. You see seven different cans – some cans are big while others are small; some are tall and thin, while others are short and wide. I'd like you to take one minute on your own to think about the following question: *What is the shape of the perfect can*?"

Mr. Marshall didn't waste any time at all. The bell had rung only a few minutes ago, and we were already assigned a task.

Not wanting to be the second victim of his angry wrath, I stared at my desk, too scared to look up and make eye contact with Mr. Marshall.

"Okay, sixty seconds are up. Now, what I want you to do is to pair up with the person beside you. Share your thoughts and ideas with your partner. You've got three minutes."

I looked at Patrick. He appeared shaken. I extended my hand.

"I'm Bethany."

He shook it. "Nice to meet you."

I dropped my voice to a whisper and lowered my head close to his.

"The teacher is so mean. You were just a few seconds late. What's his problem?"

"That's okay," he whispered back. "He's always been like that."

"Like what?"

"Concentrate!" yelled Mr. Marshall. I looked up and saw Mr. Marshall directly in front of Patrick and me. Bonnie and Breanna were close-by and they looked just as scared as I was.

"Young lady, what is your name?"

"It's Bethany."

"Bethany, are you and Patrick discussing the assigned question, or chit-chatting about something else?"

I was too stunned to answer. Thankfully Patrick came to the rescue.

"I'm sorry, sir. We're going to talk about the question right now."

"Good. Let's focus."

Mr. Marshall walked to the other end of the classroom. I took a deep breath to clear my head.

"What do you think about the question?" asked Patrick.

"What was his question?" I whispered.

Mr. Marshall introduced the next activity, to share our answers with our group of four. Thankfully, Bonnie and Breanna were paying attention: the question was to describe the shape of the perfect can.

After five minutes of discussion, Mr. Marshall got our attention.

"I'm going to point to one person in each group. When I point to you, I want you to stand up, tell us your name, and then share your group's answer with the rest of us."

I saw his finger pointed at me.

I took a deep breath and stood up, holding the tuna can in my hand.

"My name is Bethany MacDonald. Our group talked about designing a can that's easy to hold. This tuna can is great, because it's wide enough to fit in the palm of your hand. Also the can shouldn't be too tall, or else it'll be really hard to stack in a grocery store. The perfect can should be just a few inches high, like a soup can, and just wide enough to hold in your hand. That's what we think a perfect can is."

"Thank you, Bethany. Okay, moving on."

After five more people had shared, Mr. Marshall pointed towards one last student. She smiled and stood up.

"My name is Gillian Lowell. Our group's answer was completely different from the others – we didn't talk about the can's heaviness, height, or width. We don't think those things are important. We think the important thing is to reduce the amount of metal in the can. Look at our pineapple juice can – it's too tall and too thin. If you make the can shorter and wider, you can get the same amount of juice and use less material. If you make the can's *surface area* as small as possible, then you've got a perfect can."

"Excellent, thank you very much," said Mr. Marshall. "I liked all of your responses, but let me follow up on something that we just heard. Gillian talked about the surface area of the can, and how the goal is to make that surface area as small as possible. Can someone remind us what surface area is?"

Someone in the front raised his hand. "My Grade 9 teacher taught us that if you wrap the can in paper so that all the metal is covered, then the surface area is the total area of the paper you use."

"I like that," replied Mr. Marshall. "Very good. In other words, the surface area of a cylindrical can is the sum of three areas: the circular lid at the top, the circular base at the bottom, and the side area around the can. Who can tell me the formula for the surface area of a cylinder?"

Vanessa raised her hand. "It's $2\pi r^2 + 2\pi rh$, where $r$ is the radius of the can and $h$ is the height."

I nodded, following along Mr. Marshall's explanation on the chalkboard.

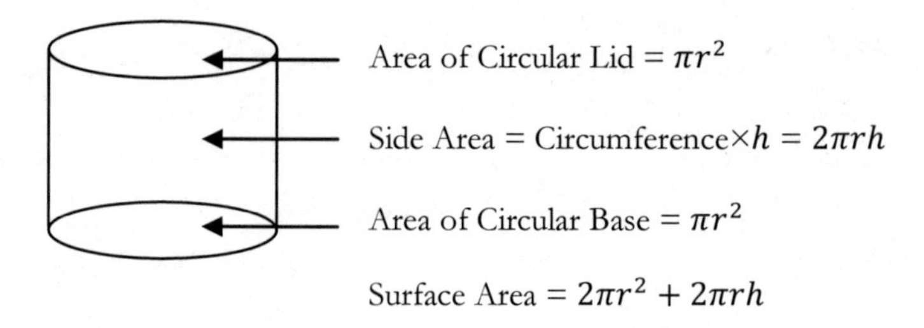


"And how do we determine the volume of a cylindrical can?"

Someone in the class raised his hand.

"The volume is $\pi r^2 h$, which is the area of the circular base multiplied by the height."

"Well done," said Mr. Marshall. "I'm now going to give you five different cylindrical cans. In your groups, I want you to calculate the volume and surface area of each can. Remember that $\pi$ is about 3.14."

Each group member got a printout, with five different shapes. Below the pictures were four questions.

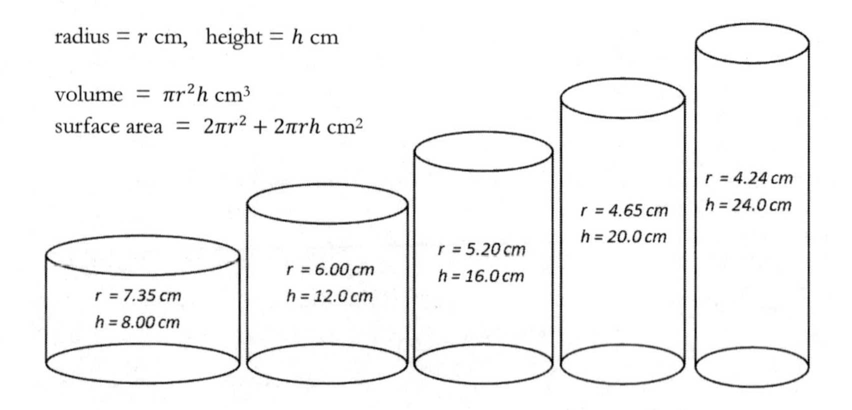


#1: What is the volume of each cylindrical can? What do you notice?

#2: Predict which of the cans has the lowest surface area. What is your guess?

#3: Confirm your prediction by calculating each can's surface area. Was your guess correct?

#4: In the can with minimum surface area, what is the relationship between $r$ and $h$?

We did the first case together on Bonnie's calculator: the radius was 7.35 cm and the height was 8.00 cm. From the formula V = $\pi r^2 h$, the volume worked out to 1357.045 $cm^3$. We rounded this to 1357 $cm^3$.

We quickly calculated the other volumes and put all the numbers in a table:

| | *Can #1* | *Can #2* | *Can #3* | *Can #4* | *Can #5* |
|---|---|---|---|---|---|
| Radius | 7.35 cm | 6.00 cm | 5.20 cm | 4.65 cm | 4.24 cm |
| Height | 8.00 cm | 12.0 cm | 16.0 cm | 20.0 cm | 24.0 cm |
| **Volume** | **1357 $cm^3$** | **1356 $cm^3$** | **1358 $cm^3$** | **1358 $cm^3$** | **1355 $cm^3$** |

"The volumes are all about the same," said Breanna, looking bored.

"That answers Question #1," replied Patrick. "All five cans have roughly the same volume."

"What's next?" asked Breanna.

"Predict which can has the lowest surface area," replied Bonnie. "I think it's the one in the middle."

"Let's check," I replied, after all of us had agreed with Bonnie's prediction.

We applied the formula SA = $2\pi r^2 + 2\pi rh$ to determine each can's surface area.

| | *Can #1* | *Can #2* | *Can #3* | *Can #4* | *Can #5* |
|---|---|---|---|---|---|
| Radius | 7.35 cm | 6.00 cm | 5.20 cm | 4.65 cm | 4.24 cm |
| Height | 8.00 cm | 12.0 cm | 16.0 cm | 20.0 cm | 24.0 cm |
| Volume | 1357 $cm^3$ | 1356 $cm^3$ | 1358 $cm^3$ | 1358 $cm^3$ | 1355 $cm^3$ |
| **Surface Area** | **709 $cm^2$** | **678 $cm^2$** | **692 $cm^2$** | **720 $cm^2$** | **762 $cm^2$** |

"I was wrong," said Bonnie. "The second can has the smallest surface area."

"Whatever," said Breanna.

Patrick ignored her and began reading the last question. "In the can with the minimum surface area, find the relationship between $r$ and $h$."

"That's easy," I replied. "The height is twice the radius: $h$ = 12.0 cm and $r$ = 6.0 cm. We're done."

After all the groups had finished, Mr. Marshall taught a short lesson where we learned how to graphically demonstrate that the minimum surface area occurred when $h$ = 12.0 cm and $r$ = 6.0 cm, when the volume of the cylinder was exactly 1356 cm$^3$.

He did this by using some algebra to express the height $h$ as a *function* of $r$, and found a simple formula for the surface area which only used the single variable $r$. He then used a graphing calculator to plot a curve, describing what happens to the surface area as the radius varies.

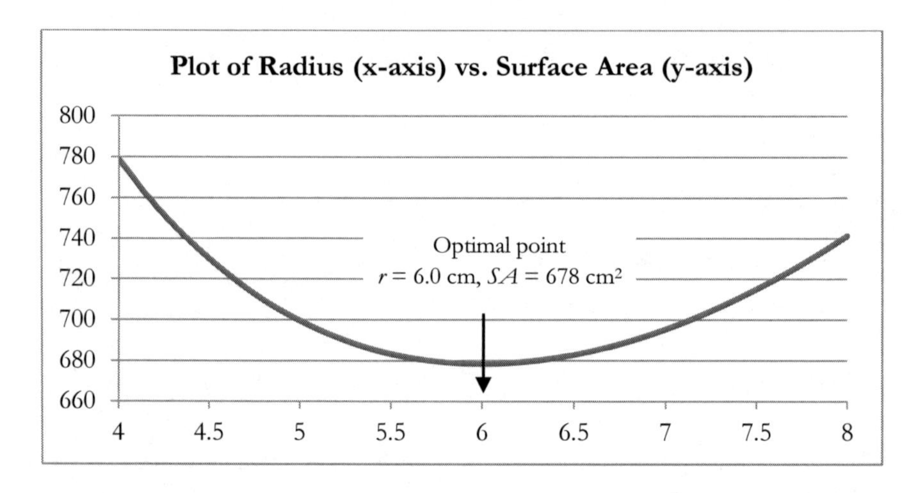


From the image, it was clear the surface area was minimized when the radius was 6.0 cm.

Gillian raised her hand. "For this question, the best option is to make the radius 6 and the height 12. Is that always true? To make the surface area as small as possible, is the height always twice the radius?"

Mr. Marshall smiled. "That is something you will determine in your assignment."

"That means the answer is yes," said Gillian.

"Perhaps," said Mr. Marshall, in a tone that invited no further comments.

"Is it easy to prove?"

"Prove what, Gillian? Please be specific when you ask a question."

"That if you want to make the surface area as small as possible, the height must be twice the radius?"

Mr. Marshall paused. "Yes, that is correct. For any cylindrical can, the surface area is minimized when the height is twice the radius. On your assignment, you will justify this result for a specific can I will give you. Unfortunately, there's no way to prove the general result formally, using Grade 10 mathematics. To do that, you will require Grade 12 calculus."

I slowly raised my hand. "You can't prove it using Grade 10 math?"

"No, it's impossible. You need calculus, which some of you will study two years from now."

"But what if you . . ."

"Like I said," replied Mr. Marshall, glancing at the clock, "without calculus, it's impossible."

I decided not to say any more, but made a note in my journal to ask Rachel and Mr. Collins about this.

Mr. Marshall reached over and took out a thin stack of papers from one of his file folders, giving one copy to each student.

"Before the bell rings, I want to explain the assignment. Each of you will be given a food product that comes in a cylindrical steel can, such as Dole's Pineapples or Campbell's Chicken Soup. As you can see, that information is on the top of your sheet. Each of you will get a different product. You will buy that item at the grocery store, and measure its radius and height to calculate its volume.

"You will also go online and find out how many cans of that product are sold in Canada each year. Now doing the same type of graphical analysis as we just did, you will determine the minimum surface area for a cylindrical can with that volume. You will write a detailed report on your findings, with an estimation of the amount of steel that the company could reduce by simply switching the shape of the can to minimize the surface area.

"Students will get bonus marks for doing additional research to estimate how much money the company would save, and the amount of greenhouse gas emissions that would be eliminated, by switching the shape of the can to the dimensions that are mathematically optimal."

Vanessa raised her hand. "You want us to write a detailed report? But isn't this math class?"

"Yes, I want you to write a detailed report," replied Mr. Marshall. "You will be marked not just on the accuracy of the mathematics, but also on the structure of your report, as well as on your spelling and grammar. You will first hand in the report to me. After I've marked it, you will make the required corrections and mail your final report to the President of the company whose product you are analyzing."

Breanna looked at Mr. Marshall with a raised eyebrow.

"Are you serious?"

"Most certainly. What is the point of studying mathematics? Is it just to memorize formulas and rules so you can get high marks for college and university? No. You are studying mathematics because it empowers you with the tools to influence and impact social change!"

I looked at Mr. Marshall, wondering what planet he was from.

"Each of you will be writing to the president of a major Canadian company, like Tim Hortons or Campbell Soup. You will explain how these companies can apply Grade 10 mathematics to save millions of dollars and make their businesses more environmentally sustainable."

I stared at Mr. Marshall. Didn't he realize that we were only fifteen years old?

Vanessa raised her hand. "It says that the assignment is due next Monday. Can we complete all this in just six days?"

"You will," said Mr. Marshall. "But only if you get started right away."

The bell rang. We walked out, shaking our heads in disbelief.

"That was insane," Bonnie said.

"Tell me about it," I replied.

Patrick walked up to us. The four of us headed towards our lockers to get ready for the next class.

I glanced over at Patrick. He was a few inches shorter than me, and wore his long black hair in a ponytail, just like our scary teacher. I turned to him.

"That Mr. Marshall guy is crazy."

Patrick shook his head. "He's strict, but he's a good man."

"What are you talking about?" I said. "You were ten seconds late for class, and he yelled at you in front of everyone."

"And look at that assignment he just gave us," added Bonnie. "Brutal."

Patrick shrugged. "He has high expectations. He's been like that ever since I was a kid."

"Well, he's a jerk," said Breanna.

"Totally," added Bonnie.

I turned to Patrick. "You've known him since you were a kid?"

Patrick sighed and shrugged in defeat. He dropped his voice to a whisper. "Mr. Marshall is my father."

# 21

"It's impossible without calculus. Mr. Marshall's right."

"That's not true, Mr. Collins," I replied. "Rachel Mullen said it could be done!"

"Did she show you how?" he responded, sipping Le Bistro's new fair-trade coffee.

"No, she didn't. But she facebooked me last night with some advice."

"She facebooked you?"

"You don't know what Facebook is?"

Mr. Collins shook his head. "I've heard of Facebook but didn't realize it could be used as a verb. So what did Rachel write when she 'facebooked' you?"

"I printed it out," I said, handing Mr. Collins a sheet of paper.

> Bethany, use *dimensional analysis* to derive a relationship between the surface area of a cylinder and its volume, to prove that for a fixed volume, the surface area is minimized precisely when the cylinder's height equals its diameter ($h = 2r$). Good luck!

I looked at Mr. Collins. "What's dimensional analysis?"

He returned the paper to me.

"You know, Bethany, I was remembering that we started these Saturday sessions exactly three years ago this week. At that time, I was still teaching at Sydney High, and you were a Grade 7 student at Pinecrest. Look at how far you've come since then."

"There's so much I still don't know," I said, impatiently waiting for him to answer my question.

"I feel like I've taught you everything I know. These Saturday sessions were such a highlight for me."

I looked at Mr. Collins. "Did you just use the past tense?"

"What do you mean?" he asked.

"You said these sessions *were* a highlight for you."

"You're right, that was a slip," said Mr. Collins, looking uncomfortable.

I could tell he wasn't his usual outgoing self today. Maybe he missed the classroom, and was sad he couldn't go back to the classroom because of the province's rule that every teacher had to retire at age sixty-five. As I pondered this, Mr. Collins brought me back to reality.

"Bethany, let's focus on the task at hand," he said. "Mr. Marshall says the cylindrical can problem can only be formally proven using calculus, but Rachel Mullen says you can justify the result with dimensional analysis. Well, let's follow Rachel's advice.

"To introduce dimensional analysis, let me give you a simple example. We measure *distance* in metres, *time* in seconds, and speed (or *velocity*) in metres-per-second."

$$\text{Distance} = m \qquad \text{Time} = s \qquad \text{Velocity} = m/s$$

"So suppose you're walking for ten seconds at a constant velocity of 3 $m/s$. What's the distance you will have covered?"

"Easy," I replied. "Thirty metres. Ten times three equals thirty."

"Good. Now I'm going to write down two equations. Which one is correct?"

$$\text{Distance} = \text{Velocity} \times \text{Time} \qquad \text{Distance} = \text{Velocity} \div \text{Time}$$

"The first one."

"That's right," said Mr. Collins. "Can you explain to me why the formula works?"

I hesitated. "It just does."

Mr. Collins wrote down some additional notes on his sheet of paper.

$$\text{Distance} = \text{Velocity} \times \text{Time} \qquad \text{Distance} = \text{Velocity} \div \text{Time}$$

$$m = \frac{m}{s} \times s \quad \text{(TRUE)} \qquad m = \frac{m}{s} \div s \quad \text{(FALSE)}$$

I saw the two equations, and understood why the first formula was correct.

"Good," said Mr. Collins, noticing my reaction. "The first equation has to be true because the *units are consistent* – we end up with metres on the left side and metres on the right side. In the second equation, the units don't make any sense, so it can't possibly be correct. Just check the units on the left side and on the right side, and see if they match: that's all there is to dimensional analysis!"

"But why do they call it dimensional analysis? You know, the word *dimension*?"

"Dimension is just a property of measurement: length, area, volume, acceleration, force, energy, distance, and so on. The dimension is the quality you're measuring. For example, say you know the volume of a cylinder is 100 $cm^3$. The 100 is the *quantity*, and $cm^3$ is the *quality*. When you do dimensional analysis, you're analyzing the units of a mathematical formula, the quality not the quantity, to check whether the equation makes sense."

"Cool."

"Okay, let's try another example. What is the correct unit for energy?"

"I don't know."

"Energy is a quality that you can measure, so it needs to have a unit of measurement. There's a famous formula for energy, discovered by Einstein. What is it?"

"$E = mc^2$."

"Do you know what that means?" he asked.

"No idea," I said, shrugging.

"$E$ stands for energy, $m$ stands for mass, and $c$ stands for speed, namely the speed of light. For many centuries, it was assumed that energy and matter were completely different, until Einstein demonstrated through his famous equation that they were in fact interchangeable. The standard unit for mass is kilograms ($kg$), and the standard unit for speed is metres-per-second ($m/s$). Using dimensional analysis, what must be the correct unit for energy?"

"If $E = mc^2$, then the unit for energy has to be $kg\frac{m^2}{s^2}$. That looks weird."

"But that's correct! Energy is measured in *joules*, and one joule is defined as one kilogram metre squared, per second squared. If that sounds confusing, here's another equivalent definition: one joule is the amount of energy expended, or the amount of work done, in applying a force of one *newton* through a distance of one metre."

"What's a newton?"

"You tell me. Force is measured in newtons. I've just told you that Energy = Force × Distance. What's one newton?"

On my notepad, I wrote down the equation

$$kg\frac{m^2}{s^2} = Newton \times m$$

"Distance is measured in metres, so one newton has to be $1\ kg\frac{m}{s^2}$. That's the only way the units work out."

"That's right. I assume you'll be taking Physics next year, where you'll learn important concepts dealing with kinematics, dynamics, waves, momentum, and energy. My advice to you is to learn the concepts and understand their significance, but don't bother memorizing all those formulas. If you use dimensional analysis, you'll be able to figure out the formulas for yourself."

"That's cool," I said. "But what does all this have to do with Mr. Marshall's question on cylindrical cans?"

"Honestly, I'm not sure. But I think we can figure it out together by simplifying the problem and finding a pattern. Let's investigate the relationship between the perimeter of a rectangle and its area, and determine how we can minimize the perimeter if we are given a fixed area."

Mr. Collins took out a sheet of paper and sketched a rectangle.

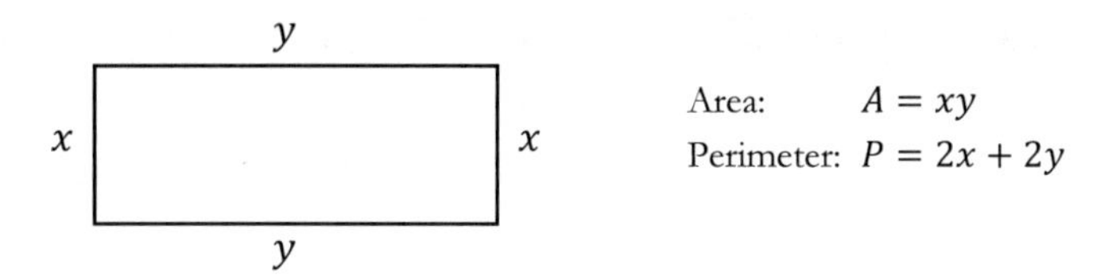


He then filled a table with some measurements.

| **Length** | 1 cm | 2 cm | 3 cm | 4 cm | 6 cm | 9 cm | 12 cm | 18 cm | 36 cm |
|---|---|---|---|---|---|---|---|---|---|
| **Width** | 36 cm | 18 cm | 12 cm | 9 cm | 6 cm | 4 cm | 3 cm | 2 cm | 1 cm |

"Bethany, here are the side lengths for nine different rectangles. I want you to determine the area and perimeter of each rectangle. And when you do, make sure you write down the units of measurement."

I quickly determined each rectangle's area and perimeter, and added this information to the table.

| Length | 1 cm | 2 cm | 3 cm | 4 cm | 6 cm | 9 cm | 12 cm | 18 cm | 36 cm |
|---|---|---|---|---|---|---|---|---|---|
| **Width** | 36 cm | 18 cm | 12 cm | 9 cm | 6 cm | 4 cm | 3 cm | 2 cm | 1 cm |
| **A** | 36 cm$^2$ | 36 cm$^2$ | 36 cm$^2$ | 36 cm$^2$ | 36 cm$^2$ | 36 cm$^2$ | 36 cm$^2$ | 36 cm$^2$ | 36 cm$^2$ |
| **P** | 74 cm | 40 cm | 30 cm | 26 cm | 24 cm | 26 cm | 30 cm | 40 cm | 74 cm |

"Good. Notice that area is measured in cm$^2$, and perimeter is measured in cm. What I want you to do now is calculate $P^2/A$, the square of the perimeter divided by the area. By dimensional analysis, you'll see the units cancel out. And so $P^2/A$ is a *dimensionless quantity* – it's a pure number."

For each of the nine rectangles, I used my calculator to determine the value of this ratio $P^2/A$.

| Length | 1 cm | 2 cm | 3 cm | 4 cm | 6 cm | 9 cm | 12 cm | 18 cm | 36 cm |
|---|---|---|---|---|---|---|---|---|---|
| **Width** | 36 cm | 18 cm | 12 cm | 9 cm | 6 cm | 4 cm | 3 cm | 2 cm | 1 cm |
| **A** | 36 cm$^2$ | 36 cm$^2$ | 36 cm$^2$ | 36 cm$^2$ | 36 cm$^2$ | 36 cm$^2$ | 36 cm$^2$ | 36 cm$^2$ | 36 cm$^2$ |
| **P** | 74 cm | 40 cm | 30 cm | 26 cm | 24 cm | 26 cm | 30 cm | 40 cm | 74 cm |
| **$P^2/A$** | **152.11** | **44.44** | **25** | **18.78** | **16** | **18.78** | **25** | **44.44** | **152.11** |

"Bethany, look at the numbers in the bottom row. What's the smallest number? What can you infer?"

"The smallest number is 16. Just from looking at the numbers, I think this ratio $P^2/A$ has to be at least 16. The ratio is equal to 16 when the rectangle's length and width are equal. And the ratio is more than 16 in all the other cases."

"Excellent," said Mr. Collins. "Do you think you can formally prove that?"

"I think so. You want me to prove that $P^2 \geq 16A$ where the perimeter is $P = 2x + 2y$ and the area is $A = xy$. It doesn't look too hard."

"Can you show me?"

Using a little bit of algebra, I proved that $P^2$ had to be at least 16A. I was also able to deduce that $P^2$ could be *exactly* 16A provided the length of the rectangle equalled its width – i.e., only in the special case when $x = y$, when the rectangle was a square.

"Wonderful," replied Mr. Collins. "Let me recap what we did. We started with a rectangle with some fixed area A. We used dimensional analysis to show that the quantity $P^2/A$ was dimensionless, and algebraically proved this pure number had to be at least 16. You showed this optimal case, when the ratio is exactly 16, could only be attained if the rectangle is a square.

"In other words, you proved that in any rectangle, the perimeter must be at least four times the square root of its area. And that the minimum perimeter can only be achieved in the case of a square."

"I completely get it," I said, nodding. "So the same method will solve Mr. Marshall's cylinder question?"

"Yes, I'm certain of it," he said. "You want to try it?"

"Totally."

"Okay, the first thing is to figure out the units. Say you have a radius of $r$ cm and a height of $h$ cm. Forget about $r$ and $h$, and just focus on the quality. What are the units for surface area and volume?"

"Surface area is measured in $cm^2$, and volume is measured in $cm^3$."

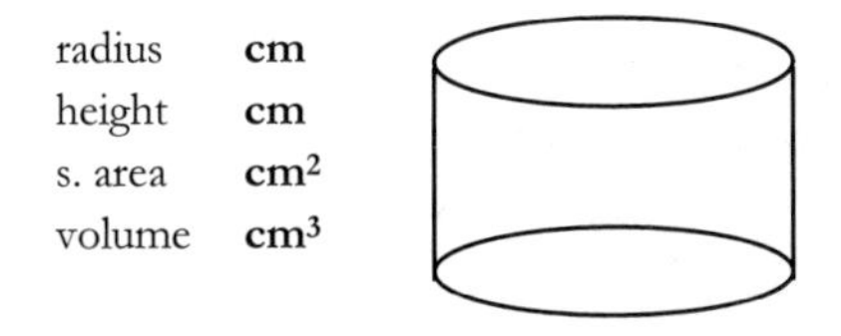


"If you wanted to create a dimensionless quantity, a pure number, what could you do?"

I stared at the paper until the answer came to me.

"I'd look at $SA^3/V^2$, the cube of the surface area divided by the square of the volume. Both the numerator and the denominator would have $cm^6$ for their units, and the units would cancel out."

"Great! So as we did before, make a table and figure out the smallest value of $SA^3/V^2$."

I spent about ten minutes considering various shapes for cylinders, and used my calculator to determine the possible values of $SA^3/V^2$, as I varied the shape of the can from really thin to really wide.

"What do you notice, Bethany?"

"Well, I've tried all these different cases, and I'm sure that the smallest possible number for this ratio $SA^3/V^2$ is $54\pi$, which is about $54 \times 3.14 = 169.56$."

"From your analysis, it appears $SA^3/V^2$ must be at least $54\pi$. When is this ratio exactly $54\pi$?"

"Whenever $h = 2r$, whenever the height is twice the radius."

"Always?"

"It seems that way. Whenever the cylinder's height is twice its radius, I always get $54\pi$."

"And what did Rachel write when she 'facebooked' you?"

I glanced at the printout of her Facebook post.

> Bethany, use *dimensional analysis* to derive a relationship between the surface area of a cylinder and its volume, to prove that for a fixed volume, the surface area is minimized precisely when the cylinder's height equals its diameter ($h = 2r$). Good luck!

Mr. Collins smiled. "I think you can take it home from there. All you have to do is use some Grade 10 algebra to show that $SA^3/V^2$ is at least $54\pi$, and you will have completed Mr. Marshall's assignment."

"Without calculus."

"That's right. All you need is a little bit of Grade 10 algebra. No calculus."

"I can definitely do that. The assignment is due on Monday, so I'll finish this tonight. And I'll show you the proof when we meet next Saturday."

Mr. Collins sighed and put down his pen.

"I'm sorry, Bethany. We won't be able to meet next Saturday."

"You're out of town?" I asked.

He paused. "Have you heard of the Halifax Preparatory School?"

"Definitely," I said. "That's the rich-kid private school in Halifax. Everyone hates that place."

"I forgot," said Mr. Collins, tight-lipped. "Miss Carvery was so proud of your perfect score on the Grade 7 Gauss contest a few years ago – because of your outstanding performance, Pinecrest Junior High came second in all of Nova Scotia, second only to the team from Halifax Prep."

"I hear they get first place in everything: science fairs, sports, spelling bees. I don't think they've ever come second in anything. Everyone calls that school *Preparation H*."

Mr. Collins shook his head. "Yes, that's been the school nickname ever since stores began selling Preparation H hemorrhoid cream – fifty years ago."

He took a deep breath.

"Bethany, my brother is the headmaster at Halifax Prep, or Preparation H, as you so eloquently call it. He called to let me know that his wife, a math teacher at that school, died of a heart attack two days ago."

I gulped.

"I'm so sorry."

"Imagine that – a sudden heart attack in the first week of school. She was one of the best teachers in the province. She's been part of our family for forty years. We're all in shock."

I felt sick to my stomach.

Mr. Collins had tears in his eyes. "My brother called me yesterday, and asked me to move to Halifax for the rest of the school year, to take over my sister-in-law's classes. Apparently, I'm the only person in the province qualified to teach the International Baccalaureate courses they offer there. Because Halifax Prep is an independent school, the province's mandatory retirement rules don't apply."

He took a napkin and wiped off his eyes before looking back at me.

"I've accepted his offer. My wife and I will be leaving Cape Breton."

My heart sank.

"We will be moving to Halifax to support my brother in whatever way we can. I need to be there for my family. It's the right thing to do."

"Mr. Collins, I'm so sorry about what I said about Halifax Prep."

"I forgive you," he said, smiling at me. "You couldn't have known. I'm just sorry we won't be able to work together anymore. I'm sure you understand."

I nodded.

*Wow, this was it.*

"So when do you leave for Halifax?"

"In a few minutes. My car's all packed. The visitation is tonight and the funeral is tomorrow morning. My wife left earlier this morning."

"You didn't go with her?"

"No. I told her I wasn't ready to go until 3 p.m. since I had an important meeting."

"With who?" I asked.

"With the most determined, capable, inquisitive, and creative student I've met in my forty-two years as an educator."

A tear fell out of my eye.

"Thank you, Bethany," he whispered. "I'll miss you."

Mr. Collins stood up and embraced me in a hug.

I sniffled. "Thank you, Mr. Collins. For everything."

"You're welcome," said Mr. Collins. "And let your mother know that Ella is enjoying life in Newfoundland, but she misses Cape Breton – and especially misses her figure skating coach."

"I'll let Mom know."

"I'm not going to say goodbye, Bethany. I'll see you again – maybe sooner rather than later."

"How do you know?" I whispered.

He wiped his eyes with his arm and looked at me.

"I'll see you in Halifax at the end of April."

"What's happening then?" I asked.

"You'll find out."

With that, Mr. Collins smiled and walked out of Le Bistro for the last time.

# 22

"You went to Vanessa's birthday party last night?"

"Yeah, we did," hissed Breanna. "You got a problem with that?"

A bad week turned even worse. On Saturday, Mr. Collins moved to Halifax. On Sunday, I had a weird phone call with Cooper where he seemed completely distracted and unhappy to talk to me. And I just learned my two closest friends were at the bowling alley on Monday night, having fun with Vanessa and Gillian.

Mr. Marshall got our attention, and started the class with a group activity.

I solved the assigned problem right away, explained the solution to Patrick, and watched Bonnie and Breanna talking about something else, whispering and giggling to each other.

Mr. Marshall called us to attention.

"As always, I want each group to share their answers. Let's start here on the right. Breanna, please share what your group came up with."

"I don't know," replied Breanna with a shrug.

"You don't know?" replied Mr. Marshall, raising his voice. "Your group had five minutes to work together. Even if you didn't solve the problem, you should have at least come up with some ideas. So tell me, what ideas did your group come up with?"

"Ask her," she replied, pointing to me.

"No, Breanna. I am asking you. What ideas did your group come up with?"

Breanna was silent. Patrick raised his hand. Mr. Marshall waved him off.

"Patrick, I did not call on you. Breanna, give me one idea from your group."

She shrugged and stared at her desk. Bonnie looked away, avoiding eye contact with our teacher who was just a few feet in front of us. Mr. Marshall walked to his desk and wrote something down.

"As all of you know from the course outline, participation counts for ten percent of your overall mark. Each member of Breanna's group has just lost one percent from their final grade."

"Why?" I asked, looking up at Mr. Marshall. "Patrick and I solved the problem. Here, let me show . . ."

"Silence!" said Mr. Marshall, giving me goose bumps. "You are sitting in a group for a reason. You do these think-pair-share activities for a purpose. You succeed as a team and you fail as a team."

He turned to the rest of the class. "Let that be a warning to all of you. Okay, let's move on to the next group. Vanessa, share with us your group's thoughts."

As Vanessa shared how her group solved the problem, I glared at Breanna and Bonnie. Breanna glared right back, while Bonnie stared blankly at the tiles on the floor.

I turned towards Vanessa to hear her group's response, and made eye contact with Gillian, who smiled mockingly.

I couldn't wait for class to end. I was barely paying attention to the responses of the other groups and scribbled mindlessly in my notepad. I glanced over at the clock, happy to see that there were only ten minutes left.

"Class, let me get your full attention. Close your notebooks, put your pens down, and turn to face me."

I reluctantly obeyed.

"Over the past week, you have looked at numerous problems involving surface areas and volumes, which has introduced you to the process of *mathematical modeling*. I have given you questions from real-life contexts, such as maximizing the volume of a waffle cone given a fixed surface area, and minimizing the surface area of a soup can given a fixed volume. You took these practical problems and modeled them in the language of mathematics, and you worked with your fellow group members to determine the optimal solution. Now that you have a taste of real-life mathematical problem-solving, I want to tell you why this is so relevant for your lives, and how mathematics empowers you with the tools to impact social change and create a more sustainable future.

"As some of you know, I belong to the Membertou First Nation. In fact, I am one of the band councillors. Those of you whose parents and grandparents grew up here in Sydney will be quite familiar with Membertou's oppressive history, including being the first aboriginal band in Canada to be forcibly relocated. For decades, the majority of us were unemployed and on welfare, with little formal education.

"In the mid-1990s, our band had an operating budget of $4 million, with nearly every dollar coming from the federal government. And that year, our band, with just thirty-seven employees, produced a $1 million deficit. Being bankrupt and fully dependent on government handouts, we worked hard to fix it – for our Cape Breton community, and more importantly, for our children."

I glanced over at Patrick, seeing him gazing proudly at his father.

"Over the past two decades, we changed the entire culture of our band. We cleared our debt, established standards for quality assurance, and launched business ventures based on our four pillars of sustainability, conservation, innovation, and success. We became the first aboriginal government in the world to receive a special certification called ISO 9001.

"Our new businesses, ranging from fishing and environmental cleanup to engineering and aerospace, have brought us millions of dollars, which we invest back into the community. We are no longer dependent on government handouts. And we now employ five hundred people, of which half are non-aboriginal."

One boy in the back raised his hand. "My dad works at the Membertou Trade and Convention Centre."

"Yes," replied Mr. Marshall, nodding. "It's a remarkable venue. How many of you have been there?"

A bunch of us raised our hands. I went to a Natalie MacMaster concert there with Bonnie and Breanna back in July, right after I came back from the Canada Math Camp. That seemed like such a long time ago.

Mr. Marshall continued: "Many of you are familiar with Membertou, and our success story. Our efforts have paid off: we have an employment rate of ninety percent and a high school graduation rate of one hundred percent."

Vanessa raised her hand. "Sir, no offense, but what does all this have to do with math?"

"A few years ago, we started a seafood company where we honour our environment, just as our ancestors did, and combined it with a mathematical approach to quality assurance and sustainability."

He opened up his briefcase to show us two small cans of fish, with Membertou's logo on the labels.

"Our canned tuna and lump crabmeat come in cylindrical-shaped cans. They're sold in grocery stores all throughout the Maritimes. Look at the cans. What do you notice?"

Gillian raised her hand. "The height is twice the radius."

"Correct," replied Mr. Marshall. "We took a manufacturing problem, modeled it in the language of mathematics, solved it, and figured out the most sustainable solution that reduced our overall expenses and minimized our environmental impact. In your assignments, you have written letters to the Presidents of two dozen Canadian companies, encouraging them to do the same thing.

"Over the next few months, I will teach you the Grade 10 Mathematics curriculum, but I will do this through the lens of environmental sustainability and social justice. You will work on real problems and develop real solutions so that you can affect real change. I am going to teach you the entire curriculum, but frame it around actual applications and situations. As a result, you will learn the material much more deeply, and you will develop a far greater and deeper appreciation for the subject of mathematics."

I nodded silently, seeing how Mr. Marshall's demeanour was a lot stricter than that of Mr. Collins, but his end goal was the same – to help us learn and love math. Those cylindrical fish cans were cool.

"And one last thing. I have marked all your assignments, and will now hand them back to you."

"You marked them already?" asked Vanessa.

"Yes," he replied. "How can I have high standards for all of you if I don't have even higher standards for myself?"

None of us could respond to that. Mr. Marshall continued.

"Please make the requested corrections and hand me a clean copy for the next class. As discussed, you will be mailing these assignments out to the company presidents. I will bring the stamps and envelopes."

I was surprised to see I hadn't received a mark – just a post-it note from Mr. Marshall with a single sentence written on it: *Bethany, see me in my office after class.*

Confused, I followed Mr. Marshall to his office, where he asked me to wait a few minutes. While I was waiting for him to come back, I opened up my assignment and re-read what I had written.

**TO: Howard Campbell, President, Campbell Company of Canada**

Dear Mr. Campbell,

My name is Bethany MacDonald and I am a Grade 10 student at Sydney High School in Sydney, Cape Breton, Nova Scotia. I am writing to tell you that your company can reduce manufacturing costs while saving the environment at the same time. In this letter, I will explain the results of my analysis.

Your company's condensed Cream of Mushroom soup can has radius 3.3 cm and height 10.0 cm. The formula $V = \pi r^2 h$ gives the <u>volume</u> of the can, measuring how much soup the container can hold. And the formula $SA = 2\pi r^2 + 2\pi rh$ gives the <u>surface area</u> of the can, measuring how much steel is used.

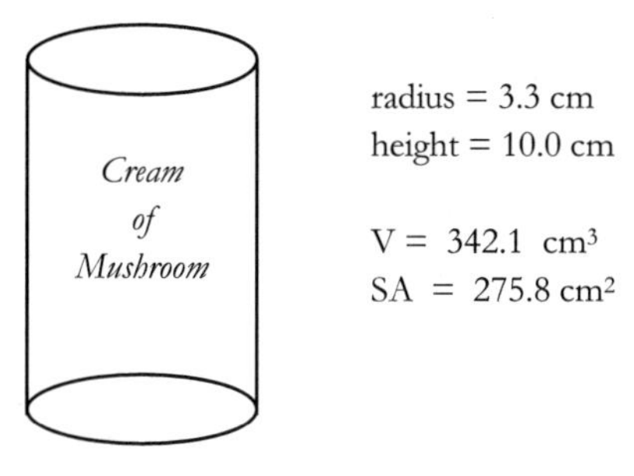


Now here's another cylindrical can that's a little wider and a little shorter, with the exact same volume, having about 2% less surface area. This soup can has radius 3.79 cm and height 7.58 cm.

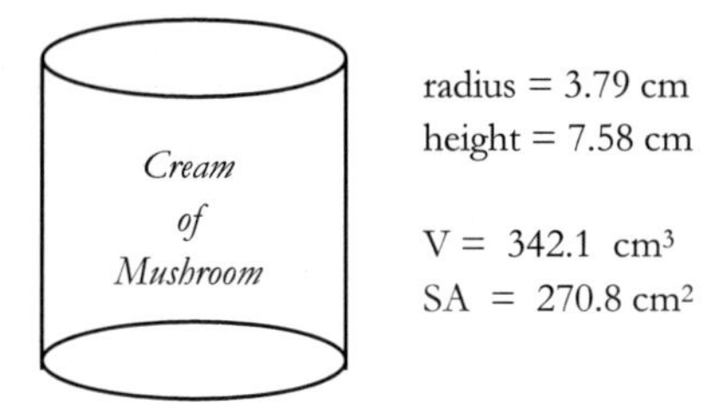


In this cylindrical can, the height is the same as the diameter (since the diameter is just twice the radius). It turns out that to minimize the surface area, what you want to do is make sure the height equals the diameter.

Whatever you put into the can, whether you put a little bit of soup or a lot of soup, this is the "perfect" shape that is most environmentally sustainable.

If you use $1,000,000 dollars of steel to produce Cream of Mushroom soups all across Canada, then by switching the shape of the can, you use 2% less steel (2% less surface area), which means that your company will save $20,000. Do this for all of your canned soup products, and you'll save a lot of money. And it's better for the environment too, since you're shipping less steel from plant to plant, and cutting down on greenhouse gas emissions – both in the amount of energy you need to produce the steel, and the number of trips your truck drivers have to make.

To close this letter, I will mathematically prove the following: that for any cylindrical can, with any fixed volume, the surface area is minimized by making the height of the can equal to its diameter.

Even though this is not part of my assignment, I'm going to prove this for you because your company produces cans of many different sizes, and my solution will work for all of your products. My teacher Mr. Marshall wanted me to explain just the Cream of Mushroom can ($V = 342.1\text{cm}^3$) by using a graph, in the same way he taught us. But I have a better solution, a rigorous proof that works for cans of *all* volumes just by using some Grade 10 math. And I don't need calculus!! ☺

I'm going to define the "happy value" and the "magic ratio". Call these two quantities $x$ and $y$, respectively.

Let $x$, the happy value, be the height of the cylinder divided by its diameter.

Let $y$, the magic ratio, be the cube of the cylinder's surface area divided by the square of its volume.

For example, in your Cream of Mushroom can with height 10.0 cm and radius 3.3 cm, the happy value is $x = \frac{h}{2r} = \frac{10.0\ cm}{6.6\ cm} = 1.5$ and the magic ratio is $y = \frac{SA^3}{V^2} = \frac{(275.8\ cm^2)^3}{(342.1\ cm^3)^2} = 179.176 = 57.034\ \pi$.

And in my "perfect" Cream of Mushroom can with height 7.58 cm and radius 3.79 cm, the happy value is $x = \frac{h}{2r} = \frac{7.58\ cm}{7.58\ cm} = 1$ and the magic ratio is $y = \frac{SA^3}{V^2} = \frac{(270.8\ cm^2)^3}{(342.1\ cm^3)^2} = 169.646 = 54.000\ \pi$.

Notice that my can has a magic ratio of $54\pi$. I'll prove this is the lowest possible value for this magic ratio.

Step 1: we determine[1] formulas for $x$ and $y$.

Step 2: we prove[2] that $y = \frac{2\pi\,(1+2x)^3}{x^2}$. In other words, $y$ can be written as a function of $x$.

Step 3: we prove[3] that the magic ratio is $54\pi$ when the happy value is equal to 1, and the magic ratio is greater than $54\pi$ when the happy value is not equal to 1.

Step 4: we conclude that $SA^3/V^2 \geq 54\,\pi$, which is mathematically equivalent to $SA \geq \sqrt[3]{54\pi} \cdot V^{2/3}$.

So for any can with volume V, the minimum surface area works out to $\sqrt[3]{54\pi} \cdot V^{2/3}$ which occurs precisely when the can's "happy value" is 1, i.e., when the height of the cylindrical can equals its diameter.

Yours sincerely,
Bethany MacDonald
Grade 10 Student
Sydney High School

---

[1] The happy value is $x = \frac{h}{2r}$ and the magic ratio is $y = \frac{SA^3}{V^2} = \frac{(2\pi r^2+2\pi rh)^3}{(\pi r^2 h)^2}$.

[2] This holds since $y = \frac{SA^3}{V^2} = \frac{(2\pi r^2+2\pi rh)^3}{(\pi r^2 h)^2} = \frac{(2\pi r^2+2\pi r\cdot 2rx)^3}{(\pi r^2\cdot 2rx)^2} = \frac{(2\pi r^2+4\pi r^2 x)^3}{(2\pi r^3 x)^2} = \frac{(2\pi r^2)^3\,(1+2x)^3}{(2\pi r^3)^2\,x^2} = \frac{8\pi^3 r^6(1+2x)^3}{4\pi^2 r^6 x^2} = \frac{2\pi\,(1+2x)^3}{x^2}$.

[3] We have $(1-x)^2 \cdot (1+8x) > 0$ for any positive $x$ not equal to 1. Expanding, we get $1 + 6x - 15x^2 + 8x^3 > 0$, which is the same as $1 + 6x + 12x^2 + 8x^3 > 27x^2$, implying that $(1+2x)^3 > 27x^2$. Thus, $\frac{(1+2x)^3}{x^2} > 27$. Since $y = \frac{2\pi\,(1+2x)^3}{x^2}$, this proves that $y = 2\pi \cdot \frac{(1+2x)^3}{x^2} > 2\pi \cdot 27 = 54\pi$.
And if $x = 1$, then we get $y = 2\pi \cdot \frac{(1+2x)^3}{x^2} = 2\pi \cdot 27 = 54\pi$.
This completes the proof. See, no calculus!! ☺

As I finished reading the last line, a nervous smile crossed my lips.

Mr. Marshall returned. He closed the door to his office.

"Bethany, your assignment was unacceptable."

"But what about . . ."

"Don't interrupt me," replied Mr. Marshall. "I felt my instructions were clear. You were asked to model the problem by graphing the relationship between the radius and surface area, just as we did in class. Twice you mentioned you didn't need calculus. What purpose did you achieve in writing that?"

I didn't answer. I didn't want to interrupt him again. After a few seconds of silence, he looked at me.

"I would like an answer, Bethany."

"Sorry, sir," I replied, staring at the floor. "I shouldn't have done that."

"Why did you not follow my instructions for the assignment?"

I paused.

"Last Saturday, I worked with Mr. Collins, and he taught me this technique called dimensional analysis. We came up with an idea we thought would work for any can of any volume, and I figured out the rest. I was proud of myself for coming up with the proof, and I thought you'd be impressed."

"I most certainly was impressed, and I commend you for your efforts. However, what was the point of mentioning you didn't need calculus, let alone saying it twice?"

"Because you insisted it was impossible to come up with a general proof using Grade 10 math, that we needed Grade 12 calculus. You said it was impossible . . . twice."

"You have a serious attitude problem, Bethany."

"I'm sorry."

"What motivated to you to behave like this?"

"I was just following Rachel Mullen's advice, who told me to always question authority and to constantly challenge traditional thinking, so that we can replace incorrect ideas with new knowledge."

"And who is Rachel Mullen?"

"She won a gold medal for Canada at the International Math Olympiad. She taught me at the Canada Math Camp in Ottawa a few months ago."

Mr. Marshall hesitated. "Yes, I heard you were invited to that camp. Clearly, it had a profound impact on you."

"So what I did was wrong?"

"It was wrong, Bethany. Asking me a question is acceptable; questioning me is unacceptable. Informing me that I made a mistake is acceptable; taunting me on an assignment is unacceptable. There is a right way and a wrong way to do things."

"I understand, Mr. Marshall."

"As you mature, you will learn that success is measured not by what you do, but how you do it. I hope today's experience has taught you a valuable lesson."

"Yes, sir. I'm sorry for what I did. Did I fail the assignment?"

"You can hand in a new assignment, where you will write the letter in the way I asked. If you submit it before next class, I will remark it without penalty."

"Thank you, sir."

I stood up to leave.

Mr. Marshall paused. "I received a memo from the principal this morning, informing me of a new math outreach event taking place in Cape Breton. Given your obvious interest and aptitude for the subject, I think you would enjoy participating."

"What is it?"

He handed me a flyer for the "Nova Scotia High School Math League".

I quickly skimmed the details: it was a Saturday-morning event open to all high schools with the emphasis on cooperative problem-solving rather than individual competition; there would be regional games taking place on the first Saturdays of October, December, and February, with the top three schools in the Cape Breton regional school board being invited to participate in the Provincial Finals.

My heart skipped a beat when I read where and when the finals would be taking place: *in Halifax at the end of April.*

"Sounds great, Mr. Marshall. I'm in."

# 23

I logged into Skype for my 3:00 p.m. call with Cooper.

It was the first Sunday of October and our relationship had grown cold.

In June, right after we came back from the math camp, we talked on Skype every day. In July, we spoke every other day and sent Facebook messages back and forth. In August, we spoke twice a week and cut back on the Facebook messaging. When school started in September, Cooper said he would be so busy he could only talk once a week, every Sunday afternoon. The calls were getting shorter, both in frequency and duration.

*Were we no longer girlfriend and boyfriend?*

Two weeks ago, Cooper just gave one-word responses to my questions. He stared at the screen like he was playing a computer game while talking with me. Last week, he wasn't even logged into Skype. When I sent him a text asking where he was, he wrote back and said he was busy.

There was no apology, not even a note saying he'd call me back the next day.

It was 3:03 p.m. . . . a palindrome.

There was no green check-mark next to Cooper's face to indicate he was logged in.

*Were we going to break up? Had we already broken up?*

I stared blankly at the screen, focussing my eyes on the empty box next to Cooper's face, and waiting for that box to be replaced with a check-mark. All of a sudden I saw a message display on the bottom right corner of my screen: *Grace is online.*

The empty box next to Grace's face was replaced by the green check-mark.

I sighed. Where was Cooper?

My computer made a loud ringing sound. It was Grace, dialling me. I hesitated for a second before clicking the green icon marked *ACCEPT.*

Grace's face appeared on my computer screen. She didn't seem as cheery and happy as usual.

"Hey, Bethany, how's it going?"

"So so," I replied. "How about you? You don't look so good."

"I had a big argument with my dad this morning."

"Oh no. That sucks."

"I'll tell you all about it in a sec. You just got off the phone with Cooper?"

"He hasn't logged in yet. We didn't speak last week either. He said he was busy. No apology, nothing."

"What's his problem? He's so lucky to be with you. Boys – so clueless and so stupid sometimes."

"Only sometimes?" I replied, trying to force a smile. "How about all the time?"

"Life sucks. Cooper's acting like a jerk and my dad is even worse. I want to hear some good news. Hey, wasn't your Math League thing this weekend?"

I told Grace about the Math League event last Saturday. She was curious to know all the details, so I filled her in: there were a half-dozen schools from all over Cape Breton, and schools competed in teams of four. There were no individual questions; instead, we got ten team questions, given to us one at a time. Each group had five minutes to decide on an answer, and we could talk as much as we wanted.

"How did your team do?" asked Grace.

"We got eight out of ten. Not bad."

"Who was on your team?"

"Patrick, the quiet guy who sits next to me in math class. His dad's the teacher I was telling you about. And we were paired up with two Grade 11s: Tommy the computer geek and Logan the hockey player."

"Are any of them cute?"

"No," I said, laughing. "But they're really cool."

I got along well with Patrick, Tommy, and Logan. They were all keen, but none were competitive. It was cool to write the contest as a team. I didn't feel the tension and anxiety I sometimes did when writing exams. And the Math League wasn't like Mr. Marshall's math class where I had to sit with Breanna and Bonnie and solve everything for them.

"And what happens after the team questions?" asked Grace.

"We end with two relays. We line up single file and each person gets a different question. To answer your question, you need the right answer to the previous question. It's like a relay race, where you pass the baton, from *A* to *B* to *C* to *D*."

"Cool. So how did it go?"

"Well, we bombed since Patrick couldn't solve the first problem on either relay, so we got zero. But we got forty points for the eight team questions we solved, so we ended up in fourth place."

"Congrats, that's awesome."

"Yeah, there were fifteen teams yesterday, four from my school. We were the top team at Sydney High. You know that girl Gillian I told you about? She was there too, on a team with her three best friends. Their team finished dead last."

"I'm sure you were happy about that."

"I was just happy Mr. Marshall didn't put Gillian on my team."

"That's cool you did the Math League," said Grace. "Brings back good memories from camp. Too bad they don't have something like that in Vancouver."

"Who knows, maybe someone will start it up there too," I said. "Hey, I just noticed – it's like 11:00 a.m. your time. Aren't you supposed to be at church?"

"I stayed home," replied Grace. "That's why my dad and I were arguing. I told him last night I didn't want to go to church anymore, and he got mad. He called me rebellious; I called him closed-minded. This morning we had the same argument, and he drove off saying I'd regret this decision, accusing me of betraying the community that had been so good to us. Remember Rachel telling us about questioning authority and challenging traditional ways of thinking? Well, why doesn't that apply to our parents?"

"Or our teachers," I added.

"My dad has a point," admitted Grace. "The church community has been good to us, especially after the accident."

I nodded, remembering Grace telling me that her mom and brother died in a car accident twelve years ago, when Grace was three years old. After the accident, Grace and her father received a lot of support from their church in Vancouver, where Grace's father was the pastor. Grace mentioned the people at their church were still helping them out, with several well-off families contributing to pay her tuition each year. She told me that was the only reason she was able to attend the Vancouver

Independent School, and felt so out of place among her classmates who had two parents and a lot of money.

Grace continued: "But even though all the church people are really nice, if they're believing in something that's not true, what's the point?"

I nodded. "Yeah."

"My dad thinks Raju brainwashed me."

"What are you talking about?"

"Well, it was Raju who got me thinking about all this. It started at the camp. You remember a bunch of us were talking about this over dinner one night – the day before the banquet . . ."

"Yeah, the Thursday night."

"Remember Raju asking me why I believed in God? He gave us a whole bunch of reasons why God couldn't exist."

"I remember. All the earthquakes and hurricanes that God failed to prevent; all those starving children dying – not just in Africa but here in Canada too; all the scientific evidence proving the earth evolved over billions of years instead of being created in just six days; all that stuff about hell . . ."

"Yeah," interrupted Grace. "That makes no sense at all. My dad says you go to heaven as long as you believe in Jesus. But if you don't believe in Jesus, then you automatically go to hell when you die. That means if Hitler believed in Jesus one minute before he died then he's dancing in heaven now? That's stupid."

"For sure," I said. "I remember Raju using the same example. What did your dad say about this?"

"He said we shouldn't question God. He said God's ways are better than our ways."

"Brutal," I said. "So that happened yesterday?"

"It happened a few months ago, but I'd been going to church anyway to make my dad happy. He doesn't like that I have questions about God. Especially after what we've gone through, he thinks I've become ungrateful. Raju and I were debating till 3:00 a.m. on the last night of the camp. At the time, I was defending Christianity but the more I think about what Raju said, the more I realize he was right."

"You and Raju were debating till 3:00 a.m.? How come you never told me about this? Where was I?"

Grace laughed. "You were with Loverboy."

I fell silent, remembering my relationship with Cooper was so much different now. I sighed, longing for a return to the way we felt about each other back in June.

"Tell me," said Grace, "what's going on with you two? Isn't he going to Cape Breton for Christmas?"

"I don't know. He still hasn't bought his plane ticket. I mean, he promised me three months ago."

"Do you still like him?"

"Yeah," I replied with a slight hesitation. "Yeah, I do. For sure. Yes. Definitely."

"Then you should call him. If he's not on Skype, then call his cell."

"But we only ever talk on Skype. Cooper says he's not allowed to call long-distance on his cell phone. His mom says his phone is just for emergencies."

"Give me a break," said Grace.

"Tell me about it."

"Well, I should get going. Thanks for listening to my problems. You're my best friend, you know."

"You're my best friend too. Thanks, Grace."

"See ya."

I hung up. Cooper still wasn't logged into Skype. I stared at my computer screen, hoping a green check-mark would suddenly appear next to Cooper's name.

But after a few minutes, I realized there was no point.

I went into the hallway and got a glass of water from the kitchen, and glanced at Mom, who was watching something on TV.

"How's Cooper?" asked Mom, turning to face me.

"He's fine," I said, walking back towards my room. "I have to make one more phone call."

I closed the door. I took a deep breath and called Cooper's cell phone.

*Ring, ring, ring, ring, ring, ring.*

No answer. Not even a voice mail. I tried again. Same result.

I needed to know what was going on.

Thanks to Google, I found *canada411.ca*, the online Yellow Pages guide that matched home addresses to phone numbers. I found Cooper's phone number in Hamilton, Ontario.

After I rehearsed what I was going to say to Cooper, I slowly dialled the ten digits.

A woman answered the phone. "Robertson residence."

I took a deep breath. "Yeah. Hello. Uh, is Cooper there?"

"I'm sorry he's not in right now," replied the woman. "This is Cooper's mother. May I ask who's calling?"

"It's Bethany," I replied.

"Hello. And are you Rosalind's friend from Elmdale?"

I paused. Elmdale College was the all-girls school in Hamilton, a short drive from Lakewood Academy, Cooper's all-boys school. Cooper hated the girls at Elmdale, saying they were all snobby and stuck-up. At least that's what he told me back in June.

*Who was Rosalind?*

"Excuse me, are you still there?" asked Mrs. Robertson.

"Sorry, I'm still here," I quickly replied. "Cooper and I met at the math camp in Ottawa."

"Oh yes, the math camp. Cooper had a wonderful time there. I'm sorry, what was your name again?"

"My name is Bethany MacDonald," I said, my stomach tightening. "Did Cooper mention anything about buying a plane ticket to Nova Scotia?"

"No, he didn't. What's in Nova Scotia?" she asked.

*His girlfriend.*

I wanted to say those two words, but couldn't make out the words.

"If you'd like, I can tell him you called. Cooper should be home soon. He's at the movie theatre with Rosalind right now."

"Who's Rosalind?" I blurted out.

"Cooper's new girlfriend."

My throat went dry. I nearly dropped the phone. I tightly held the receiver to my ear but no words came out.

Mrs. Robertson spoke. "Would you like to leave your phone number so I can get him to call you?"

My body went numb.

"Hello, are you still there? Hello? Hello?"

"Yes, I'm still here," I said, snapping out of it. I took a deep breath. "No, there's no need for Cooper to call me back. Sorry I disturbed you."

"No problem. Say, what was your name again? Stephanie? Melanie?"

Without replying, I hung up the phone and collapsed into my bed.

What just happened?

My mind went blank. I just laid there, staring at the white paint on the ceiling. I tried to cry but the tears wouldn't come out. I couldn't feel a thing.

Of course Cooper was too busy to talk to me. He had a girlfriend.

But wasn't I his girlfriend?

In July, when we were talking on the phone, Cooper mentioned he had once dated a girl from Elmdale, but things ended really badly. Did he get back together with her?

Or was it somebody else?

I reached over to the side of my bed, and grabbed the framed picture of Cooper and me at the math camp banquet. I loved that picture: me in my elegant purple dress, Cooper in that fancy suit with the matching purple tie. His arm was wrapped around my shoulder. We both had huge smiles on our faces. That was just a few months ago.

How could he have lied to me?

I sat up. With all my strength, I flung the picture frame against the wall, not caring about the loud crack of glass shattering the frame.

Things started to make sense. Cooper always talked to me in his room via Skype so his parents would never find out about his Nova Scotian girlfriend, the same way I was private and shy about Cooper's identity, telling only Breanna and Bonnie and making them promise to keep it a secret. I waited until mid-July, when I finally told Mom about the special boy who would be visiting me in Cape Breton over Christmas.

Obviously that visit wouldn't be happening anymore.

There was a knock. I heard Mom's voice, asking if she could come in.

I sighed. I knew I was going to get yelled at.

"Yeah, Mom. Come in."

Mom opened the door and walked in. She carefully avoided the broken glass near the door and sat down on the edge of my bed.

Mom didn't appear mad. Surprised, I rolled over and placed my head in her lap, the way I used to as a kid.

"Bethany, darling," whispered Mom.

She didn't need to ask what was wrong. The broken picture frame said it all.

I began sobbing.

# 24

The bell rang.

Geography class was finally over. I packed my stuff and headed towards my locker to get ready for Physical Education class. Our Phys. Ed. teacher, Mr. Campbell, wanted us to meet outside, by the main entrance to the school. He had told us that we would be running five kilometres today, and he would be recording our times.

I was in no mood to run. Especially after the break-up with Cooper the day before, I wasn't in the right frame of mind. But there was nothing I could do.

Breanna headed towards me. She walked past me, deliberately avoiding eye contact and marched towards Bonnie's locker which was just a few metres from mine.

"What the hell?" said Breanna. "Campbell's making us run in this weather?"

"Yeah," replied Bonnie. "It's so unfair this counts for marks."

Early October in Cape Breton was usually quite sunny, but today it was cold, foggy, and windy. Mom had the weather channel on this morning: 4 degrees, overcast, and 30 km/h wind gusts. I grabbed my gym bag from my locker, thankful I had packed an extra layer for my run today.

Just as I closed my locker, I saw Patrick walk past me, head down.

"Patrick," I called out. "Wait up."

I saw Breanna and Bonnie turn and look towards me.

"Hey, what's up?" I asked, catching up to Patrick. "You okay?"

"Sorry about Saturday," said Patrick, barely looking at me.

"Sorry about what?"

"For choking in the Math League," mumbled Patrick. "We were tied for first before the relay. Because I screwed up the relay, we finished fourth."

"Don't worry about it," I said, the Math League the least of my concerns. "No big deal, okay?"

"It is a big deal. I let the team down. You probably want someone better on the team."

"Definitely not," I replied. "The next Math League is in January. I'm going back, and I want us to be on the same team again."

"You really mean that?" asked Patrick.

I put my hand on his shoulder, and smiled at him.

"Yeah, definitely."

As Patrick walked away, I heard Breanna snicker.

Ignoring Breanna, I walked into the girls' locker room to change into my running gear. Breanna and Bonnie followed me into the locker room, and placed their bags close to mine. Both were laughing and whispering. I leaned closer.

"Hey, you think she likes him too? You know, double-dipping with her math league boyfriend and her math camp boyfriend?"

I angrily pulled my sweater over my head.

Breanna continued: "I wonder if her math camp boyfriend knows she's cheating on him."

I turned around, and glared at Breanna.

"Cooper broke up with me yesterday. He was dating another girl the whole time. The whole time. Are you happy?"

Breanna didn't answer, and gazed towards the showers in the other direction.

"I'm single," I continued, face flushed. "Go ahead and cheer. Cooper dumped me, okay? He dumped me."

"I'm sorry," said Bonnie.

Breanna continued staring in the other direction.

"What's wrong with you, Breanna? We used to be good friends. What happened?"

She turned to face me.

"You became a stuck-up jerk," Breanna said, raising her voice. "Every time we hung out, you always talked about yourself. The math camp. The boyfriend. Cooper this, Cooper that. Him coming over for Christmas, you going to Ontario next summer. You, you, you."

"I thought you'd be happy for me," I mumbled.

"Sure, at first. But then it got really annoying because that's all you ever talked about. Then school started. You come back from math camp, get all smart, and then make the rest of us feel stupid. When do you ever let us get a question in math class? Why do you always have to solve everything for us as if we can't do anything ourselves? Huh?"

I had no answer. Breanna still wasn't done.

"And when have you ever asked how I'm doing? Or how Bonnie's doing? Why does everything always have to be about you?"

A lump caught in my throat.

Bonnie tapped Breanna on the shoulder. "Hey, cool it. Let's go, or we're going to be late."

Breanna angrily got up and walked out the door, followed by Bonnie who cast a sympathetic look over at me before following Breanna.

I grabbed a few tissues to blow my nose, and ran outside where Mr. Campbell was waiting. The wind chill was a lot worse than I had expected, and I cursed, realizing I didn't have gloves on.

"Thank you for finally joining us, Bethany," said Mr. Campbell. "Now that we're all here, let me remind you of the instructions."

As I cupped my hands to my mouth, I saw Breanna facing the front and ignoring me. Glancing to my right, I saw Gillian smirking at the sight of tears rolling down my cheeks.

I tried to block out Gillian and Breanna as I listened to Mr. Campbell explain how we would run 2.5 kilometres towards the Canadian Forces base, and head straight back.

Just to ensure none of us were cheating, Mr. Campbell said that he too would be running, and waiting for us at the halfway point. He informed us that another Phys. Ed. teacher, Mrs. MacEachern, would greet us at the finish line to record our times.

"Look, I know it's really cold and windy. Just do your best. As an extra incentive, the faster you finish, the faster you can eat lunch. Okay, ready girls? Go!"

With that, he started running and the rest of us followed him. A girl named Jennifer quickly ran past Mr. Campbell and opened up a large gap between herself and the rest of us.

Through my sniffles, I found a steady pace near the back, jogging a few steps behind Bonnie and Breanna.

A huge gust of wind propelled us forward. Especially because we were running downhill, I was going a lot faster than normal.

As much as I tried to block it out, I found myself replaying my argument with Breanna in the locker room, realizing to my dismay that everything she accused me of was true.

All I did was talk about myself. She was right. I had become a self-centered jerk.

I had a hard time concentrating. Every part of me wanted to cry.

I suddenly felt an intense stabbing pain just below my right ribcage, and slowed down my pace. I continued to jog, step by step, hoping that the cramp would just go away. I sniffled.

I took some deep breaths, remembering a tip I learned a few months ago . . . from Cooper.

Cooper once told me he got "side-stitches" whenever he was running too fast and hadn't properly stretched beforehand. Remembering his method for making the pain go away, I concentrated on my breathing pattern, and instead of inhaling every fourth step when my right foot hit the ground, I held my breath for an extra stride and inhaled every other time my left foot hit the ground.

Within a minute, the pain had subsided. I continued taking deep breaths and adjusted my breathing patterns every few minutes until the side stitch was completely gone.

I hit the dead end by the Canadian Forces base, and saw Mr. Campbell waving at me.

"Great job, Bethany. Sixteen minutes. You're halfway done."

I was about twenty steps behind Breanna and Bonnie. As they turned around, I heard Breanna complaining about having to run back uphill. A few seconds later, I too made the turn and was immediately hit by a giant gust of wind.

As I struggled to run forward, I pumped my arms a bit harder and tried to focus on my breathing. Because of the wind, I found it difficult to keep running. Ahead of me, I could see Bonnie and Breanna slowing down.

It was frustrating to have my thoughts keep coming back to Cooper. As much as I tried to block out any thoughts involving him, Cooper's words kept replaying in my head.

Cooper got animated whenever he talked about team sports, and how everyone had to play a specific role to produce the intended result: that the key to basketball was moving when you didn't have the ball, to create open space for your teammates; that football was about eleven people each doing one specific thing, and how a play broke down if even one person failed to complete their task; that the optimal strategy in cycling was to have teammates bike behind each other to reduce drag and conserve energy; that in auto racing, each team had two different cars, so that the two drivers could apply a technique called *drafting*, with the lead car's bumper touching the nose of the rear car, propelling both cars to a faster speed by cutting down the wind resistance.

An idea hit me. I ran to catch up with Breanna and Bonnie, whose jog had turned into a brisk walk.

I overtook them and turned around. "Hey guys . . . run right behind me."

"What are you talking about?" said Bonnie.

"Trust me," I shouted, making sure they could hear me over the loud wind. "Run right behind me."

Bonnie shrugged her shoulders and started jogging again, while Breanna reluctantly followed behind her. I looked over my shoulder.

"Right behind me, Bonnie," I said as loud as I could. "Make sure . . . there's no space . . . between us . . . or between . . . you and Bre . . ."

"Why?" she yelled back.

"I'll explain . . . after!"

The big hill approached. I looked over my shoulder again, seeing Bonnie right behind me, and Breanna just behind her. "Stay with me!"

I pumped my arms as hard as I could, shielding Breanna and Bonnie from the wind, and allowing them to draft right behind me. After a full minute, we reached the top of the hill.

I stopped with my back turned away from the wind, hunched over with my hands on my knees. Bonnie and Breanna joined me at the top, and looked a lot less exhausted than me.

Bonnie spoke up, "You okay?"

I shook my head as I hyperventilated. "Give me . . . one . . . sec . . ."

Breanna looked at me. "You blocked the wind."

We paused to catch our breath.

Breanna shook her head in disbelief. "Here, let me lead."

"Thanks," I said, nodding. "Ready when you are."

"Let's go."

We ran in a straight line, one right behind the other, me being careful to match strides with Bonnie and Breanna. It was so much easier running behind than in the front, and I could barely feel the wind in my face.

After a few minutes, Breanna got tired and moved to the side. Bonnie took the lead. Breanna slipped in right behind me, so that I was sandwiched between the two girls.

After Bonnie got tired, I took the lead.

We passed at least five girls running by themselves, struggling to deal with the wind in their faces. I couldn't help but notice that one of them was Gillian.

With about five hundred metres to go, Breanna took the lead again.

"Great job, girls!" called out Mrs. MacEachern as we reached the finish line. "Thirty-three minutes exactly!"

We collapsed in exhaustion, lying on the concrete next to the entrance of the school.

After my breathing returned to normal, I sat back up and stretched out my legs. The three of us watched as Gillian and several others finished the course.

Bonnie turned to me. "That thing we just did . . . what's it called?"

"Drafting," I said. "They do it in cycling and auto racing."

"Something from Math Camp?" she asked.

I paused. Cooper was now part of my past.

"Not anymore."

Breanna looked confused by my response. She started walking towards the entrance.

"Let's talk inside. It's way too cold."

"Wait for me, I'm coming too," said Bonnie, getting up to follow her.

I quickly stood up and followed my two best friends from Sydney High into the locker room. As we sat down on one of the benches, I realized there was something important I needed to say.

"Hey, Breanna?"

"Yeah, what?"

"I was a complete jerk. I hope I can be your friend again."

Breanna smiled and nodded, and I felt a huge release of tension. There was something else I needed to say.

"I'm sorry, Breanna."

She came over and gave me a hug.

"It's all good."

# 25

I flipped open my cell phone and re-read Rachel Mullen's text message.

*Your self-worth is not defined by being somebody's girlfriend.*

Even though it was now April, six months after the break-up, there were times when I still missed Cooper. Whenever I started to relive those memories of Ottawa and long to be reunited with Cooper, I'd look at Rachel's message to remind myself of what was most important.

Shortly after Cooper broke up with me, Rachel called me on Skype a couple of times. She helped me get over him, reminding me that Cooper and I had only known each other for six days, and that his inability to commit to a long-distance relationship was the only reason for the break-up. Rachel encouraged me not to blame myself because somebody else was insecure and immature.

Rachel texted me again today, wishing me luck in the High School Math League final. I still couldn't believe that the provincial championship would take place on Saturday, less than forty-eight hours from now.

I walked into the living room, and sat down to eat dinner with Mom.

I was excited and nervous. Sydney High had squeaked into the finals, coming in third overall among all the schools in Cape Breton.

"Are you all packed?" asked Mom.

"Yeah," I replied, mouth full of chicken casserole. I swallowed.

"Mr. Marshall says we're leaving right after lunch, so we can skip our last two classes and make it to Halifax around dinner time."

"That's great, Bethany. And you'll get to see Mr. Collins too."

"Yeah, I'm so excited about that! Of course, Halifax Prep made it to the provincial championship, so I'll definitely see him there."

Mom nodded. "I'm sure Sydney High will give Halifax Prep a run for its money. Your team keeps getting better and better."

"Yeah. We were fourth in October, third in December, and second in February. So if the pattern continues, we'll win on Saturday and beat those losers from Preparation H."

"Now, now," replied Mom. "You know it's not about winning."

"Yeah, yeah, I know," I said.

I knew our team was really good, especially since we started practicing twice a week ever since the start of the semester. We were peaking at just the right time.

"I really like how you stuck together and found a way to work together as a unit – it's such a change from what you told me in October when you were carrying the team, solving all the questions yourself."

"Yeah," I nodded. "Patrick's confidence is way up, especially since he got both relay questions last time. And the two Grade 11s, Tommy and Logan, are really good."

"Remind me," said Mom. "Logan was the one who lives computer games, and Tommy was the muscular boy who plays hockey?"

"Other way," I said. "Logan's the hockey player. Oh yeah, speaking of hockey, the principal was so happy about us going to provincials. She said it's been years since Sydney High sent a team to Halifax for a tournament that wasn't hockey-related."

Mom smiled. "Yes, I'm sure she's happy about that. Making it to the Math League championship is great for the school's reputation. And also I'm sure she's relieved not to have to spend all that money on twenty hotel rooms and an expensive school bus."

"Well, she did, since the hockey team made it to provincials again this year. But yeah, a math team is a lot cheaper than a hockey team. Just three hotel rooms and the cost of gas for Mr. Marshall's van."

"You've got your own hotel room?"

"Yeah, since I'm the only girl. Mr. Marshall and Patrick are pairing up, and so are Tommy and Logan. I've got a hotel room all to myself."

"Your first time sleeping in a hotel. And you've worked so hard to get there."

The phone rang. Mom leaned over and picked it up.

Mom turned to me, and cupped her hand on the phone. "Bethany, it's for you."

I put down my spoon and washed down the last morsel of casserole with my juice. I walked over and grabbed the phone from Mom.

"Hello?"

"It's Patrick. I've got bad news."

"What's wrong?" I asked.

"Logan's really sick. Remember he was coughing a lot at lunch today, and his voice was all screwed up? Well, he went to the doctor right after that. My father just got a call from his father. Logan has laryngitis. The doctor told him to stay home this weekend."

"Oh no," I said. "Logan can't come?"

"No, he can't," replied Patrick. "The team will be just you, me, and Tommy. There's no way we can get a replacement on such short notice. We can't do the relays, but at least we can do the team questions."

"What do you mean we can't do the relays?"

"Well, remember what happened in February? There was a team with only three players. They had to sit out the relay, since the organizers said the same person couldn't answer two questions. Such a stupid rule."

"We're screwed then," I said, leaning against the wall with my eyes closed.

"Well, we can still play," said Patrick. "But we wanted to stick it to Preparation H. That's impossible now, since we'll automatically get zero on the relays."

"Damn it," I muttered.

I looked down to see Mom sitting at the table with a frown on her face, wagging her finger at me in disapproval.

"My father tried calling a bunch of people in Grade 11 and Grade 12 – you know, the ones that played in previous regional games. But no one was available. Logan's really disappointed."

"We can't get anybody?" I asked. "Even if we could get someone half-decent, we'd just switch the relay order. Put that person first, then you second, Tommy third, and me last. All they'd have to do is get the first question and we'd be fine."

"How about Bonnie or Breanna?" suggested Patrick.

"They're going to the Natalie MacMaster concert tomorrow night at the Membertou Centre. We were invited too, remember?"

"Oh yeah. Okay, my father and I will keep calling people, hoping someone says yes, but it doesn't look good. Thought you should know. I should go, and let Tommy know what's going on."

"Thanks, Patrick. Bye."

I hung up the phone, a bit more forcefully than I meant to. I sat down opposite Mom, shaking my head.

"We practiced so hard all year, and then Logan gets sick the day before. It's so unfair."

Mom paused and spoke slowly. "You know, I was worried something like this would happen. It's so risky when you put all your eggs in one basket."

"Mom, stop," I said, cutting her off. "How is this helping?"

"You're right, Bethany. I'm sorry. You'll just have to do your best as a team of three. Forget the relay."

"But Mom, you don't understand. We worked for months to train for the relay. We practiced so much. All that hard work down the toilet. It sucks."

"Not true. All of you are in Grade 10 or Grade 11. You can do this again. And you'll be even better next year."

"But we were ready this year! If Logan didn't get sick, we would have done really well. It's not fair."

"Life isn't fair, Bethany. Maybe this is a reminder that you shouldn't put all your hopes on one thing. That you need to spread yourself out, focussing more on your other subjects . . ."

"Give me a break, Mom," I interrupted. "I don't need to spread myself out. I'm already getting an A or A-plus in everything. The classes are so boring, and the teachers don't show us any application to the real world. History is just memorizing dates and events. Business is just memorizing definitions and formulas. Science is just memorizing concepts and facts. It's so shallow. I hate it."

"It wasn't so bad when I was there," she said, defensively.

"Did I tell you what happened yesterday? Mrs. Finley was teaching us the formula for photosynthesis – you know, where plants convert carbon dioxide and water into oxygen and sugar. Someone asked why we needed to know this. You know what her response was? She said it was useful because if we were walking down the street and someone asked us to describe the chemical equation for photosynthesis, we could explain it to them. And she said it with a straight face too!"

"I really liked Mrs. Finley when she was my Science teacher."

"She's a nice lady, but her class is too easy. Why can't she make her class practical and challenging, like Mr. Marshall did with math? Remember all the cool stuff I told you about? Writing letters to politicians about social justice issues like poverty; approaching business leaders to show how they could make their companies more environmentally sustainable; doing public outreach presentations to show how people lie and deceive using statistics. Even the mayor came! Why don't the other teachers do stuff like this?"

"Because not every teacher is as passionate as Mr. Marshall. Because not every teacher is as knowledgeable as Mr. Collins. But what about English? Isn't your English teacher amazing?"

"Yeah, she is. But I'm ahead on the readings, and there's nothing to stretch me. It's not like there's an English Olympiad or something."

Mom leaned back in her chair and sighed. "You're right, Bethany."

I looked at her, stunned.

"Did you just say I'm right?"

Mom nodded. "You're someone who *needs* to be stretched. Not every teacher knows how to push their students; besides, most students don't want that. But for you, regular school just isn't challenging enough. You need things like math leagues and camps and contests to drive you, especially after having teachers like Mr. Collins and Mr. Marshall. You need challenge. You thrive on it."

She hesitated. "I know where you got that from."

"Let me guess. It was from my dad," I said. "Like everything else you blame him for: my height, my nose, my big feet, my deep voice, my stubbornness?"

"You got it from me."

"Yeah, right," I responded.

Mom looked away. I could tell she was hurt, and I knew I shouldn't have said that.

"I'm sorry," I said, leaning over and putting my hand on her arm. "I didn't mean what I said."

After a long silence, Mom looked at me.

"Bethany, did I ever tell you how I got into figure skating?"

"No," I replied. "You never tell me anything about figure skating. I didn't even know you made it to the Olympic Trials. It was Miss Carvery who spilled your secret, remember?"

"Yes, I remember," said Mom, staring at the wall.

She took a deep breath, and turned to face me.

"I was six when your grandmother, my mother, took me to the rink for my first figure skating lesson. I loved it. It came so easily to me. A few years later, I met this tough coach who saw my potential and pushed me like crazy. Every day, I had to wake up super early to practice at the rink before school. It was crazy. But the coach taught me all these spins and jumps. I could do a triple toe loop at age eleven."

"What's that?" I asked.

"A special type of jump. Anyway, I became really good really young. I qualified for the junior provincials when I was fourteen, matching up against people who were four years older. It was my first major competition. There was so much pressure, and I was nervous. I choked in the short program. I fell on all three of my jumps, and landed really badly on my last fall, nearly breaking my ankle."

"Seriously?"

"Yeah, it was bad. I somehow finished the routine, and ended up in last place. My ankle was killing me, and I wanted to quit and go home. But my coach said no, and insisted I skate in the long program the next day. I was crying and told her there was no point, since I was in dead last and had no chance of winning. But she forced me to continue. I could barely move during the warm-ups the next day, but somehow I went out there, and skated a flawless long program. I hit all my jumps, since I could plant and land on my other leg, and got the highest score of all the girls. I went from fourteenth after the short program and finished fifth overall, skating on one leg."

"That's crazy, Mom. So that's when you knew you were good."

"That's when I knew I was good. That's where the dream began, that I'd go to the Olympics someday."

"And you almost made it."

"Yes, Bethany. I almost made it."

Mom reached her hands across the table to grab mine. She squeezed tightly.

"I was hobbling on one leg after that short program. It was the provincial finals – my first big event – and I had a terrible day. I was dead last. I went out there the next day, with no pressure. I mean, how could I do any worse? So I skated the long program, just thinking about landing the next jump. I hit one jump, then two, then three, and so on, until I nailed them all. The final result didn't matter to me. That fifth place finish is the result I'm most proud of in all the years I did figure skating."

"That's awesome, Mom. You never told me that before."

"You're right," she said. "But I wanted to tell you this because you're not going to win the Math League on Saturday. So there's no pressure on you. Just do the team questions, and enjoy them. Get as many as you can. Forget about the results. And you can stick it to Preparation H next year."

I laughed out loud, and rose up from my seat to give Mom a hug.

"Thanks, Mom, you're great."

The phone rang.

I let go of Mom and picked it up. It was Patrick.

"Hey, Bethany, I've got great news, good news, and so-so news."

"Tell me the great news."

"We've got Logan's replacement. My father called around and got someone to say yes."

"Awesome! So we have a full team now. That's super. So what's the good news?"

"Remember you said if we could just get someone half-decent, all they'd have to do is get the first question of the relay, and then we'd be fine? Well, our replacement hasn't practiced at all, and I know you two don't get along, but she's really smart. If we put her first on the relay, she'll do fine."

*You two don't get along. But she's really smart.*

I felt a knot tightening in my stomach.

"It's Gillian, right?" I asked.

"Yeah."

# 26

As the van pulled up to the Cambridge Suites Hotel in Halifax, I relaxed my shoulders and let out a deep breath.

Our four-and-a-half drive from Sydney was finally over.

Mr. Marshall checked us into the hotel and gave us our room keys. He was by himself in Room 303, between the two boys in Room 301 and the two girls in Room 305. I sighed, knowing I would be sharing a room with Gillian tonight.

As we took the elevator up to the third floor, Mr. Marshall asked us to quickly get ready and meet in the lobby in five minutes. Gillian and I didn't say a word to each other as we entered our room. She silently claimed the bed closest to the washroom, quickly removing some clothes from her backpack and dumping them onto the bed.

*Whatever.*

Deciding to unpack later, I grabbed my wallet from my backpack, put it in the right pocket of my jeans and made sure my room key was safely in my pocket. I turned the doorknob, ready to go.

"I need to use the washroom first," said Gillian.

"Fine. See you downstairs," I replied, not bothering to look back at her. I walked down two flights of stairs and popped into the washroom in the lobby.

Knowing Mr. Marshall, no one dared to be late. We followed Mr. Marshall as he walked briskly along Brunswick Street, and turned down the big hill on Duke Street. A few minutes later, we entered Scotia Square Mall.

"Here is dinner money," said Mr. Marshall, handing each of us a $10 bill. "Dinner is courtesy of Sydney High. The food court has plenty of choices, so choose anything you want and we'll all sit down and eat together."

Tommy turned to us, pointing at a big store on our left. "Hey, check it out! There's an electronics store. Mr. Marshall, can I go there after dinner?"

As Mr. Marshall nodded, I laughed, reminded of how much Tommy loved everything involving computers and electronics. He always dressed the same, a crisp dress shirt to go with his faded blue jeans. He was six feet one, even taller than me, and wore thick glasses that covered some of his pimples.

During last week's practice session, Tommy mentioned how comfortable he felt with his Math League friends, and thanked us for being the only people at the school who never made fun of his appearance.

As we stepped into the food court, I held back, trying to decide between Lebanese and Indian. I noticed Gillian walk towards Taste of India, so I decided to walk in the other direction and order my dinner from the short balding man standing behind the counter of Ray's Lebanese Cuisine. I had no idea what Lebanese food was, but the smiling man named Ray assured me that I would love his famous falafel sandwich, and made a special order just for me.

Mr. Marshall found an empty circular table in the middle, and we sat down to eat our dinners. Tommy glanced over at my Lebanese sandwich.

"Eeew, what's that? Looks gross."

"Not as bad as that," I said, motioning towards the greasy onion rings on his tray, and the bacon double cheeseburger an inch from his mouth.

I took a bite of my falafel. Mr. Ray was right.

"What should we do for the relay?" asked Patrick.

Gillian put down her spoon. "Like I told you in the van, I'm not going first. Don't insult me."

Tommy shook his head. "But Gillian, the rest of us have been practicing for months. We've gotten really good at the relay."

Gillian glared at Tommy. "What was your mark in Grade 11 Math last semester?"

"I got an A," replied Tommy.

She turned to Patrick. "We were in Grade 10 Math together. What did you get?"

"B," replied Patrick.

Gillian didn't bother asking what my mark was. She knew.

"Mr. Marshall, I got an A-plus in your class," said Gillian. "Can you honestly say that I'm the weakest student on this team?"

Mr. Marshall put down his fork. "This is not about who is the strongest or weakest student. All four of you are excellent, and that is why you were chosen to represent the school. The order of the relay is not as important as you make it out to be."

"But Mr. Marshall, I don't want to go first!"

"I would like the four of you to discuss and decide this amongst yourselves."

"But Mr. Marshall . . ."

"Like I said, you need to decide this amongst yourselves."

With that, Mr. Marshall stood up and walked towards the other end of the food court, where he sat down by himself at a table by the window, and took a newspaper out from his briefcase. His lips were pursed, and I could tell he was annoyed.

Tommy folded his arms. "Gillian, I hate to say it, but the math league is a lot different from class. You came to one math league event and your team bombed."

"I had a bad day. It was my first time at the Math League, and it was new for me."

"It was new for us too," I snorted. "What's your point?"

"The point is that I don't want to go first."

I stared back defiantly. "Okay, what do you want then?"

"I want to go last," said Gillian. "To be the anchor."

Tommy laughed. "You've got to be kidding."

"If I'm not the anchor, I'm going home. The three of you are on your own."

I rolled my eyes in disgust. Patrick was about to say something, but Tommy cut him off.

"You're bluffing," said Tommy, smiling at Gillian.

"No, I'm not."

Tommy shook his head. "When you play as much poker as I do, you can tell when someone's lying. Gillian, you're not going to sit out tomorrow. Want to know why?"

Gillian folded her arms and squinted at Tommy, before looking away. Tommy leaned closer.

"If you sit out tomorrow, you're going to have to explain that to Mr. Marshall, and I know he scares you just as much as he scares the rest of us. And if you quit on us, you're going to have to explain that to your mother when you go home. From what I've heard about your mom . . ."

"Shut up," she snapped, cutting him off. "Okay, fine, I'll compete. But I'll get the questions wrong on purpose."

Tommy laughed. "That's an even bigger bluff. I guarantee you won't."

"You want to bet?"

"Everyone in the entire school knows how competitive you are. If you get your questions wrong, you'll just confirm what everyone already knows – that when it comes to math, Bethany's just a lot better than you."

Gillian's face flushed. She stared defiantly at me before turning to Tommy.

"Unlike Bethany, I can actually handle pressure. It's not about how many practice sessions I've attended. It's not about how many math camps I've gone to, or whether I used to have my own personal math coach. It's about who can best handle the pressure of a big event."

Gillian turned to me and flashed a wicked smile. "Remember the Spelling Bee?"

I shrugged. "Sure. You beat me. What was that, three years ago?"

"That's right," she said. "I beat you. Case closed."

Gillian folded her arms and took a sip of her drink.

"Look guys," interjected Patrick, "this isn't going anywhere. Let's just vote and decide. Otherwise this is never going to end."

During elementary school, I felt bullied because it was always Gillian's clique against me. Four against one. The clique was on her side. This time, Tommy and Patrick were part of my clique, and the numbers were on my side. It was time to put Gillian in her place.

"I agree with Patrick. If you want to end the discussion and have a final vote, raise your hand."

Patrick and Tommy quickly put up their hands, while Gillian looked away. Feeling confident, I continued.

"Okay, three votes to one, we agree to end the discussion and make a decision. So it seems that we have four options to choose from, based on whether Gillian goes first, second, third or fourth. From how we've done the relays during practice, Tommy should go after Patrick, and I should go after Tommy."

I looked at Gillian. "I assume you have no problems with that."

She didn't reply.

I took out a sheet of paper and scribbled the four options with my pen:

| | 1st | 2nd | 3rd | 4th |
|---|---|---|---|---|
| **Option A** | **Gillian** | Patrick | Tommy | Bethany |
| **Option B** | Patrick | **Gillian** | Tommy | Bethany |
| **Option C** | Patrick | Tommy | **Gillian** | Bethany |
| **Option D** | Patrick | Tommy | Bethany | **Gillian** |

"Raise your hand if you want Option A," I said.

I put up my right hand, and was joined by Tommy.

"How about Option B?" One hand went up; it was Patrick's.

"Option C?" Silence.

"Option D?" No answer. Gillian looked away, staring at the floor.

I cleared my throat. "No matter what Gillian chooses, Option A gets the most votes. So we've decided. Let's go tell Mr. Marshall."

Gillian stood up and knocked her empty cup in my direction.

"Do whatever you want," she yelled. "I don't care!"

Seeing the commotion, Mr. Marshall came back to our table and demanded what was going on. Gillian complained, saying we were unfairly ganging up on her, and announced that she was going back to the hotel room. Mr. Marshall was about to respond but Gillian started walking away, and we stared at her as she turned the corner to exit Scotia Square.

Mr. Marshall told us that he was going back to talk to Gillian. He sighed, and gave us permission to stay inside the mall.

We were all relieved to finally have some Gillian-free time. Patrick and Tommy went straight to the electronics store, while I headed to the bookstore. An hour later, we walked back up the hill towards the Cambridge Suites Hotel. Walking up to the third floor, I went straight past my room and followed the boys into their room, and flipped on their TV.

The room was freezing. I told the boys I'd pop into my room, get my sweater, and come straight back. I pulled my room key out of my pocket and prepared to swipe it, when I heard Gillian crying and shouting inside our room. I leaned closer, putting my ear against Room 305.

"Mom, why did you force me to come here?"

I heard sniffling, followed by silence.

"Look, I told you. I don't even like math. Why don't you ever think about what I want?"

I heard a sob, followed by a loud honk of Gillian blowing her nose.

"Mom, stop calling me a failure! It's not my fault! I tried to be the relay anchor, but they outvoted me three to one. I did my best. What else am I supposed to do?"

Another loud honk.

"I don't care about that stupid scholarship . . . I don't care what Victoria and Caroline did! Mom, stop comparing me to them. It's my life!"

I knew that Gillian had two older sisters, who had both graduated with the highest marks in their years. Both were chosen as the class valedictorian. I assumed Gillian was referring to them, especially since Bonnie told me that both of Gillian's sisters got full entrance scholarships to St. Francis Xavier in nearby Antigonish, Nova Scotia, one of Canada's top universities.

"So what if Bethany gets the scholarship? Why do I have to be the best in everything, anyway?"

I heard a loud crack, probably Gillian slamming the phone against the table. She started to yell.

"I hate you, Mom. I hate you even more than I hate Bethany. I hate your guts, Mom. Stop ruining my life!"

I was startled by the sound of a door opening behind me. Patrick leaned out of Room 301. I looked at him and put my finger to my lips.

He got the hint. I tiptoed back into Room 301 and closed the door gently. My head was spinning, trying to process what I had just heard.

*What scholarship was Gillian talking about?*

Patrick looked at me. "What was that?"

"Oh, nothing," I said, staring blankly at the ceiling.

I felt sympathy for Gillian. I never thought I would ever feel that way about her. But overhearing that, I did.

"Didn't you say you were going to get your sweater?"

"I changed my mind. I'm not cold."

I sat down on the bed, and folded my arms across my chest to keep my body warm. Tommy glanced over at me.

"Hey, remember how I'm good at catching people bluffing?"

"Yeah, so?"

Tommy reached into his bag and threw an oversized black sweater in my direction. He laughed.

"You're welcome, Bethany."

# 27

We arrived at Dalhousie University, the site of the Math League provincial championship.

The seats and tables were arranged in groups of four, just like Mr. Marshall's classroom. We sat down at an unoccupied table, with Gillian and Tommy on one side, and Patrick and me on the other. I looked at Gillian, who was directly across from me, and forced a smile. She pretended not to notice. Like it or not (and neither of us did), we were going to be teammates for the next three hours.

A few teams had taken their seats, and I noticed a half-dozen students walking around and chatting. A tall slim brunette, probably in her late twenties, was standing at the front and arranging some envelopes. I had heard that the Nova Scotia High School Math League was started by a graduate student at Dalhousie; surely this was her.

Just then, a team of four boys walked in, each wearing a navy blue suit with the letters HPA stitched on its left front lapel. So this was the infamous Halifax Preparatory Academy. Two of the boys had a striking resemblance to Albert and Raju.

A few seconds later, I saw a friendly face. He leisurely strolled in, hands in his pockets, eyes scanning the room. He found our table, flashed a big smile, and started walking towards us.

"Mr. Collins!"

"Bethany!" said my favourite teacher, wrapping me in a huge hug. "How are you?"

"I'm great."

Mr. Collins walked over to my teammates. No introductions were necessary, since he already knew everyone. He had taught Tommy's father and Gillian's older sisters, and knew their families quite well. Of course, he knew Patrick through his father. When Mr. Marshall returned to our table, he gave Mr. Collins a friendly handshake and handed him a coffee. Mr. Collins joked that he had joined the dark side, but his true loyalty was still with the school where he had taught for over forty years.

I looked up at these two men, the stern father and gentle grandfather in my life, and realized how lucky I was. I was already motivated to do well, but seeing my mentors here gave me an additional boost.

The tall brunette called us to attention, invited all the students to take their seats, and asked the teachers to join her at the front of the classroom. Mr. Collins and Mr. Marshall wished us luck and told us they'd come back to chat later. The brunette switched on her microphone and smiled at us.

"Good morning, everyone! My name is Paula and I will be your host for today's Math League provincial championship. I'm currently finishing my Ph.D. in combinatorial game theory here at Dalhousie, and I am delighted to welcome you to our campus. We have sixty-four students from sixteen schools all across the province. Welcome and good luck to you all."

Paula asked each teacher to be the "proctor" for a team from a different school. I smiled when I realized who our proctor would be.

It was time. I got out my pens and paper, and tried to relax.

Mr. Collins reached into his envelope and handed each of us a small sheet of paper containing the first question:

> Question #1: Sarah looks at her watch at exactly 8:59 a.m. What is the measure of the angle formed by the hour hand and minute hand?

"You have five minutes," said Paula. "Starting right now."

"Let's draw a picture," suggested Tommy.

Whenever we could, we tried to draw a picture. Especially for a question involving a clock, a diagram made a lot of sense. As Gillian silently looked on, Patrick drew a picture with the time showing nine o'clock.

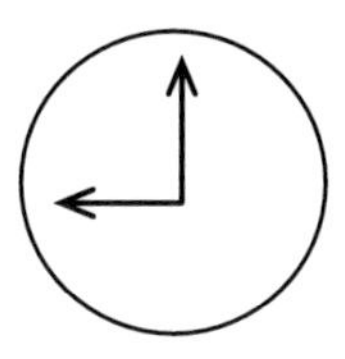

"Bethany always tells us to consider a simpler case. 8:59 is really close to 9:00. So the answer has to be around ninety degrees."

Tommy nodded, and picked up his pencil. "Since 8:59 is just before 9:00, the answer has to be a bit less than ninety degrees. Here, let me draw a bigger picture."

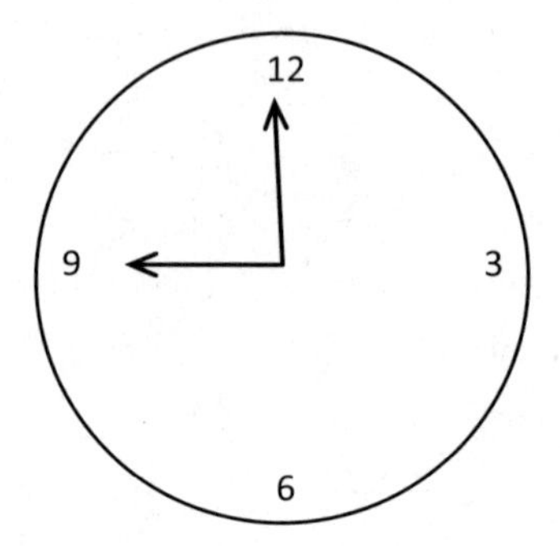


I pointed to the minute hand on Tommy's diagram.

"I have an idea. The minute hand makes one full 360° revolution every hour. That works out to 6 degrees every minute. And now look at the hour hand . . ."

Gillian cut in. "I got it! Since 9:00 a.m. has an angle of 90°, 8:59 a.m. has an angle of 84°, since it's 6 degrees less. That's it."

She grabbed the official answer sheet and wrote down 84° with a thick black Sharpie pen.

I glared at Gillian. "It's not a race to see who solves it first. We don't write on the answer sheet until everyone on the team agrees."

Gillian looked at me. "What the hell's your problem? We've got the answer."

"No, we don't," I softly replied.

Patrick looked at me strangely. "We don't?"

I shook my head, and pointed to Tommy's diagram. "The picture's wrong. As the minute hand is moving, so is the hour hand. Here, let me fix it."

I grabbed an eraser from the table and re-drew the hour hand a bit darker, and marked the 84° that Gillian had just established.

"The diagram looks like this," I said.

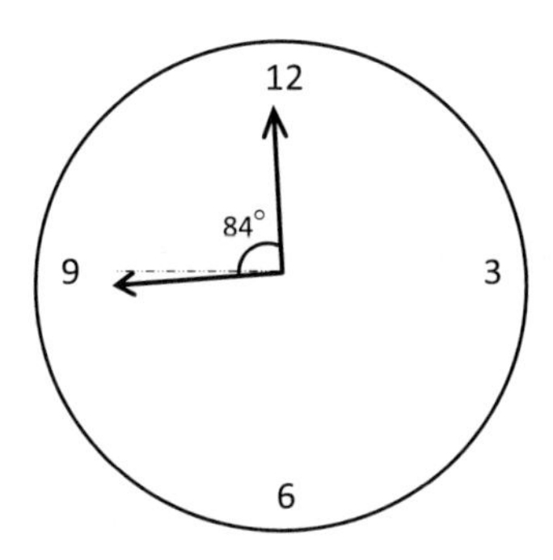


"Shoot," said Tommy. "Bethany's right. The hour hand's moving too. From 6:00 to 9:00, the hour hand makes a quarter turn. That's a quarter turn every three hours, or 90 degrees every 180 minutes."

Patrick nodded. "I get it. The hour hand travels clockwise half a degree every minute."

Tommy grabbed his pencil and pointed to the half degree space between his incorrectly-positioned hour hand and my correctly-positioned hour hand. "So if it's 8:59 a.m., one minute *before* nine o'clock, then we've got to *add* 0.5° to our answer. The right answer is 84.5°."

I looked at Patrick. "Do you agree?"

Patrick paused, and after a few seconds, he nodded. "Yeah. I agree."

I turned to Gillian. "How about you?"

She bit her lip and nodded.

"We have consensus," I said. "Gillian, you have the answer sheet. Could you please write down the correct answer?"

Gillian glared at me and muttered something under her breath. She crossed out her earlier answer and wrote down 84.5° in its place.

I took a deep breath, relieved we had avoided a careless mistake on the first problem.

"Time's up!" said Paula.

Mr. Collins came to our table to pick up our answer sheet. Paula asked a team from Truro to come to the front and present their solution, which was identical to ours. The scores went up on the big screen. Thirteen of the sixteen teams had gotten the full five points for the first problem.

Mr. Collins gave us the next question:

Question #2: There are $x$ books in the Old Testament, where $x$ is a two-digit number. If you multiply the digits of $x$ you get $y$, which is the number of books in the New Testament. Adding $x$ and $y$, we get 66, the number of books in the Bible. Determine $x$ and $y$.

I have an idea," said Patrick. "Set up a two-variable equation. Let $x = 10a + b$ for some digits $a$ and $b$."

"Yeah," I continued. "By definition, $y = ab$, since it's the product of the digits of $x$. Then you just add up $x$ and $y$, set it equal to 66, and solve for $a$ and $b$. Should be easy."

$$10a + b + ab = 66$$

Tommy laughed. "I have a better idea."

Gillian looked at him. "What's that?"

"Just write down the answers: $x = 39$ and $y = 27$."

I stared at Tommy in amazement. "How'd you figure that out so fast?"

Tommy grinned. "My parents made me go to Sunday School as a kid."

"Yeah, so?"

"Well, Sunday School was the biggest waste of time. The only thing I remember from that old hag Miss Barton was her telling us there were 39 books in the Old Testament and 27 in the New Testament. She said it was easy to remember because you could multiply the digits in the first number to get the second number: $3 \times 9 = 27$."

Patrick laughed. "That's thc only thing you remember?"

"Yeah," said Tommy. "One useless piece of trivia in five years. Thanks for everything, Miss Barton."

We quickly agreed that Tommy's answer was correct, since $39 + 27$ added to 66. Patrick wrote down $x = 39$, $y = 27$ on the official answer sheet. Everyone exchanged high-fives except for Gillian and me.

Since we had so much time left, Patrick and Tommy started to chit-chat about something they saw last night at the electronics store. Just to satisfy my curiosity, I double-checked the answer using some algebra, showing that $(x, y) = (39, 27)$ was the unique solution.

"Okay, time's up!" said Paula. A team from Bridgewater came to the front and presented their solution, which was nearly identical to my algebraic approach.

We got the next three questions without any difficulty, and were clicking as a team. After five questions, only four of the sixteen teams had a perfect score, including our team from Sydney High.

| SCHOOL | SCORE |
|---|---|
| Halifax Preparatory Academy | 25 |
| Lunenburg High School | 25 |
| Parkdale High School | 25 |
| Sydney High School | 25 |

It was time for the break, and I went to the back of the classroom to grab a snack and some juice. Mr. Collins came by to congratulate me on our team's performance so far.

It was so nice to catch up with Mr. Collins, and see him again after such a long time. He told me how much he missed Cape Breton but was loving the teaching at Halifax Prep.

"How is your Mom doing?"

"She's fine," I said. "But her job is getting worse and worse, and she complains about it all the time. Her new boss is horrible."

Mr. Collins was about to say something, but Paula announced the end of the break. As we walked back towards our seats, Mr. Collins dropped his voice to a whisper.

"I know you and Gillian Lowell have never gotten along. But you two are teammates today. Mr. Marshall told me she only came because she was a last-minute replacement, and I know she's not as strong as you or the two boys. But I overheard what you said to her after she made that mistake on the first question, and I could tell how embarrassed she was. She knows you're far better in math than she'll ever be. This is the one subject where she's unable to compete with you. I encourage you to make her feel welcome, and give her a chance to contribute something to the team."

"But Mr. Collins, she's a . . ."

"Bethany, please. Give her a chance to contribute, and let her regain some confidence. You'll need her on the relay. You always said that cooperation is better than competition. Well, Gillian needs to hear that from you."

I nodded and sat back down. Rejoining my teammates, I looked at Gillian.

"I'm sorry for what I did," I said, crossing my fingers under the table.

"Sorry for what?" scoffed Gillian.

"For criticizing you, and not making you feel like an equal member of the team. I'm sorry."

"Yeah, fine. Whatever."

*Whatever to you too.*

That apology didn't go over so well, but at least I tried. Fortunately it didn't matter, as we solved the next four problems without any contribution from Gillian. Lunenburg and Parkdale had missed Question #9, which left just two teams with a perfect score.

| SCHOOL | SCORE |
|---|---|
| Halifax Preparatory Academy | 45 |
| Sydney High School | 45 |
| Glace Bay Education Centre | 40 |
| Lunenburg High School | 40 |
| Parkdale High School | 40 |

I saw that Glace Bay had moved up into a tie for third. We knew that school was good; they had won each of the three regional games in Cape Breton.

Mr. Collins gave out the last team question:

> Question #10: Each of 16 rooms has a "point value", as indicated in the diagram below. Your team's task is to collect as many points as possible. Enter at the start, exit at the finish, and pass through each room no more than once. What is the greatest number of points your team can collect?

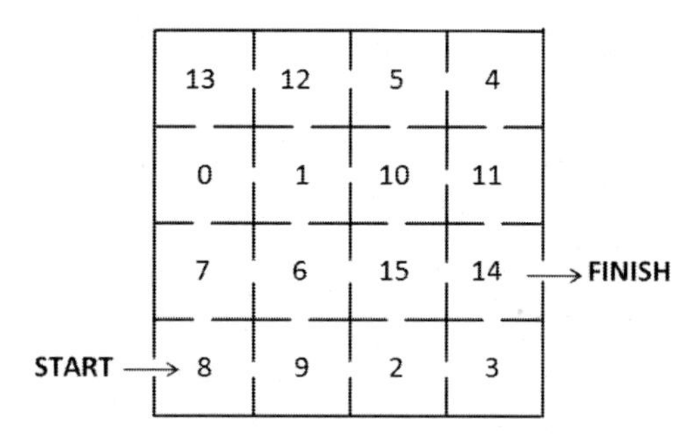


I looked at Gillian. "All I can see is that the numbers from 0 to 15 are all there. Any ideas, Gillian?"

"Are you patronizing me?" she asked.

"No," I lied. "I'm not sure where to start. I want to know your ideas."

While Gillian looked away, Patrick excitedly pointed to the diagram on his question sheet.

"If we can find a way to go through all the rooms exactly once, then we're done. That'll give us the maximum number of points, 0+1+2+3+4+5+6+7+8+9+10+11+12+13+14+15, whatever that is."

Tommy did a quick calculation. "It's 120, by Bethany's Staircase."

Out of the corner of my eye, I saw Gillian scowl. Clearly, she remembered that Grade 5 class.

Patrick nodded. "So the best we can do is 120 points. Is that possible?"

Tommy drew a path on his question sheet. "I can't do 120, but I can do 117. All but the 3-point square in the bottom right corner."

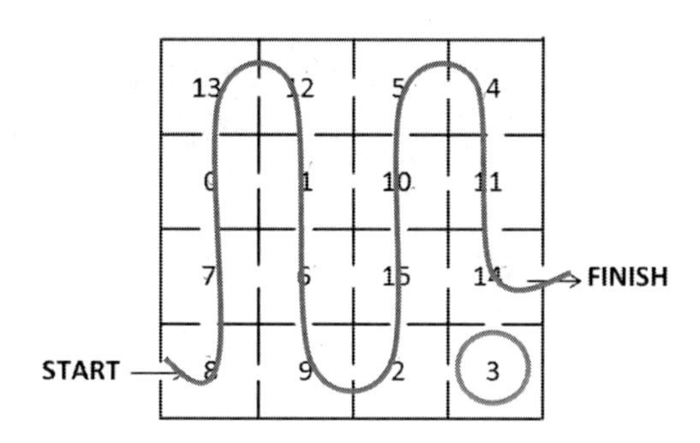


"Cool," I said. "Do you think that's the answer?"

"No, that's too easy," said Tommy. "This is the last question on the provincial final."

Patrick pointed to the diagram, and started drawing paths lightly using his pencil. "I'm sure there's a clever way to go through all 16 squares. We just have to find it."

Gillian nodded and started drawing some paths, seeing if she could find a single path that went through all the squares. Tommy and I joined in, but none of us were able to do it.

Patrick tapped his pen on the table. "How about a path that goes through 15 squares, all except for the one marked 0 points? Then you still get the maximum score."

"Yeah, great idea!" said Tommy.

A minute later, we still had nothing. Something wasn't right. I glanced over at the Halifax Prep table, and saw them working frantically. The boy who looked like Albert Suzuki was scratching his head, thinking.

I suddenly remembered how the real Albert had solved a hard grid problem at the Canada Math Camp by considering a *checkerboard colouring*. He coloured the squares black and white, like on a checkerboard, and presented an elegant argument that counted the number of squares of each colour.

I realized the same idea would work on this problem.

"Hey, check it out," I said, leaning across the table. "I'm going to colour every other square black, like on a checkerboard. I'll draw a random path from the start to the finish. Look at the colours!"

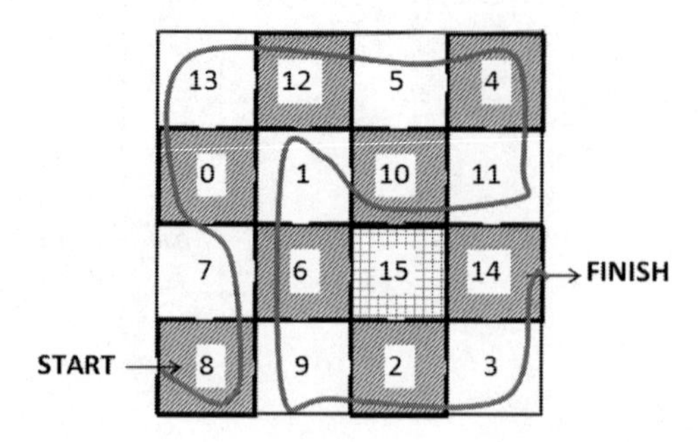

"I don't see it," said Patrick.

"They alternate! Look at my path – it starts at 8 then goes to 7, then 0, then 13, and so on. The colours alternate. All the even numbers are on Black and all the odd numbers are on White!"

"Yeah, so?"

"So we're done!" I said, nearly shouting. "A path starts with Black and ends with Black, so for every path, there has to be one more black square than white square. A path can only hit 7 white squares!"

B W B W B W B W B W B W B W B

"One minute left," announced Paula.

I kept talking. "Any path can go through at most 15 squares! The square you miss has to be a white square. Yes, yes, yes!"

"I don't get it," said Gillian. "You're making no sense."

"Look! The square marked 0 is black – it's an even number! So it's impossible!"

"Hurry, Bethany!" shouted Tommy. "What's the answer?"

"I don't know if it can be done, but the best path has to go through all the squares except for the lowest *odd* number, which is the *white* square marked 1."

"Okay, let's try it!" said Patrick. We frantically tried to draw a path that hit every square except for the 1-point square.

Gillian pumped her fist. "I got it!"

She flipped her diagram towards us.

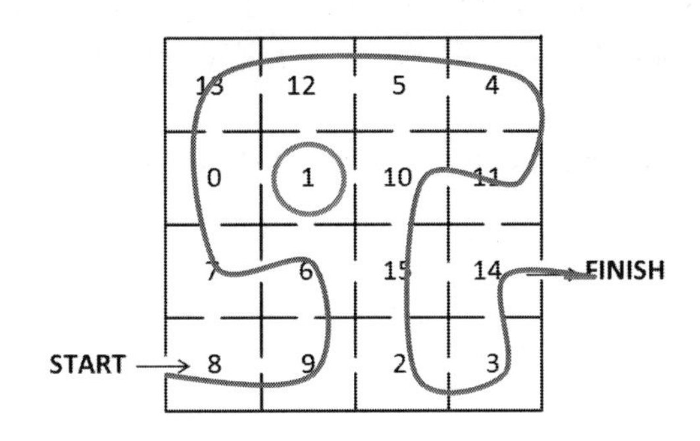

I quickly grabbed the answer sheet and wrote down 120 points, the sum of the numbers from 0 to 15.

"No!" shouted Gillian. "It's 119! We skipped the 1-point square!"

I quickly crossed out my incorrect answer and replaced it with Gillian's correct answer of 119.

"Time's up!" announced Paula.

We collapsed in our seats and sighed in relief.

I turned to Gillian. "Thank you."

She looked shocked. After a long pause, she nodded at me.

"You're welcome."

Paula updated the scores. Not only did Parkdale High miss Question #10, so did Halifax Prep. My heart skipped a beat, realizing that we were all alone in first place.

| SCHOOL | SCORE |
|---|---|
| Sydney High School | 50 |
| Glace Bay Education Centre | 45 |
| Halifax Preparatory Academy | 45 |
| Lunenburg High School | 45 |

Paula looked at us. "Would someone from Sydney High please present their solution to us?"

Tommy turned towards me. "Bethany, it's all yours."

I shook my head. "Gillian, could you do it?"

She stared at me in disbelief. "Seriously?"

I nodded, as did Patrick and Tommy. Gillian hesitated before slowly standing up from her seat and walking to the front of the room to present our team's solution.

I looked towards Mr. Collins, and saw him smiling back at me.

# 28

We were in first place.

I splashed cold water on my face and stared at my reflection in the bathroom mirror. I closed my eyes and tried to get my heart rate down.

Shortly after my perfect score on the Grade 7 Gauss contest, Mr. Collins told me about the infamous Halifax Preparatory Academy. They hadn't lost an academic competition of any kind, in any subject, at any grade level, in ten years. As I dried my face and hands with some paper towels, I tried to erase the memories of that conversation from three years ago.

It was time for the relay. I was dreading it, especially after Gillian walked out on us last night. Would she whine and complain about having to go first, and break all our confidence and momentum?

I slowly walked back into the main classroom, hands in my pockets, staring at the ground. I accidentally bumped into the short Asian boy from Halifax Prep. He didn't apologize as he continued walking forward, looking tense and frazzled. Maybe he too was thinking about his school's undefeated streak.

When I approached my table, I let out a sigh of relief. Gillian, Patrick, and Tommy were sitting in a row, in that order. There was an empty chair on Tommy's right. I quickly sat down.

Paula got our attention, and we turned to face her.

For the benefit of the half-dozen people participating in the Math League for the first time, Paula briefly summarized the format of the relay. Each team would get an answer sheet containing four boxes labelled *A*, *B*, *C*, *D*, and each student would get one of the four relay questions. The first student would solve her problem, enter it in the box marked *A*, pass the answer sheet on to the second student, who would need to use that value of *A* to solve his problem, and write his answer in the box marked *B*. This process would continue down the line, until all four students had answered their question.

If all four boxes were filled in with the correct answer, the team would be awarded five points with up to five bonus marks for being one of the first teams to finish. If at least one of the boxes was incorrect, the proctor would simply hand back the answer sheet to the final student, without indicating which answers were wrong. Paula reminded us of three rules: no talking was allowed; the answer sheet could only be passed back and forth with no eye contact; and any form of cheating such as looking at our teammates' questions would immediately lead to disqualification.

As the relay team's anchor, my job was to take the most difficult question, solve as much of it as I could without knowing the value of *C*, and try to get a general formula for *D* in terms of this unknown variable *C*. As soon as Tommy passed me the sheet with his value of *C*, I could substitute his answer into my formula, and immediately write down *D* on the answer sheet. If I did my job properly, there would be just a few seconds between Tommy finishing his problem and Mr. Collins getting the completed answer sheet.

Paula called each of the proctors to the front and gave them the envelope containing the problems for the first relay. Mr. Collins walked back to us and placed our questions in front of us, face-down, and handed the answer sheet to Gillian.

"You have ten minutes," said Paula. "Starting right now."

<u>RELAY #1, Question #4</u>

A group of *C* friends sit around a circle, of which two are boys and the rest are girls. Each girl writes down the number of seats from her to the nearest boy. These numbers are then added together to produce one final "score". In the example shown (with $C$=9), the total score is 11.

The *C* friends take their seats so that the final score is as large as possible. Let *D* be this value, the maximum possible score. Write down *D*.

For example, this girl is <u>2</u> seats from one boy and <u>4</u> seats from the other boy. So she is 2 seats away from the <u>nearest</u> boy.

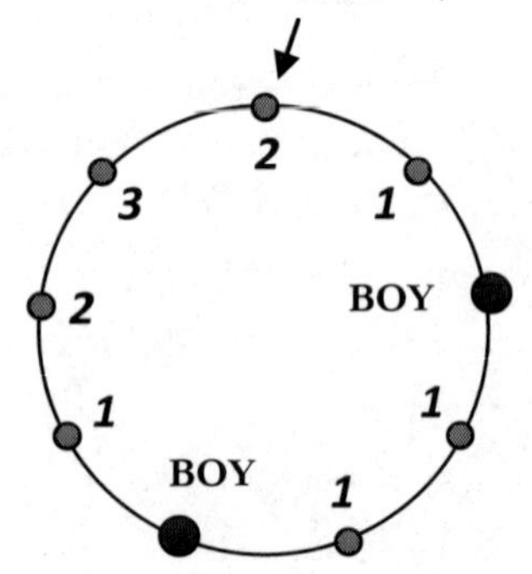


**Score = 1+2+3+2+1+1+1 = 11**

To get a feel for the problem, I began with simple scenarios. The cases $C$=1 and $C$=2 made no sense since there were two boys around the circle, and there would be no girls to produce a final score.

The case $C$=3 was easy since the one girl had to be sandwiched between two boys, and be one seat away from each of them. The total score had to be 1.

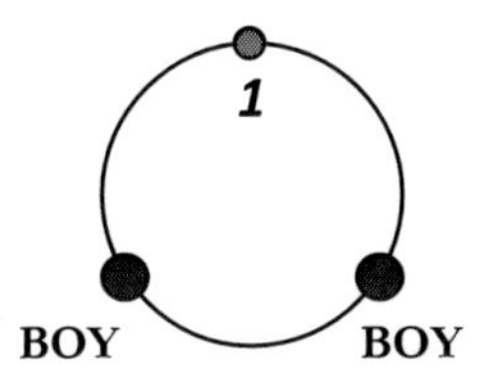


The next case was $C$=4. Either the two boys were sitting opposite each other, or next to each other. In both scenarios, each of the two girls would write down the number 1, since each girl would be seated next to a boy. No matter what, the final score would have to be 1+1 = 2.

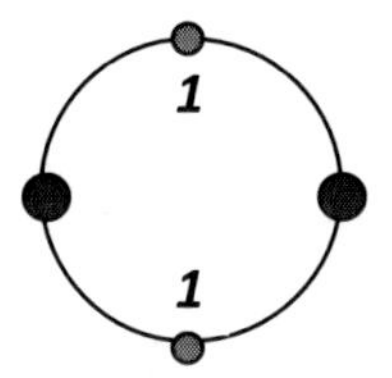


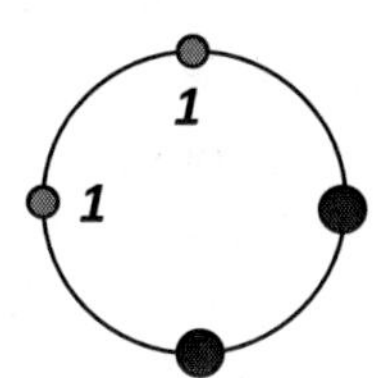


The next case was $C$=5. There were two possible cases: either the boys were sitting beside each other, or they were not.

In the first case, one girl would be sitting two seats away from both boys.

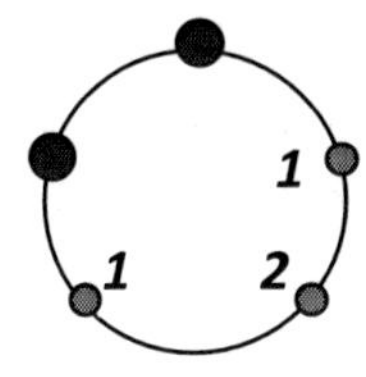


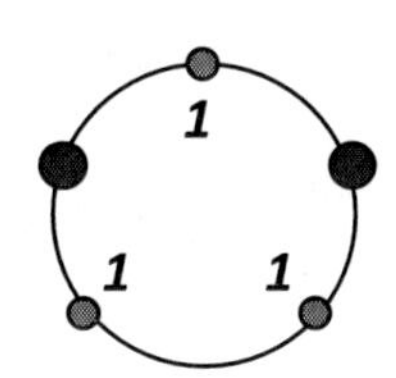


So the first seating arrangement would produce the maximum total score of 1+2+1 = 4.

The goal was to *maximize* the total score. To do this, I realized I needed to sit the two boys beside each other, which would make the total score as high as possible.

To confirm my intuition, I looked at the example given in the question with $C$=9, which had a total score of 11. I found a better seating arrangement which produced a total score of 1+2+3+4+3+2+1 = 16.

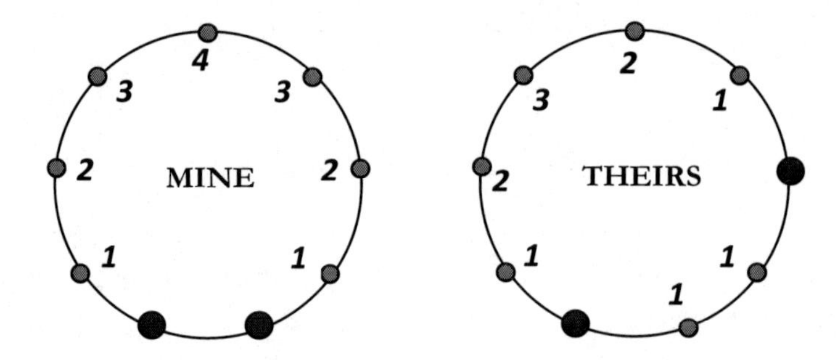


The symmetry was too beautiful. This had to be correct. Without hesitation, I starting calculating the maximum possible score for each value of $C$, by placing the two boys beside each other. This maximum score was the value of $D$. For example, I showed that if $C$=9, then $D$=16.

For each $C$ up to 12, I calculated $D$, and wrote the information in a table:

| Friends ($C$) | 3 | 4 | 5 | 6 | 7 | 8 | 9 | 10 | 11 | 12 |
|---|---|---|---|---|---|---|---|---|---|---|
| Max Score ($D$) | 1 | 2 | 4 | 6 | 9 | 12 | 16 | 20 | 25 | 30 |

I stared at the numbers, trying to find a pattern. After a while, I saw it.

The key was to consider the odd and even cases separately. Once I did this, the pattern was completely clear, with perfect squares (1×1, 2×2, 3×3, 4×4, 5×5) for odd values of $C$, and the product of consecutive integers (1×2, 2×3, 3×4, 4×5, 5×6) for the even values of $C$.

| $C$ | $D$ |
|---|---|
| 3 | 1×1 = 1 |
| 5 | 2×2 = 4 |
| 7 | 3×3 = 9 |
| 9 | 4×4 = 16 |
| 11 | 5×5 = 25 |

| $C$ | $D$ |
|---|---|
| 4 | 1×2 = 2 |
| 6 | 2×3 = 6 |
| 8 | 3×4 = 12 |
| 10 | 4×5 = 20 |
| 12 | 5×6 = 30 |

I determined a general formula for each case:

If $C$ is odd, then $D = \frac{C-1}{2} \times \frac{C-1}{2}$ If $C$ is even, then $D = \frac{C-2}{2} \times \frac{C}{2}$

All I had to do was wait for my teammates to complete their questions. As soon as Tommy passed me the answer sheet, I'd glance at his number $C$, use one of my formulas to calculate $D$, and then we'd be done the relay.

"Five minutes left," said Paula in a booming voice.

I glanced to my left, concerned Tommy was anxiously tapping his pen, not writing anything down. The answer sheet was not in front of him.

"Halifax Preparatory!" shouted a teacher, indicating that Halifax Prep had just completed the relay.

*Shoot.*

Halifax Prep now had the maximum score of 10 for the first relay, which included the five-point bonus for being the first team to finish. They turned to each other and slapped high-fives.

*Come on, Patrick. Hurry up!*

"Glace Bay Education Centre!" shouted another teacher.

Our rivals from Cape Breton had just gotten 9 points, including the four-point bonus for being the second to finish. They too started celebrating.

Tommy received the answer sheet from Patrick. I sighed in relief, knowing a third-place finish would keep us in the lead.

Tommy looked at Patrick's answer, rapidly filled in his box, and threw the answer sheet towards me.

| | Sydney High School |
|---|---|
| *A* | ***8*** |
| B | ***11*** |
| *C* | ***9*** |
| *D* | |

From Tommy's answer of $C$=9, I knew my answer was $D$=16. I hastily wrote down 16 in the last box and handed the completed answer sheet to Mr. Collins.

Mr. Collins took it, rapidly scanned the four numbers and shook his head. He wordlessly handed the sheet back to me.

*Oh no.*

I wasn't sure which of the four answers were wrong, but I knew that if Tommy's answer was $C=9$, my answer of $D=16$ had to be correct. I quickly passed on the answer sheet back to Tommy, who immediately passed it back to Patrick. I saw Patrick lean forward with his head down, putting his hands to his face.

"Parkdale High School!" bellowed Mr. Marshall, who was proctoring that school.

*Come on Patrick, you can do it!*

"Lunenburg High School!" shouted another teacher, less than ten seconds later.

Patrick wrote furiously, repeatedly stopping to tap his pen against the table.

"Charles Tupper High School!" announced another teacher, meaning that five of the teams had completed the relay, ensuring that we wouldn't get any bonus points. Our five-point lead had evaporated.

"Thirty seconds!" yelled Paula.

Shaking his head, Patrick scribbled something on the answer sheet and passed it on to Tommy. If we could complete the relay in time, we would get five points and be tied for first with Halifax Prep going into the final relay.

Tommy immediately crossed out his answer and tossed me the sheet. From his new answer $C=17$, I used my formula to conclude $D=64$. I quickly handed the completed sheet to Mr. Collins.

| | Sydney High School |
|---|---|
| *A* | ***8*** |
| B | ***~~11~~ 13*** |
| C | ***~~9~~ 17*** |
| D | ***~~16~~ 64*** |

Mr. Collins looked at the four numbers and shook his head, giving the answer sheet back to me.

I winced.

"Time's up!" yelled Paula.

Patrick tossed his pen up into the air, and stretched forward so his forehead was touching the end of the desk.

"Sorry, guys," he whispered, his voice barely audible.

Tommy patted him on the shoulder. "It's fine. Don't worry."

"I couldn't solve my question. I suck."

Gillian looked over at the slip of paper containing Patrick's question. She didn't say anything but her eyes told the whole story. *I would have solved it.*

Patrick looked up with tears in his eyes. "I can't believe this is happening."

Paula updated the standings, showing that Sydney High got just one point for the relay, from the first question Gillian solved correctly.

| SCHOOL | TEAM | RELAY 1 | TOTAL |
|---|---|---|---|
| Halifax Preparatory Academy | 45 | 10 | 55 |
| Glace Bay Education Centre | 45 | 9 | 54 |
| Lunenburg High School | 45 | 7 | 52 |
| Sydney High School | 50 | 1 | 51 |
| Parkdale High School | 40 | 8 | 48 |

We were now in fourth place. Gillian looked furious.

Patrick turned to Gillian, his confidence shattered. "I think we should switch for the next relay. You're much better than me."

"Didn't I tell you that yesterday?" snapped Gillian.

"Hey," said Tommy, turning towards Gillian. "He tried his best. Cool it."

We were four points behind Halifax Prep going into the final relay. The deficit was too big to overcome, especially against them. Whatever we did now, it would make no difference. I looked at Gillian.

"You can be the anchor."

"Fine," said Gillian. "It doesn't matter now, anyway."

Gillian stood up, allowing Patrick to move into the leadoff position, with Tommy now in second, me in third, and Gillian in fourth. I glanced over at Patrick.

"Hey, Patrick. It's okay. We'll get the next relay."

Tommy patted him on the shoulder again. "You're going to get this question. I know it."

I looked over at Gillian, and saw how annoyed she was. She was bitter and angry: bitter for being stuck in the leadoff position in the first relay, and angry we didn't listen to her last night.

Maybe Gillian was right. If we had listened to her, we would be in first place instead of fourth going into the final relay.

"It's now time for the last relay," announced Paula. "You are about to get the most interesting yet most difficult relay in the history of the Nova Scotia Math League. I gave this relay to four math professors at Dalhousie, and even they couldn't complete it in ten minutes. Especially that last question – well, I'll just say it's the most challenging problem we've ever given. If any of you want to change the order of your relay team, you've got thirty seconds to do that, starting right now."

No one from our table moved. Part of me wanted to be the anchor, but the other part of me didn't care, realizing Halifax Prep would win no matter what. If the question was as hard as Paula said it was, none of us would be able to solve it anyway.

Gillian tapped me on the shoulder. "Let's switch."

I shook my head. "It makes no difference. It's over."

"No, it's not. If we get this and they don't, we win."

"Halifax Prep is up by four. It's impossible."

Gillian stood up, and motioned for me to take the anchor position. "It's not impossible."

"But you can be the anchor," I replied. "I don't care."

Gillian put her face inches from mine and clenched her jaw. "I want to win."

Stunned by Gillian's intensity, I switched seats just as Paula told us the relay would begin.

I scanned the room and saw Mr. Marshall. My mind flashed back to Mr. Marshall's first words to me back in September.

I smiled at the memory. Sure, it was a long shot. We'd have to solve the relay, get a bunch of bonus points, and have the top three schools all mess up.

It was improbable . . . *but not impossible.*

Mr. Collins distributed the questions for the final relay.

"You have ten minutes," said Paula. "Starting right now."

RELAY #2, Question #4

For each integer $n$, we make a list of all its divisors less than $n$, and add up the total. If the total is equal to $n$, we say that the number is *perfect*. For example, 28 is a perfect number because its divisors are {1, 2, 4, 7, 14} and these divisors add up to 1+2+4+7+14 = 28. The first three perfect numbers are 6, 28, and 496.

There is only one perfect number with exactly $C$ digits. Let $D$ be the sum of the digits of this $C$-digit perfect number. Write down $D$.

I looked up from my paper. This didn't look that bad.

To get a feel for the problem, I wrote down all the divisors of 6, 28, and 496. I verified that all three numbers were indeed "perfect":

$$6 = 1 + 2 + 3$$
$$28 = 1 + 2 + 4 + 7 + 14$$
$$496 = 1 + 2 + 4 + 8 + 16 + 31 + 62 + 124 + 248$$

Not noticing any pattern, I tried some other values for $n$, seeing that sometimes the sum of the divisors was larger than $n$, and sometimes it was smaller than $n$.

$$12 \neq 1 + 2 + 3 + 4 + 6 \qquad \text{(the right side is too large)}$$
$$45 \neq 1 + 3 + 5 + 9 + 15 \qquad \text{(the right side is too small)}$$

I tried a few other things, but made no headway.

I thought about factoring the three perfect numbers, seeing if I could find any pattern. But there was nothing obvious here either.

$$6 = 2 \times 3$$
$$28 = 2 \times 2 \times 7$$
$$496 = 2 \times 2 \times 2 \times 2 \times 31$$

"Five minutes," announced Paula.

*What, five minutes already?*

From my periphery, I saw Tommy quickly scribble his answer and pass the sheet on to Gillian.

I looked over to the far end of the classroom and noticed the Raju-lookalike from Halifax Prep sitting in the #3 position, passing the answer sheet to the Albert-lookalike, the anchor.

*It's over.*

I tried a few more things on the perfect number question, but nothing came to mind. I was all out of ideas. If Gillian gave me a large number like $C$=10 or $C$=20, how could I determine what the 10-digit or 20-digit perfect number was supposed to be? There were just too many possibilities.

I glanced down at my question sheet, and my eyes moved down to the last line: "Let $D$ be the sum of the digits of this $C$-digit perfect number. Write down $D$."

Why would they ask for the sum of the digits, rather than for the number itself? What difference would that make to the problem?

It made no sense. Once you found the $C$-digit perfect number, it was a simple exercise to just add up the $C$ digits to get the final answer.

As I thought about why they worded the question in such a strange way, I realized that by asking for the sum of the digits, it would be easier for our proctor to check whether the answer was correct. Instead of looking at some big number like 913887024 to see if all the digits matched up, Mr. Collins could just verify a simple sum like 42. In other words, the correct answer for $D$ had to be small.

I nearly jumped out of my seat.

*Hurry up, Gillian! Finish your problem!*

I wasn't sure if I had enough time to carry out my plan. But as long as Patrick, Tommy, and Gillian correctly solved their problems, I knew my "problem-solving strategy" would give us full points on the relay, since there was no limit to the number of times I could hand in the answer sheet to Mr. Collins.

*Come on, Gillian! Hurry up!*

I glanced over at the anchor for Halifax Prep, who was furiously scribbling some notes on his notepad, trying to find that elusive perfect number.

The goal was to get the correct answer, which didn't mean I had to actually determine this *C*-digit perfect number.

For example, if *C* = 6, then there were nine hundred thousand possibilities for this 6-digit perfect number, everything from 100,000 to 999,999. However, there were only *54* possibilities for the *sum* of these digits, ranging from 1 (from 1+0+0+0+0+0) to 54 (from 9+9+9+9+9+9). While I couldn't guess and check 900,000 possible answers, I could with 54.

Gillian pumped her fist. She filled in the answer sheet and threw it on my desk.

| | Sydney High School |
|---|---|
| *A* | ***5*** |
| B | ***17*** |
| *C* | ***8*** |
| *D* | |

The answer *C*=8 meant the perfect number was at most 99,999,999, implying *D* was at most 72. How quickly could I get the answer?

Starting from the top-left corner of my box, I immediately wrote down 1 and passed it to Mr. Collins, who had positioned himself to my right.

He looked surprised to get the answer sheet so quickly.

Mr. Collins scanned the four boxes, shook his head, and handed the answer sheet back to me.

I crossed out 1 and wrote 2 next to it, giving the sheet back to Mr. Collins.

He looked puzzled, wondering what I was up to. He silently returned the sheet. I rapidly crossed out 2 and wrote down 3.

I handed the sheet up to him, and he gave it straight back.

Mr. Collins chuckled.

He figured out what I was doing. With a huge grin on his face, he knelt down right in front of me, to make the transition between us as efficient as possible.

I wrote down the numbers in turn.

4, 5, 6, 7, 8, 9, 10.

Each time Mr. Collins gave the sheet right back to me.

"What the hell are you doing?" whispered Gillian.

I ignored her. I kept crossing off the numbers as fast as I could.

11, 12, 13, 14, 15.

Mr. Collins passed the sheet back five times.

I kept writing.

16, 17, 18, 19, 20, 21, 22, 23, 24.

My hand and shoulders were throbbing in pain.

"Thirty seconds left," announced Paula.

I was getting exhausted.

25, 26, 27.

As Mr. Collins handed the sheet back to me for the twenty-seventh time, he stood fully straight and turned his body towards Paula.

I crossed out 27, quickly scribbled down 28, and handed it back.

"Sydney High School!" bellowed Mr. Collins.

*Yes!*

We started high-fiving each other. I even slapped hands with Gillian.

"Time's up!" shouted Paula.

The Math League was over.

Gillian's jaw dropped. She silently pointed to the anchor from Halifax Prep, who was pounding his fist against the table.

Halifax Prep didn't finish the relay. That meant they got three points, since the last person didn't get the answer. We got five points for finishing the relay, and five bonus points for being the first team to finish.

And that meant . . . *oh my God.*

"Way to go," whispered Mr. Collins.

I glanced over and saw Patrick lay back in his chair, smiling in obvious relief.

"Hey," I said, with tears rolling down my cheeks. "We did it."

Paula got our attention.

"Congratulations to all of you. This concludes the provincial finals for the Nova Scotia High School Math League. Give yourselves a round of applause."

I glanced over and through my tears, locked eyes with Mr. Marshall. He smiled and gave me the thumbs-up, as Paula introduced a well-dressed gentleman from Dalhousie University.

I was so wrapped up in my thoughts I didn't catch the man's name or title, and couldn't process a single word he said. After he finished his remarks a few minutes later, I joined others in the applause.

"Before we eat lunch together," said Paula, "I'd like to announce the winning teams. In third place, with a final score of fifty-seven points, is the team from Glace Bay Education Centre in Cape Breton. Could the team please come up with their teacher to collect their plaque."

The four students from Glace Bay came up and posed for a picture with their teacher. The teacher stepped off the stage and shook hands with Mr. Marshall and Mr. Collins, two fellow Cape Bretoners.

"In second place, with a final score of fifty-eight points, is the team from the Halifax Preparatory Academy."

The four boys from Halifax Prep stood up and walked to the front. They looked dejected. As the team posed for a picture, I saw that the only person smiling was their grey-haired teacher.

"And with sixty-one points, please join me in congratulating the champion of this year's Nova Scotia Math League: Sydney High School in Cape Breton."

The four of us bounced to the front of the classroom. Paula leaned over to hand me a huge trophy.

Mr. Marshall stood over to the side, but Patrick pulled him in for our group photo.

I held the trophy high in the air as the camera clicked. I passed the trophy to Gillian, who asked the photographer to retake the picture.

The students and teachers applauded us as we walked off the stage. As we sat back down in our seats, Mr. Collins gave me a high-five.

Patrick tapped me on the shoulder.

"Bethany, what were you doing in the last relay?

"What do you mean?"

"Passing answers back and forth with Mr. Collins. What the heck was that about?"

"I had no clue how to find the eight-digit perfect number. So I had to solve it a different way."

"How did you do it?"

I explained how I first guessed $D=1$, and repeatedly incremented the answer by one until I finally arrived at the correct answer.

Patrick shook his head in disbelief. "You guessed twenty-eight times?"

I smiled and shrugged.

He dropped his voice to a whisper. "Isn't that cheating?"

"No," I said, laughing. "It's problem-solving."

## The Canadian Mathematical Olympiad, Problem #3

**Twenty-five men sit around a circular table. Every hour there is a vote, and each must respond *yes* or *no.***

**Each man behaves as follows: on the $n^{th}$ vote, if his response is the same as the response of at least one of the two people he sits between, then he will respond the same way on the $(n + 1)^{th}$ vote as on the $n^{th}$ vote; but if his response is different from that of both his neighbours on the $n^{th}$ vote, then his response on the $(n + 1)^{th}$ vote will be different from his response on the $n^{th}$ vote.**

**Prove that, however everybody responded on the first vote, there will be a time after which nobody's response will ever change.**

# Solution to Problem #3

"It's impossible," I mumble.

I'm on the toilet, head down, rubbing my temples with my fingers.

I bolt up in shock, recalling Mr. Marshall saying those words on my first day at Sydney High School. I then remember what happened after he said that.

I solved that cylinder problem – without calculus.

Then I remember what happened after I said those words to Gillian. We won the Math League.

*The Math League!*

A few minutes ago, as I was pondering the Olympiad problem on Circle Voting, I knew I had solved a similar problem before. Recalling that magical day in Halifax from two years ago, I remember.

It was the Math League relay. Not the perfect number question – the one before that.

*Yes, yes, yes!*

I flush the toilet, rinse my hands, and spring back to the boardroom while wiping my hands on my pants. The blonde lady supervising my exam appears surprised that I've run back to my seat.

I grab my pen and sketch two diagrams from that Math League relay. Even though it was nearly two years ago, the images are still so clear.

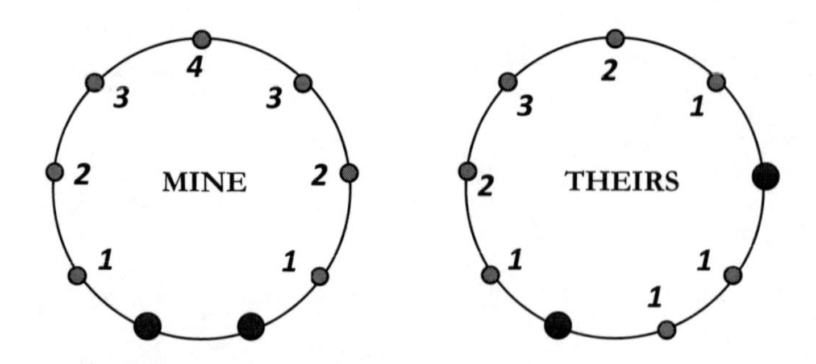


In both my scenarios, there were two boys and seven girls, where each girl wrote down the number of seats from her to the nearest boy.

*The number of seats to the nearest boy.*

Just before going to the washroom, I came up with the key insight of "coalitions", which happens whenever two or more people sit in consecutive seats and vote the same way. Once a woman is in a Yes-coalition, she will vote Yes forever; similarly, once a woman is part of a No-coalition, she will vote No forever. I realize how to explain my coalition idea by relating it to this Math League problem.

Every woman not in a coalition will switch her vote on the next round, to conform to the response of her neighbours. I want to show that eventually every woman will join a coalition, which will rigorously prove that after a certain point, no one's vote will ever change.

From my examples, I know any woman not in a coalition will eventually get sucked into one, and now I know how to justify that.

I mark all the coalitions at the very beginning and assign a specific number to every woman not in a coalition: *the number of seats from her to the nearest woman in a coalition.*

I make up a scenario with nine women, where I create two coalitions: a No-coalition at the top with three women and a Yes-coalition at the bottom with two women. I mark the two coalitions with a highlighter. For each woman not in a coalition, I write down the number of seats from her to the nearest woman in a coalition.

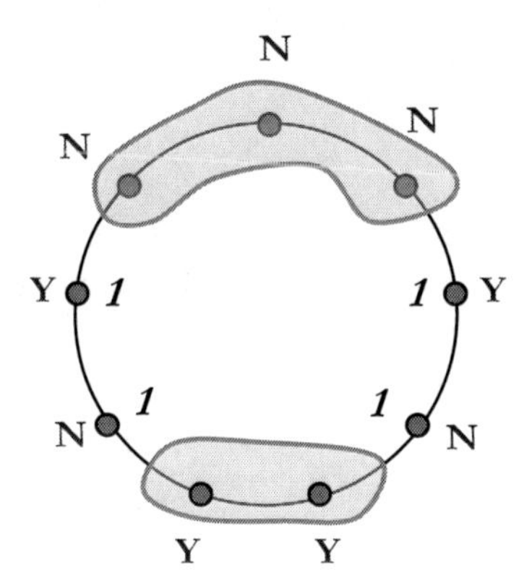


It's clear what happens in the following round. All the women who are not part of a coalition have the number 1, which implies they will get drawn into a coalition one round later.

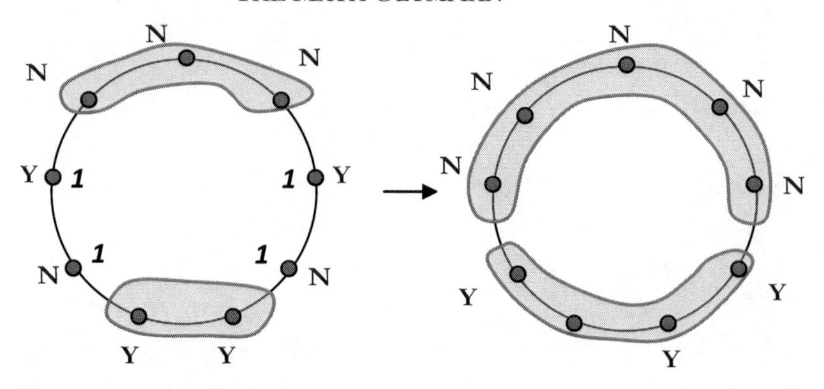


I see this pattern will continue: everyone who has the number 2 will get sucked into a coalition two rounds later, everyone who has the number 3 will get sucked into a coalition three rounds later, and so on. As the coalitions draw in new people, a woman not in a coalition finds herself one seat closer to someone in a coalition, and will eventually get drawn in too.

Just to make sure, I look at another scenario from my earlier work, where I had the votes arranged in a straight line.

Round 1: N N N N N **Y N Y N Y N** Y Y Y Y Y
↓ ↓ ↓ ↓ ↓ ↓
Round 2: N N N N N N **Y N Y N** Y Y Y Y Y Y
↓ ↓ ↓ ↓
Round 3: N N N N N N N **Y N** Y Y Y Y Y Y Y
↓ ↓
Round 4: N N N N N N N N Y Y Y Y Y Y Y Y

For each woman in each round of voting, I write down the number of seats from that woman to the nearest coalition.

Round 1: 0 0 0 0 0 **1 2 3 3 2 1** 0 0 0 0 0
↓ ↓ ↓ ↓ ↓ ↓
Round 2: 0 0 0 0 0 0 **1 2 2 1** 0 0 0 0 0 0
↓ ↓ ↓ ↓
Round 3: 0 0 0 0 0 0 0 **1 1** 0 0 0 0 0 0 0
↓ ↓
Round 4: 0 0 0 0 0 0 0 0 0 0 0 0 0 0 0 0

Yes, my idea is correct! The two people marked **3** after Round 1 get sucked into a coalition three rounds later, in Round 4. In every column, the numbers go down by one until it reaches zero.

I know how to explain it.

But something's still missing. What if no one belongs to a coalition in the first place? This was what made my head hurt and made me panic earlier.

Didn't I find a counterexample with four women? I scan the sheets of paper lying on the table, and find the diagram I'm searching for.

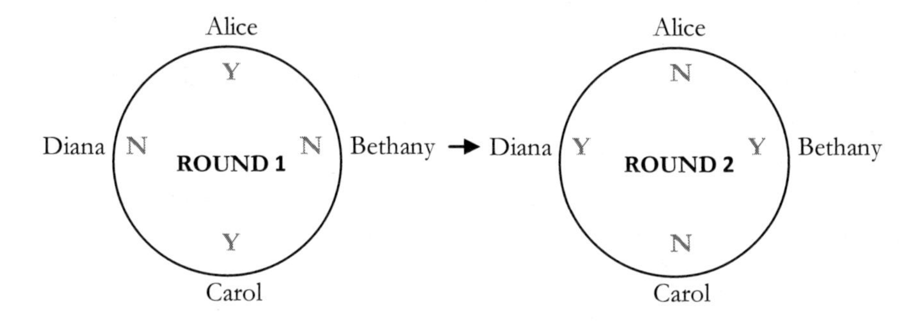


In this example, no woman would ever be assigned a number, since no woman would ever belong to a coalition.

The argument doesn't hold – technically, each woman is infinitely many seats away from a coalition.

What am I missing? I re-read Problem #3: "Twenty-five men sit around a circular table . . ."

*Twenty-five! An odd number!!*

Because twenty-five is odd, I can show that at least one coalition must exist at the beginning. Just to get the details right, I look at a simpler odd scenario with five people.

If there does not exist a coalition after the first round of voting, Alice and Bethany have to vote differently. So if Alice votes YES in Round 1, then Bethany has to vote NO. That means Carol has to vote YES, which means Diana has to vote NO, and this forces Ellie to vote YES. That's a contradiction because Ellie and Alice form a coalition.

I see immediately that the same technique can be extended to twenty-five women, with the last step being a forced coalition between Woman #25 and Woman #1. No matter what, at least one coalition has to exist after the first vote. My solution works.

Because twenty-five is odd, my approach is completely correct, with no gaps or errors. I take a deep breath and begin writing my solution, remembering to change “women” to “men”. To explain my solution, I prove two small results, called *lemmas*, which I use to justify my proof. The lemmas correspond to the two key discoveries I’ve made: there exists at least one coalition after the first vote; and once a man enters a coalition, he will never change his vote. I use these lemmas to show that eventually every man will enter a coalition.

After twenty-five minutes, I let out a sigh of relief, and put down my pen. My solution is perfect.

*11:00 a.m.*

The Canadian Math Olympiad finishes in exactly one hour, and I still have two problems left. The hardest two problems. The two problems that separate the Math Olympians from everyone else.

Solving both problems in such a short amount of time will be hard. Really hard. And highly improbable.

But not impossible.

I take a deep breath, and open the envelope containing Problem #4.

**Problem Number:** 3
**Contestant Name:** Bethany MacDonald

Label the men $M_1, M_2, M_3, M_4, \dots, M_{24}, M_{25}$, starting with any man and going clockwise around the circle. For each $1 \leq x \leq 25$ and $r \geq 1$, define $v_{x,r}$ to be the vote of the $x^{\text{th}}$ person $M_x$ in the $r^{\text{th}}$ round of voting.

For any integer $n \geq 2$, we say that a group of $n$ men form a *coalition* after the $r^{\text{th}}$ vote if these men are sitting in $n$ consecutive seats around the circle, and all respond the same way on the $r^{\text{th}}$ vote.

Lemma #1: There exists at least one coalition after the first vote.

Proof: Suppose on the contrary that no coalitions exist. Then $v_{x,1} \neq v_{x+1,1}$ for each $x = 1, 2 \dots 24$. By symmetry, we can assume that $v_{1,1}$ is YES. Then $v_{2,1}$ is NO, $v_{3,1}$ is YES, $v_{4,1}$ is NO, and so on. Specifically, each odd-numbered person ($M_1, M_3, M_5, \dots, M_{25}$) votes YES and each even-numbered person ($M_2, M_4, M_6, \dots, M_{24}$) votes NO. This is a contradiction, as $M_1$ and $M_{25}$ are two neighbours who both vote YES. We have therefore justified that at least one coalition must exist after the first vote.

Lemma #2: Once a man enters a coalition, he will never change his vote.

Proof: Suppose $M_x$ enters a coalition after the $r^{\text{th}}$ vote, i.e., his vote that round agrees with (at least) one of his neighbours. By symmetry, we can assume that $v_{x,r} = v_{x+1,r}$. Since $M_x$ and $M_{x+1}$ agree, they vote the same way the following round, so that $v_{x,r+1} = v_{x+1,r+1}$. Thus, $M_x$ stays in the same coalition after the $(r+1)^{\text{th}}$ vote, by voting the same way as he did on the $r^{\text{th}}$ vote. Repeating the same argument, $M_x$ stays in the same coalition with $M_{x+1}$ in every subsequent round. Therefore, he will never change his vote.

We now use these lemmas to prove that eventually, every man will enter a coalition.

Suppose on the contrary that there is some man $M_x$ who never enters a coalition, i.e., he disagrees with both his neighbours in every vote. Among all the men belonging to some coalition after the first vote (which must be a non-empty set by Lemma #1), let $M_y$ be the man *nearest* to $M_x$.

This implies the existence of a block of consecutive people, say $M_x, M_{x+1}, M_{x+2}, \ldots, M_{y-1}, M_y, M_{y+1}$, for which on the first vote the last two people agree (i.e., $v_{y,1} = v_{y+1,1}$), and the rest disagree with both their neighbours, since none of these people belong to a coalition.

Person $M_{y-1}$ will switch his second vote to conform to $M_y$, who will retain his first-round vote by Lemma #2. Thus, $v_{y-1,2} = v_{y,2} = v_{y+1,2}$, implying that $M_{y-1}$ joins that coalition after the second vote. This reduces by one the distance between $M_x$ and the nearest coalition member.

Repeating the argument, if there are $k$ seats between $M_x$ and $M_y$, then after $k$ rounds of voting, $M_x$ must join the same coalition as $M_y$, thus establishing our desired contradiction.

This proves that each man will eventually enter a coalition, and once he enters a coalition, he will never change his vote. Thus, we have proven that after the last of the twenty-five men has entered some coalition, no man's response will ever change.

Our proof is complete.

## The Canadian Mathematical Olympiad, Problem #4

Let $\Delta ABC$ be an isosceles triangle with $AB = AC$. Suppose that the angle bisector of $\angle B$ meets $AC$ at $D$ and that $BC = BD + AD$. Determine $\angle A$.

# Problem #4: Measuring Angles

*I can do this.*

This doesn't look hard. I sketch a diagram, translating the words of the question into a picture I can visualize.

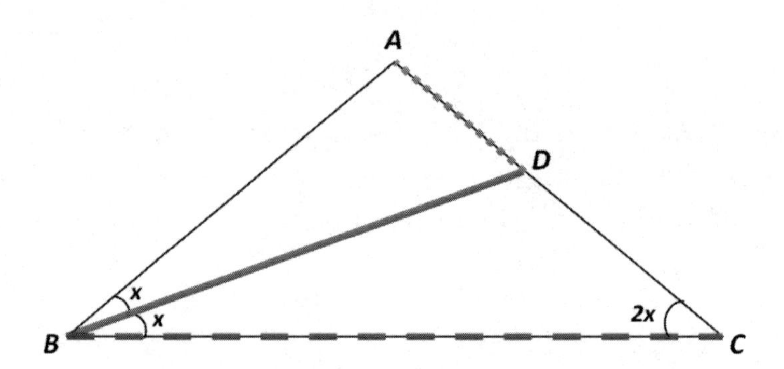


Underlining all the important words in Problem #4, I note:

- The angle bisector of $\angle B$ meets side $AC$ at point $D$. So I highlight this angle bisector $BD$ with a thick marker, ensuring that the line cuts angle $\angle B$ into two *equal* parts. I set both parts equal to some unknown angle $x$.
- Triangle $\Delta ABC$ is isosceles, with sides $AB$ and $AC$ of equal length. In other words, $\angle B$ and $\angle C$ must be equal. Since $\angle B = x + x = 2x$, I know that $\angle C = 2x$.
- The lengths of $BD$ and $AD$ add up to the length of $BC$.

From these three pieces of information, the measure of angle $\angle A$ can apparently be determined. That is what I need to find.

I have an idea. By using *trigonometry*, I can express the lengths of $BD$, $AD$, and $BC$ in terms of sines and cosines and then solve the equation to find the unknown angle $x$. If I can find $x$, I can easily calculate the measure of angle $\angle A$.

For example, if $x = 10°$, then $\angle B = 2x = 20°$ and $\angle C = 2x = 20°$, implying that $\angle A$ has to equal 140° since the sum of the angles of any triangle is 180°. I see immediately that $\angle A = 180° - 4x$, and therefore the key to the problem is finding $x$. Once I determine this angle, I will have my solution.

I use the "Sine Law" to find expressions for $BD$, $AD$, and $BC$, and from the given equation $BD + AD = BC$, I obtain an ugly trigonometric equation:

$$\frac{\sin 2x}{\sin 3x} \cdot \frac{\sin x}{\sin 4x} + \frac{\sin 2x}{\sin 3x} = 1$$

Clearing the common denominator, I produce something that looks even more scary and intimidating.

$$\sin 2x \cdot \sin x + \sin 2x \cdot \sin 4x = \sin 3x \cdot \sin 4x$$

I shake my head. Trigonometry isn't my strength, and there are too many pitfalls to solving an equation like this. First, I need time to reproduce the double-angle, triple-angle, and quadruple-angle formulas necessary to solve this equation. And if I just mess up one small calculation, like accidentally writing a plus sign instead of a minus sign, that will lead to a chain of futile computations.

And the IMO dream will be over.

Of course, Albert won't have any problems with Problem #4. He's already made the IMO team twice and won a gold medal both times. And I know Raju would also breeze through a trigonometric equation like this. He was selected to the six-member IMO team last year, and will definitely represent Canada again this year. I'm sure Grace will solve this problem too.

*11:05 a.m.*

With fifty-five minutes left, I need to solve the two hardest Olympiad problems. I take a deep breath.

It's too bad this is a question in triangle geometry. I'm so much better at circle geometry, since there are all these cool things I can deduce if I know the points are lying on the circumference of a circle.

For example, if four points $P$, $Q$, $R$, and $S$ form a *cyclic quadrilateral*, then I know that opposite angles (e.g. $\angle P$ and $\angle R$) have to add up to 180°. If angle $\angle P$ equals 80°, then I can immediately deduce that angle $\angle R$ has to equal

100°. Circle geometry problems often involve deductions like this, using the measure of one angle to determine the measure of another.

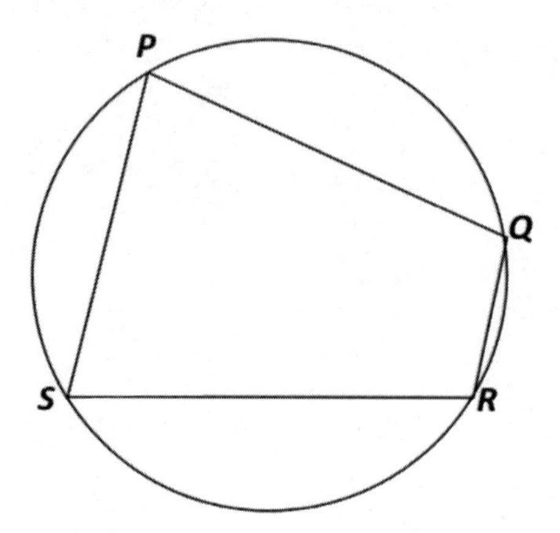


I look at my diagram again, disappointed there are no circles to be found.

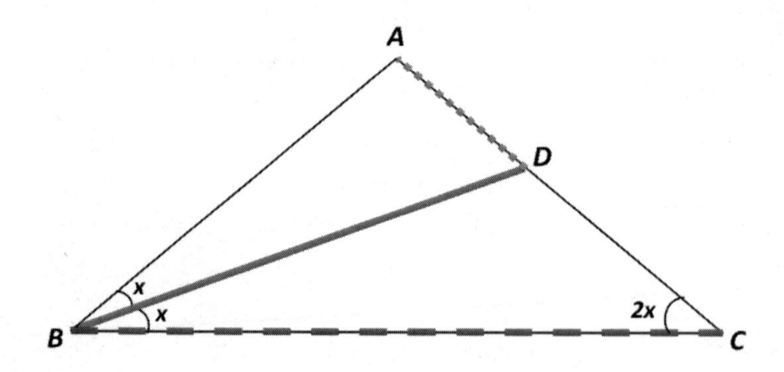


All I have is the useless equation $BD + AD = BC$.

From just this one piece of information, I have to figure out $x$.

I've already abandoned the messy trigonometric approach, which would take far too much time and have no guarantee of success. But I can't think of anything else to try.

Two summers ago, Grace and I had a week together in Vancouver, where we spent all of our time working on Olympiad-level geometry problems.

Come to think of it, that wasn't quite right. We spent *most* of our time doing geometry.

If Grace and I could work together, we'd solve this problem. But today, separated by four time zones, we're on our own.

I visualize the two of us sitting at Grace's kitchen table in her home in Vancouver, with sheets of paper scattered all over the place, frantically racing against time to solve the second-last question on the Canadian Mathematical Olympiad. I imagine myself getting a flash of inspiration that makes it clear how to solve for $x$. I see the two of us celebrating as we realize the key idea that unlocks the entire problem.

*But what's the key idea?*

I sigh and close my eyes. I know the insight is there – I just don't know where to find it.

Or how to find it.

# 29

I rubbed my eyes and rolled over to steal a glance at the alarm clock perched on top of the bedside table.

*4:33 a.m.*

I groaned, realizing that it would take days to adjust to the four-hour time difference between British Columbia and Nova Scotia. But I'd only be here for one week, and I wanted to make every moment count. I desperately tried to get some extra sleep, to match my body clock to Vancouver time.

After a while, I realized it was no use, and turned on the light to finish *Sophie's World.* I loved the last eighty pages, seeing how all the different types of philosophy connected together, with Hilde finally discovering the secret of her world on her fifteenth birthday, on June 15.

I smiled. That was my birthday too.

My stomach growled. That made sense, since it was 11 a.m. in Nova Scotia. Instead of waiting for Grace and her father to get up, I opened the door and carefully tiptoed down the stairs, walking through a dark hallway and into the kitchen. I turned on the light and looked around for an open box of cereal.

Grace reminded me that her house was my house, and that I should be free to help myself to whatever I wanted during this week we'd have together. I grabbed a bowl from the top of the cupboard and filled it to the top with Raisin Bran and milk.

I looked forward to all that we had planned together over the next six days, supporting each other to reach the goal both of us longed for: to have, for the first time in Canada's history, two girls on the IMO team.

Grace had just come back from a week-long seminar at the University of Waterloo, her reward for getting a perfect score on the Cayley contest, a competition written by over twenty-five thousand Grade 10 students from all over Canada. I almost made the seminar myself, but missed out by two questions, since I made a careless mistake on Question #21 and ran out of time on Question #25.

Grace promised to teach me everything she had learned at Waterloo.

With nothing to do or read, my eyes scanned the kitchen. There were a few photos and cards on the fridge, and several plants by the door leading to the backyard. Along one wall was a giant calendar filled with appointments and notes, alongside a framed photograph.

I got up from the table and walked towards the photo, seeing a happy couple with their two young children, a little girl holding the hand of her father and an infant boy in the arms of his mother. From the intense and focussed look in the little girl's eyes, I could easily tell that it was Grace.

The woman in the picture, holding her son, looked so happy.

As I stared at the picture, my heart sank, remembering that two of the four people in the picture were no longer alive. Grace told me during the National Camp that her mother and brother died in a car accident when she was only three.

"Good morning."

I jumped back and turned around.

A balding man stood by the kitchen entrance, his pear-shaped face covered by a thick pair of glasses. He was barefoot, wearing a plain white T-shirt and running shorts.

"You must be Bethany. I'm Grace's father, Raymond Wong."

"Nice to meet you."

I reached over and shook his hand, surprised that he was almost as tiny as Grace. Mr. Wong spoke gently, with a slight accent.

"Sorry I didn't get to introduce myself last night," said Mr. Wong. "I came home just a few minutes after you went to bed. Did you sleep well?"

"Yes," I lied. The bed was too soft and way too short. My feet dangled over the edge all night.

"I see you've met my family," he said, pointing to the picture.

I nodded, not sure what to say.

"As you can tell, I'm the one in the middle, though I've got a lot less hair than I used to. As for Grace, she looks exactly the same, doesn't she?"

I laughed. "Definitely."

"And that's my wife Carmen, and our son Jesse. Grace may have told you that Carmen and Jesse passed away thirteen years ago."

"Yes, she did," I replied. "I'm sorry."

"Thank you," he said. "We took that photo a week before the accident. It's my favourite picture of the four of us."

"It's beautiful."

Grace's dad walked towards the kitchen, and topped his cereal with some milk.

"Mind if I join you?"

"Sure, Mr. Wong," I said, motioning towards the empty seat across from me. I looked up at the clock, seeing it was 7:15 a.m. Knowing Grace, she wouldn't be waking up for a while.

"There's no need to call me Mr. Wong," he said. "You're not in school, and I'm not your teacher. Please call me Ray."

"You want me to call you by your first name?"

"Absolutely," he replied. "You're Grace's best friend. You're here on your summer vacation."

"Thank you. Ray."

"Do people call you Beth?" asked Ray.

"Some classmates used to call me Beth, but I never got used to that. And whenever I introduce myself to new people, half the time I have to say my name again. They always think it's Stephanie or Melanie."

"It's a beautiful name. Was it your mother who named you?"

"Mom's friend suggested it. Bethany apparently means *New Beginnings*."

"Oh, I didn't know that," said Ray. "I like that very much. New beginnings. There's a young couple at my church who had their first child a few months ago. They named their daughter Bethany."

"Really?" I said, surprised. "I've never met another Bethany."

"Did you know Bethany was the name of a city in Israel, just a few kilometres from Jerusalem? At Bethany, Jesus raised a man named Lazarus from the dead. Also, it was from the city of Bethany that Jesus ascended into heaven."

"Bethany is a city in the Bible?"

"Yes, it most certainly is," he replied, before suddenly pausing. "Forgive me. I promised Grace I wouldn't mention anything relating to church or Christianity. I'm sorry if I offended you."

"No problem," I replied, with a shrug. I knew Grace and Ray had a lot of tension relating to issues of religion, especially over the past year after

Grace stopped going to her dad's church, and quit believing in God. Grace told me that she got her dad to promise he wouldn't pester me about this topic, or ask me questions about why Mom and I had never gone to church.

From Grace's descriptions of her dad, I had pictured Pastor Raymond Wong as an angry and intimidating know-it-all, not this gentle soft-spoken man slurping milk from his cereal bowl.

"Grace is so happy you're here this week," said Ray. "She's been so excited, waiting for your arrival. And now you're here."

"Thank you so much for the air miles."

"You're welcome. Grace's uncle donated the air miles as a gift, and I'm so glad you could be here, with you and Grace teaching each other."

"It's more Grace teaching me than the other way around."

"Not at all," he said. "You're teaching and encouraging and supporting each other. Grace says you're a lot stronger than she is in writing up formal solutions. By working with you, she knows she has a much better chance of making the Olympiad team."

"I feel the same way," I replied.

"As I'm sure Grace told you, she's so bored in math class. She goes to an excellent school, but she finds math so boring. Her teacher is insistent that math is nothing more than a 'tool' to get the right answers to problems in engineering and physics, that if you can memorize the right formulas and learn the right steps, you can do well."

"My former math coach, Mr. Collins, taught me that math answered the *why*, not the *how* or the *what*, and that math was the art of explanation, not about memorizing stuff."

"Unfortunately, Grace's teacher doesn't agree. And that's why I'm so happy Grace heard about the Math Olympiad, because it gave her something concrete to work towards. She has so much excitement about going to the Olympiad someday. What makes it even better is that she's doing this with you, Bethany. You're her best friend, and she talks about you all the time."

"Thanks," I said. "Mr. Wong, can I ask you something?"

"Of course you can. And please call me Ray."

"Sorry, I forgot. I'm curious if you support Grace doing this – you know, pursuing the IMO."

"I'm absolutely supportive. I'm so proud of Grace for choosing to pursue her dream. Yes, it's costly *and* it's completely worth it. Why do you ask?"

"Because Mom isn't always so sure. Especially at the beginning, she said it was way too risky."

"Your mother's right."

That wasn't the answer I wanted to hear.

"She is?"

"Yes, what you're going for is a huge risk. Of course, there's no guarantee you or Grace will make the Olympiad team. But that doesn't mean it's not worth doing! You remind me of the jewellery collector who owns a massive collection of small pearls and decides to sell everything he owns just to have enough money to buy the one pure pearl that's dearest to his heart.

"Most people in life are happy with having a bunch of small pearls – but you and Grace have chosen to go all out and reach for the one big pearl that has so much more value, sacrificing everything you have with the hope of getting it. That's what the Math Olympiad represents for both of you."

"Aren't you worried Grace might not make it?"

"Not at all."

"Really?"

"If Grace doesn't get chosen for the Math Olympiad, she'll of course be tremendously disappointed. But she'll bounce back. I know she will – she's been that way her whole life. And this experience will strengthen her character for the next big goal she chooses to pursue. I'm delighted that Grace has found something so meaningful to strive for, and that she's chosen to go with you on this unpredictable up-and-down journey."

"The roller-coaster," I said, remembering Rachel's favourite metaphor.

I had a thought.

"Do you think Grace and I need to have a backup plan? You know, maybe keep a few small pearls in case we can't get the big one?"

Ray shook his head and pointed to the picture on the wall.

"Carmen and I were married for six amazing years. She was the love of my life. Neither of us had any knowledge of how long we would be on Earth, but we were completely committed to one another, giving our all to our marriage every day we were together. With anything in life, you either do it or you don't do it. There's no point being eighty percent married, putting some of your eggs in 'other baskets'. If you insist on keeping a bunch of your small pearls, then you'll never be able to reach for the big one.

"I'd argue it's the same situation for you, Bethany, with your dream. You can't control the results, but you can control your effort and your attitude. If you want this, you should pursue it with everything you've got. If not, don't bother, because you'll never obtain the pure pearl if you're striving for it half-heartedly, balancing that prize alongside a bunch of backup plans."

I heard approaching footsteps and then a familiar high-pitched voice.

"Dad, are you preaching at Bethany?"

Grace popped into the kitchen, wearing her pajamas. She looked annoyed.

I smirked. "Hey, sleepy head. Your dad and I are having a great conversation. No preaching."

"Good morning, Grace," said Ray. "You want to join us?"

"Sure," said Grace, casually. She grabbed some cereal and plopped down in the chair next to me. She flipped her hair back and adjusted her glasses.

"What were you two talking about?"

Ray handed her the soy milk. "We just covered the first half of the New Testament."

Grace glared at her father. After a short pause, Ray burst out laughing, and I joined in at Grace's expense. Grace rolled her eyes and resumed eating her cereal.

"So what's the plan for today?" asked Ray, looking at Grace.

Grace had her mouth full with cereal, so I stepped in.

"Grace told me she wanted to show me something she learned at the Waterloo Seminar, this special geometric proof. That's the first thing we'll do. For most of this week we'll be studying geometry together, since that's our weakest subject and the hardest to figure out on our own."

"Well, I'd suggest you work outside. It's a beautiful day, no wind, and you can use the picnic table in the backyard."

"That's a great idea," I replied. "So when do you have to get to work?"

"Monday is my day off," replied Ray. "I worked hard all weekend, and didn't get home until late last night. Today I'm just going to relax. I'm going to do all the things I enjoy doing most – and that includes cooking a great lunch for the two of you."

Grace gave me a thumbs up. "He's an awesome cook."

I looked at him. "Thank you so much. And thanks for the chat, Ray."

"You're most welcome." He stood up and took a few paces towards the hallway before turning to me with a broad grin.

"Now enjoy your day, and go get that pearl."

# 30

"Ready?"

"Yeah," I replied, picking up my pen.

Grace opened her notepad and used a thin ruler to create a triangle, labelling the three vertices with the letters *A*, *B*, and *C*. She took out a marker and added a vertical line.

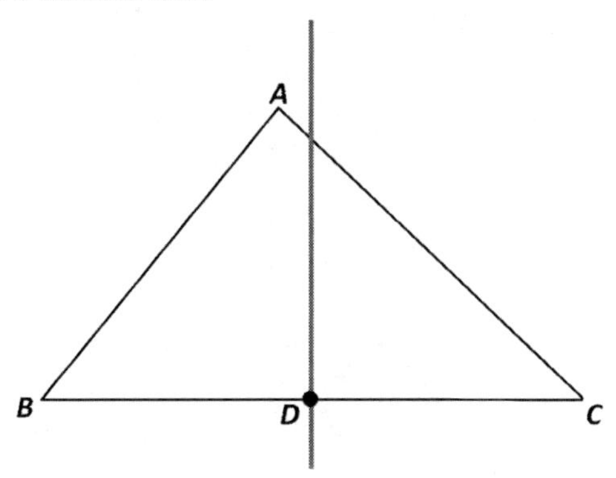


"What's that?" I asked, pointing to the vertical line passing through the point labelled with the letter *D*.

"It's the *perpendicular bisector*."

I nodded, seeing that the point *D* divided side *BC* into two equal parts, so that $BD = DC$.

Grace took out another marker to add a second line to her diagram. She used the letter *P* to mark the intersection of her two lines.

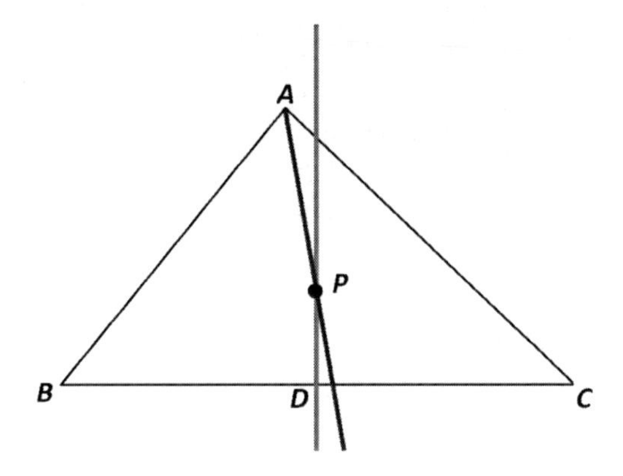

Grace's new line divided angle $A$ into two equal parts.

"That's the *angle bisector*," I said, jumping ahead of Grace.

She nodded. "Do you agree with my diagram?"

"What do you mean?"

"That if you start with a triangle $ABC$, then draw the angle bisector of $\angle A$ and the perpendicular bisector of side $BC$, the picture looks like this."

"Yeah," I replied, confused by her question. "Of course it does."

Grace proceeded to draw a whole bunch of lines and labels before handing me the sheet.

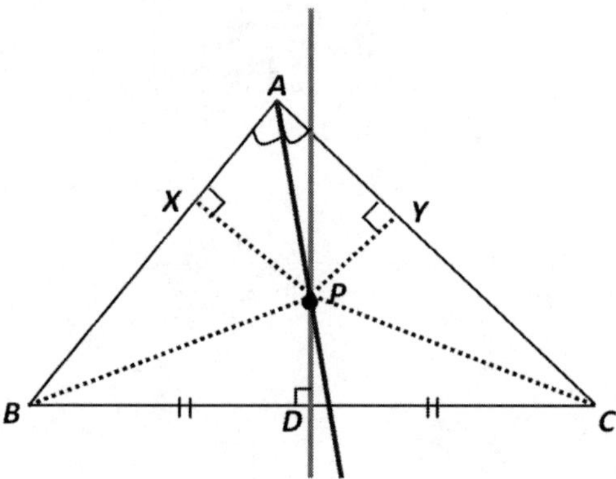


I looked at her completed diagram, noting that she had drawn in the line segments $PB$ and $PC$, as well as two other line segments $PX$ and $PY$. From the picture it was clear that $PX$ was perpendicular to $AX$, and $PY$ was perpendicular to $AY$.

"Okay, you do the rest."

"Do what?" I asked.

"Can you figure out the rest of the picture? What's equal to what?"

I started at the top of the diagram, at triangles $AXP$ and $AYP$.

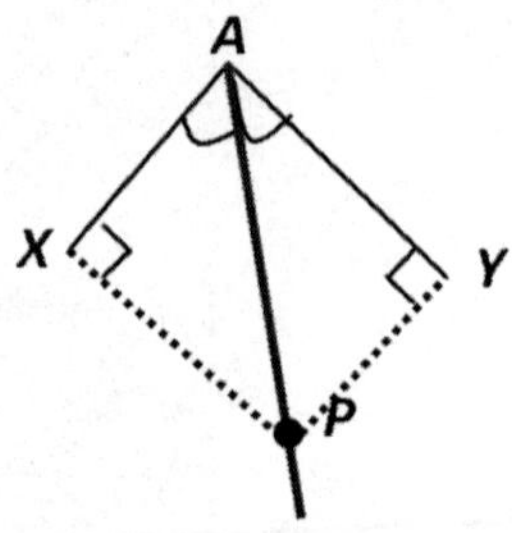

It was clear to me that the two triangles were *congruent*; they had the exact same shape and size. By viewing the line *AP* as a mirror, if we reflected triangle *AXP* in the mirror, the resulting image would be triangle *AYP*. So the angles matched up, as did the side lengths.

I looked at Grace. "*AX* equals *AY*, and *PX* equals *PY*."

"Yup," she nodded. "The Angle-Angle-Side rule. Anything else?"

I then turned my attention to the bottom part of the diagram, looking at triangles *PBD* and *PCD*.

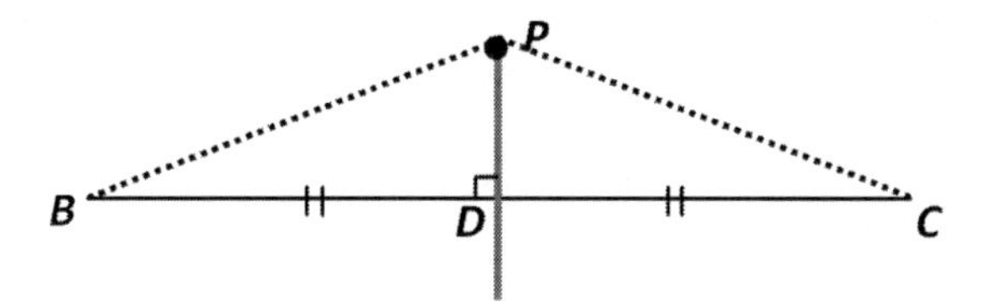


I saw instantly that these two triangles were also congruent. By viewing the vertical line *PD* as a mirror, if we reflected triangle *PBD* in the mirror, the resulting image would have to be triangle *PCD*.

I looked at Grace. "*PB* equals *PC*."

"Yup," she nodded. "The Side-Angle-Side rule. Here, let me make a new diagram."

She took out a fresh sheet of paper and re-drew triangle *ABC* without all the other information, using different coloured pens to mark down the three properties I discovered: *AX* = *AY*, *PX* = *PY*, and *PB* = *PC*.

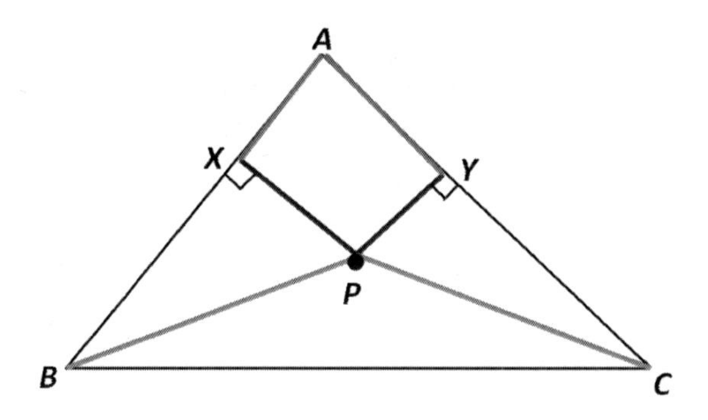


"Notice anything interesting?" she asked casually.

I stared at the diagram until I saw what Grace was referring to.

"That's impossible," I whispered.

"What's impossible?" asked Grace, with a twinkle in her eyes.

"Look at $PXB$ and $PYC$," I said, outlining those two triangles with my pen. "You've got two right-angled triangles, with two pairs of equal sides. So the third pair has to match up too. $XB$ has to equal $YC$."

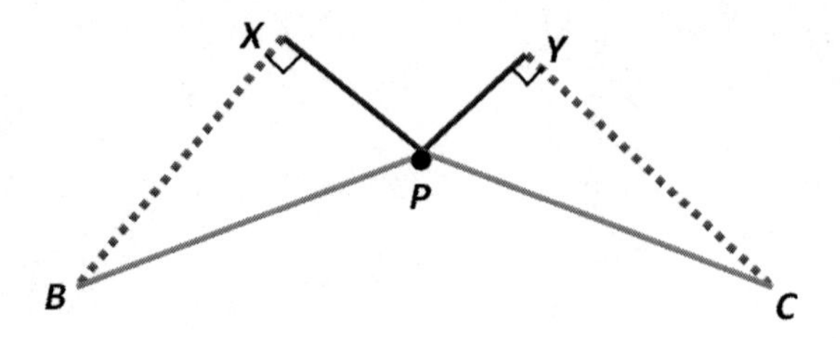


"You sure?"

"Yeah, definitely," I replied. "It's the Pythagorean Theorem."

To illustrate my point, I made up some numbers for the two pairs of equal sides.

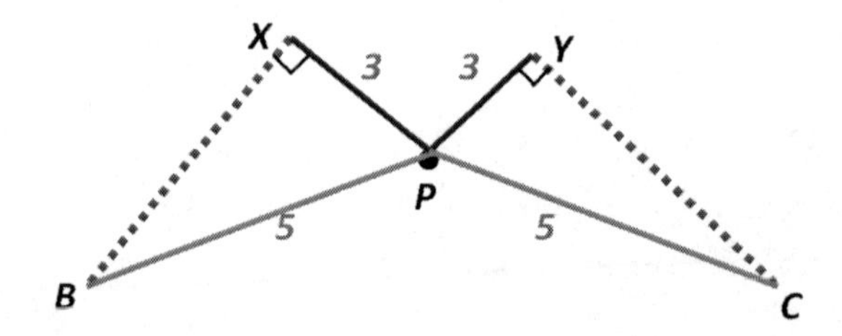


If $PX = PY = 3$ and $PB = PC = 5$, then both triangles *must* be 3-4-5 triangles, by the Pythagorean Theorem. So $XB = YC = 4$. I saw that in general, no matter what the lengths are, $XB$ has to equal $YC$.

I looked at Grace. "Something's wrong."

"Everything looks fine to me," she replied with a shrug.

"No, it's not," I said, pointing to the sheet. "Look! If $AX = AY$ and $XB = YC$, then if you add up the two equations, you get $AX + XB = AY + YC$, which means $AB = AC$. That's impossible!"

"Why?"

"Because that would prove *all* triangles are isosceles – which isn't true!"

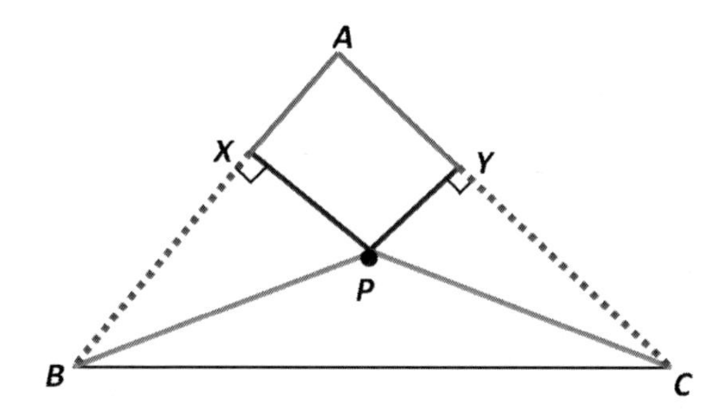


She grinned. "But your proof is right! You just proved that no matter what triangle you start with, you *have* to get $AB = AC$. That means every single triangle, no matter what it looks like, has to be isosceles! And if you just rotate the diagram, and apply the same argument to sides $AC$ and $BC$, then you have to conclude that $AC = BC$. So that means, for *any* triangle, $AB = AC = BC$. Oh cool, we just proved that all triangles are equilateral!"

I shook my head. "Okay, you've got me. What's the flaw?"

"There's no flaw. Every triangle, no matter how you draw it, has to be equilateral."

"Grace, shut up. What's the flaw?"

She stood up and stretched.

"Washroom break. Be right back."

I stared at the sheets of paper for several minutes, going through each step line by line, checking the proof several times to make sure I didn't made a careless error.

I even made a little table, using the same method I was taught back in Grade 9 Geometry, just to be extra careful.

| Step | Result | Reason | Justification |
|---|---|---|---|
| 1 | $AX=AY$ | $\Delta APX$ congruent to $\Delta APY$ | Angle-Angle-Side |
| 2 | $PX=PY$ | $\Delta APX$ congruent to $\Delta APY$ | Angle-Angle-Side |
| 3 | $PB=PC$ | $\Delta PBD$ congruent to $\Delta PCD$ | Side-Angle-Side |
| 4 | $XB=YC$ | From Steps 2 and 3 | Pythagorean Theorem |
| 5 | $AB=AC$ | From Steps 1 and 4 | $AX+XB = AY+YC$ |

Yes, everything was correct. I had proven an impossible result, that all triangles were isosceles.

Grace returned and noticed me staring at my table. I knew there was a flaw somewhere, but I couldn't find it.

"I give up."

Grace glared at me. "I thought we weren't going to say those three words. Ever."

"Well, I don't see it," I snapped.

"You'll see it," said Grace. "Just look carefully. It's there."

I bit my tongue and stared at the diagram. Eventually my eyes focussed on the point *P*, and how it was found by marking the intersection of the perpendicular bisector of *BC* and the angle bisector of $\angle A$.

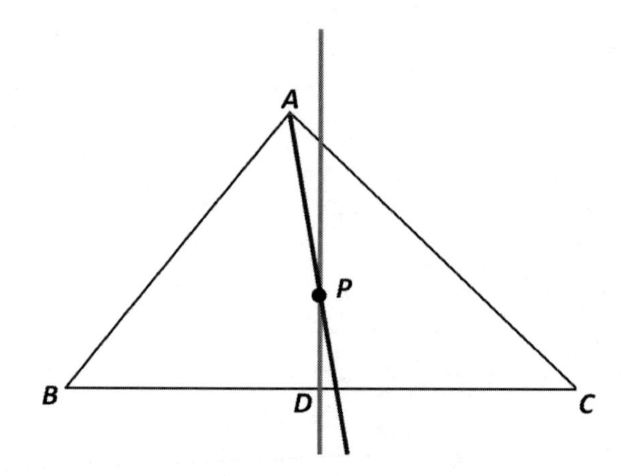


Something looked funny. I had a hunch, but wasn't completely sure.

I looked at Grace. "Hey, you have a protractor?"

"Yes, I do," she replied with a broad grin, opening up her pencil-case to hand me a familiar-looking tool in the shape of a semicircle.

I carefully measured the angle $\angle BAC$. It was just a bit more than 85°.

Since *AP* was the angle bisector, both $\angle BAP$ and $\angle CAP$ had to be half of that, or 42.5°.

Using the protractor, I measured the two angles in Grace's diagram. Angle $\angle BAP$ was around 48° and angle $\angle CAP$ was around 37°. A huge difference.

I laughed and threw the protractor at her. "That's not the angle bisector. You tricked me."

"You got it," said Grace, giving me a high-five. She took out a fresh sheet of paper and put it on top of the original sheet, carefully tracing out the same triangle with the same side lengths.

She handed me the two coloured markers.

I knew what to do I carefully measured the side *BC* to draw an accurate perpendicular bisector and the protractor to construct the exact angle bisector of $\angle A$.

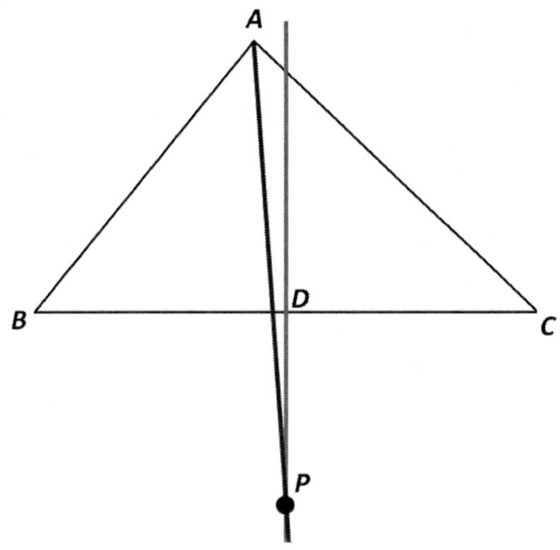


The intersection point *P*, was *outside* the triangle.

"So that's the flaw," I said. "The picture's wrong. Since *P* is outside the circle, the entire proof is bogus."

"Yup," said Grace. "When I asked you at the very beginning if you agreed with my initial diagram, you said you did. But that was a lie. So what's the moral of the story?"

"That you're full of crap?"

We burst out laughing.

"That we need to challenge *all* our assumptions," I said, turning serious. "That's what Rachel taught us."

"Exactly," replied Grace. "That was also the key point from that geometry lecture at Waterloo. If our initial assumptions are wrong, we can draw a false conclusion."

"Like all triangles are equilateral."

"Yeah," said Grace. "The prof told us a bunch of examples where people believed the wrong things for hundreds of years. Such as the sun revolving around the earth. Or the universe being stationary and non-expanding. Or that everything in geometry could be derived just from *Euclid's Elements*."

"The bible of geometry," I said, cutting in. "I remember Mr. Collins telling me about Euclid's Elements. He said all of geometry was based on five basic results – I think he called them *axioms*."

"That's right," said Grace. "One axiom is all right angles are equal to each other; another axiom is you can describe any circle from just two pieces of information – its centre and its radius."

"Wasn't there something wrong with the last axiom?"

"Exactly," she said. "Euclid's fifth axiom was that if you take any two non-parallel lines, they will eventually intersect in exactly one point."

She sketched a diagram to illustrate her point.

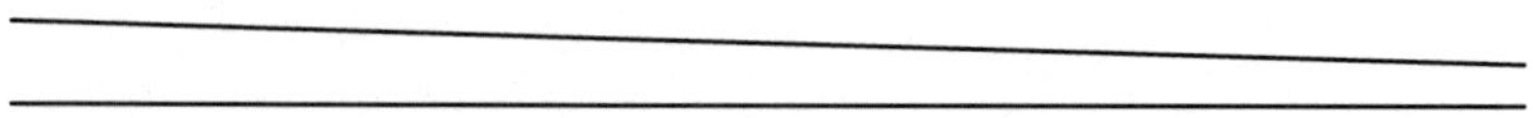

"*That's* the axiom?" I asked, vaguely recalling my lesson with Mr. Collins on this topic four years ago. "But isn't that obvious? Of course the two lines will intersect in only one point."

"Sure," said Grace. "If you assume the two lines are on a flat two-dimensional surface. But what if you have two lines on a three-dimensional surface, like on a sphere? Then it's not true."

"You can't have lines on a sphere," I said.

"Why not?"

"Because lines are straight. Spheres are curved. You can't have lines on a sphere. Obviously."

"Obviously not," said Grace, standing up. "Be right back."

Grace popped back into the house and returned holding a globe.

I suddenly had a flashback to Geography class, remembering the lines parallel to the equator were called lines of latitude, and the lines perpendicular to the equator were called meridians, or lines of longitude. I shook my head, annoyed at myself. Of course there were lines on a sphere.

Grace pointed to the equator, and was just about to say something when I cut her off.

"Yeah, yeah, you're right. Don't rub it in."

"I wasn't going to rub it in," said Grace.

"So what were you going to show me?" I asked, pointing to the globe.

"The Waterloo prof taught us about *great circles*, which cut the sphere into two equal hemispheres," said Grace, tilting her palm to simulate a chopping motion. "So the equator is a great circle, and so are all the vertical meridians. But other than the equator, the horizontal latitude lines are not great circles."

"Sure, I get it," I said.

"On the surface of a sphere, *lines* are just the great circles."

I paused to ponder this, while starting at the globe. "Yeah, I see that."

"If you take any two vertical meridians, those lines are not parallel. And these two lines do not intersect in exactly one point."

"Yeah," I responded. "They intersect in *two* points: the North Pole and the South Pole."

"Exactly," said Grace. "So all of Euclidean geometry breaks down on the surface of a sphere. All these theorems we take for granted, just aren't true."

"Such as?"

"Like the angles of a triangle always adding up to 180°. It's not true on a sphere."

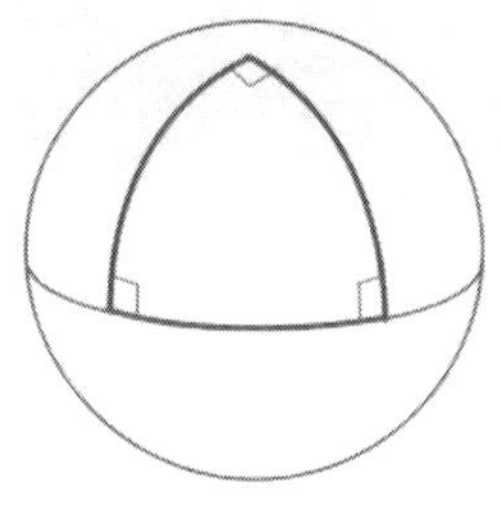

To illustrate her point, Grace used a pencil to lightly mark three points on her globe: the North Pole, and two points on the Equator. All the angles were 90°, so the angles of this triangle had to add up to 270°.

"That's cool, eh?"

I nodded. Yeah, that was cool.

"The prof showed some surfaces where two non-parallel lines intersect infinitely many times, and other surfaces where two lines never intersect. It took mathematicians two thousand years to figure out the existence of non-Euclidean geometries, that these different types of geometry were just as valid as Euclid's. It completely changed our understanding of the world; at least, that's what the prof said to us. And he also taught us the ideas behind hyperbolic geometry and elliptic geometry."

"Sounds intense."

"It was," admitted Grace. "I was lost most of the time. But the ideas were really cool, especially the stuff about spheres and great circles. Oh yeah, that reminds me. Here, check this out."

She grabbed the globe and used her fingers to point to two cities: Tokyo and Toronto.

"What's the shortest distance between these two cities?"

I paused, knowing that it was a trick question.

"Something to do with great circles, right?"

"Yup!" she replied, tracing a great circle with her finger. "There's exactly one great circle that passes through Tokyo and Toronto – it's a route that goes through the middle of Alaska. So that's why every flight from Tokyo to Toronto goes through Alaska."

"I didn't know that. Why?"

"Because it's the shortest distance. Just as straight lines describe the shortest path between two points $A$ and $B$, the great circles trace out the shortest flight route between any two cities in the world."

"The shortest flight from Tokyo to Toronto goes through Alaska?"

"Weird, eh?"

"Crazy."

I had an exciting thought.

"Grace, your prof said it took mathematicians two thousand years to figure out that Euclid's understanding of geometry was incomplete. Right?"

She nodded.

I continued: "It took two thousand years to disprove certain scientific theories that everyone believed to be true."

"Yeah, so?"

"It took two thousand years to contradict what everyone had always *assumed* – that there was only one correct type of geometry."

"Bethany, what the heck are you getting at?"

"Doesn't this story perfectly relate to something you're always talking about?"

Grace looked confused and shrugged.

"Doesn't this story completely settle the issue you're always arguing about . . . with your dad?"

I shook my head in disbelief, shocked that Grace wasn't clueing in.

After a few seconds, Grace laughed. She turned to me with a huge smirk on her face.

"I can't wait for lunch."

# 31

"Wow, this looks amazing."

"Thank you," replied Ray, handing me a plate. "And Bethany, if you think this looks good, just wait until you see what's for dinner."

I smiled and looked at all that Ray had prepared: fresh corn bread, spinach and mushroom salad with two types of home-made dressing, and a whole salmon topped with an assortment of spices and herbs.

We cleared our notes from the picnic table so that all three of us could sit down. I used my knife to cut the salmon fillet, surprised that the fish was a bright shade of pink.

"That's sockeye from the Pacific," said Ray. "Completely different from your Atlantic salmon. How is it?"

"Amazing," I replied, mouth half-full with pieces of salmon and corn bread.

"So, how was your morning?" asked Ray.

"Really good, but tiring," said Grace.

"Three hours of geometry," I added. "Grace showed me some of the stuff she learned at Waterloo, then we spent the rest of the time working on old contest problems. And before that, we talked about the five axioms of geometry."

"Axioms?" asked Ray.

"They're statements that you can't actually prove, but because they're obvious, you just assume they're true. And from these five axioms you can prove every theorem in geometry."

"Can you give me an example of an axiom?"

"Sure," I said, drawing two lines in my notepad.

I pointed to the diagram. "If two lines aren't parallel, they have to meet at exactly one point."

Ray stared at my diagram, and nodded. "Yes, that's clear."

Grace chuckled.

"That's an axiom," I said. "A simple statement you just assume so that you can prove some cool results. That's what we talked about today."

"That's great," said Ray. "Well, I'm glad you two are taking a break after lunch, and taking advantage of the weather. Grace, where are you taking Bethany this afternoon?"

"Olympic Village," said Grace. "I'll show her around False Creek and Granville Island."

Ray nodded. "That's one of the best parts of Vancouver. While you're down there, you can also take the ferry to Science World, if you're interested in that."

"Good idea," said Grace. "Yeah, Bethany, I'll take you there."

I nodded, and exchanged a knowing look with Grace.

"Can I ask you a question?" I said, turning to Ray.

"Of course you can."

"How long have you been religious?"

Ray put down his fork, and looked over at Grace. "I promised you we wouldn't talk about this."

"It's okay, Dad. Answer her question."

"Well, Bethany," said Ray, turning to me. "I've actually never been religious."

"Come on, Dad!" replied Grace, folding her arms. "You know exactly what she's asking."

"Sorry," said Ray. "There's actually a difference between religion and faith, but that's another subject for another time. You want to know how long I've believed in God. I assume that's your question?"

I nodded.

"I found Christianity when I was a twenty-year-old university student. So it's been twenty years."

"And how do you know that your religion . . . sorry, how do you know that your faith . . . is true?"

"Because when I think about the last twenty years, I've changed so much. Ever since I committed myself to following the life and way of Jesus Christ, I've been transformed into someone so unlike the person I was in the first twenty years of my life. I'm more joyful, more loving, and more generous. I'm less angry, less judgmental, and less cynical."

Grace shook her head in disgust. "So someone who doesn't share your faith isn't joyful and loving and generous? And all these people are automatically angry, judgmental, and cynical?"

"No, that's not what I meant," said Ray, shaking his head. "I know people from other religions, as well as people who don't believe in any god, who are a lot more loving and generous than I am."

I was confused by Ray's self-contradictory reasoning. I needed to ask him an obvious question.

"How do you know your faith is true?"

"I can't be certain that my faith is true," admitted Ray. "Christianity isn't some theorem you can prove mathematically. It's faith, and by definition, that's different from certainty."

"So why do you believe in God?"

"Because ever since I was a teenager, I had struggled with the answers to life's deepest questions: Who am I? Why am I here? What does my life mean? And when I was in university, I found a fellowship group on campus, and went to check it out. I discovered that Christianity gave such a satisfying response to these deep questions of life, and I was filled with this indescribable sense of peace. I tasted something I couldn't un-taste. And that's why I was so attracted to the faith."

"Give me a break, Dad," said Grace, rolling her eyes. "Tell Bethany the real reason you were so attracted to the faith."

Ray blushed. "I knew that one of my classmates attended the fellowship group. And I'll fully admit I went to the group because of her. She was the one who led me to Christ. I guess I fell in love twice: with Jesus, and with Carmen."

I looked over at Grace. "That was your Mom?"

Grace gave a light nod, and looked away.

"That's beautiful," I said, turning back to Ray. "But if I can be honest, you didn't answer my question. You said your faith gives you joy and peace, but that doesn't allow you to conclude that Christianity is true. How do you know God exists if you've never actually seen God?"

Ray paused. "You and I believe in many things that we can't physically see."

"Like what?"

"Subatomic particles. Ultraviolet rays. Black holes."

"Yeah, but that's different from God."

Ray looked at me. "How so?"

"Because those things have been scientifically proven, or logically proven, to be true."

"Do you need scientific evidence or logical reasoning in order to conclude that something is real?"

"Most certainly," I said.

"Then how would you describe love? Is it scientific? Rational? Logical? Just some random neurons firing in the brain? Or is love something that transcends scientific evidence and logical reasoning?"

I hesitated. Grace jumped in.

"Dad, don't change the subject. You still haven't answered Bethany's question."

Ray looked at me. "Okay, let me try. You want to know how I can be certain of God's existence even though I haven't physically seen him, and can't prove him scientifically or logically. Well, I believe that there is a personal loving God who is revealed in Jesus Christ – in other words, Jesus' existence proves God's existence, because I believe that Jesus *is* God. I also believe that God inspired the Bible as our means of having a relationship with him. And I believe in the Bible because there is so much evidence, both historical and scientific, that validates its authenticity. That is why I put my faith in God – not only in his existence, but in his ability to save us from the things that mess up our lives."

"But what if Jesus didn't exist, or what if the Bible isn't authentic?"

"Then my faith would be a lie – a cruel hoax that has deceived billions of people over the past two thousand years."

"Exactly," said Grace.

"I think I understand," I said, speaking slowly to make sure I didn't offend Ray. "In order for you to make sense of your Christian faith, you have to *assume* the existence of a human God named Jesus, who inspired a bunch of people to write what's in the Bible. That's what you're saying?"

He paused and took a sip of water. After considering his response, he looked at me.

"Yes. That's what I am saying."

"Those are your two *axioms*, right?"

Ray nodded. "Yes. They're the axioms of my faith, and without them I wouldn't have a faith."

"But if your axioms are wrong, then your conclusion is wrong. The Christian God doesn't exist and your faith is untrue."

"That's right," said Ray, after an even longer pause. "But based on everything I've studied and experienced over the past twenty years, I'm confident that both of these axioms are correct. That's my worldview."

"What do you mean?"

"A worldview consists of the assumptions you hold about the world in which you live. Think of a worldview as the contact lenses for your eyes – they're the prescription that brings things into focus. I have one particular Christian worldview among many possible Christian worldviews. Others hold a Muslim worldview, or a Buddhist worldview, or an agnostic worldview, or an atheist worldview, or some other worldview. You and I may have a different worldview, but no matter what, we all have a worldview. We all choose to wear some type of contact lens in order to see the world."

Grace cut in. "I don't."

"Sure you do," he replied. "We all do. Grace, you say it's arrogant to believe that there's one universal faith or religion. You say it's impossible that Jesus can provide salvation for all of human history, because truth is relative to one's culture and to one's historical time period. Grace, *that's* a worldview."

"No, it's not."

"Of course it is. It's called post-modernism. You believe truth is relative and no moral values exist. You believe that truth and reality are socially constructed, that the only absolute truth is that there is no absolute truth. Grace, you said so yourself, that what was true for Jewish people in the first century was different from what was true for Europeans in the seventeenth century, and how both are different from what's true for Canadians here in the twenty-first century."

"Yeah, that's right," said Grace. "Knowledge evolves over time, and therefore truth evolves over time."

"That's what you believe," said Ray. "That's your worldview."

"Is she wrong?" I asked.

"Yes, she most certainly is," said Ray, putting down his fork and pushing his plate off to the side.

"Why?" I asked.

"Because there *are* absolute truths. For example, right now it's 12:30 p.m. in this time zone – it's not two minutes slower for you and one minute faster for me. Just as time is absolute, so is truth. There must be an absolute standard that distinguishes right from wrong – otherwise, how would we live our lives?"

"We reason with our minds," replied Grace. "We make decisions rationally and logically, instead of following a bunch of outdated religious nonsense."

"So in *any* situation, no matter how complex, you can make the right decision just by thinking about it logically?"

"Yes!" replied Grace and me, simultaneously.

"Well," said Ray. "I did a minor in economics, a long time ago, and I remember this game that might clarify the situation. In fact, I brought this up in my sermon yesterday."

"What's the game?" I asked, a bit too eagerly.

"Say I give you two options. The first option is that you get one dollar – completely for free, no strings attached. The second option is that I give you a fair coin and you flip it: if it lands heads you get one thousand dollars, and if it lands tails you get nothing. You can pick either of the two options. Which option do you pick and why?"

"Obviously the second option," I said. "I have a fifty percent chance of getting a thousand bucks, and that's a lot better than the guaranteed one dollar."

"The *expected value* of the second option is five hundred dollars," said Grace, cutting in. "You always pick the option with the higher expected value."

"Always?" asked Ray.

"Yes," I replied.

"Okay," said Ray. "Here's a different scenario. Would you prefer ten dollars, or a fifty percent chance of getting ten thousand dollars? Still take the second option?"

"That's the same thing," I said, seeing Grace nodding in agreement. "You're just multiplying each option by ten. That doesn't change anything: the first option is $x$, the second option is a fifty percent chance of $1000x$."

"Exactly," added Grace. "The expected value of the second option is still five hundred times the expected value of the first option. Of course you take the second option."

Ray continued. "A thousand dollars, or a fifty percent chance of getting a million dollars?"

"The second option," I said instantly.

"A million dollars, or a fifty percent chance of getting a billion dollars?"

"The second option," I said, but with a slight hesitation. Grace slowly nodded.

"You'd risk one million to have a fifty percent chance of winning one billion?"

"Yeah," I replied. "That's the logical choice."

"Bethany's right," said Grace. "An expected value of one million versus an expected value of five hundred million. For sure."

Ray raised an eyebrow. "A billion dollars, or a fifty percent chance of getting a trillion dollars?"

I paused. So did Grace. I knew what option I'd pick but didn't want to admit it.

"Hey, Dad, that's not fair," said Grace.

"Why not?" asked Ray. "As you say, it's the exact same question as all the others. So, Bethany, which option would you pick – the first or the second?"

I shrugged my shoulders in resignation. "The first option. I'd take the billion."

"But why? Isn't the expected value of the second option five hundred times more? Isn't the second option the *rational* choice?"

I hesitated, unable to form a coherent reply.

Grace looked coldly at her father. "But Dad, that's just a hypothetical example, with a bunch of big numbers. In the real world, the best choice is always rational. It's always based on logic."

"Always?"

"Yes!"

"All right, then let me show you something else," said Ray. "I'll be right back."

As Ray went into the house, I finished the last bite of my lunch. He came back out a few seconds later with a roll of one-dollar coins, which he ripped open and arranged in front of me.

"Bethany, here are twenty-five dollars for you. This is not hypothetical – this money is yours to keep."

"Really? What's the catch?"

"Well, we're going to play a little game where you get to decide how to divide this money between yourself and Grace."

"Sounds good to me."

"Here are the rules. There's only one round to this game, so we're not playing this again with Grace deciding how to divide the money. And for the purposes of this game, pretend you've never met Grace before, and that after this game, you'll never see her again. You're two complete strangers, so you have zero interest or motivation to be kind to her. It's not like she's taking you to Science World."

"Okay," I laughed.

"The rule is that you have to give her at least one dollar, but you get to decide just how much money you want to give. Grace, you then have two options: you can either *accept* Bethany's proposal or you can *reject* Bethany's proposal. If you accept it, the money is split according to what Bethany has decided; if you reject it, then neither of you get anything and I get my twenty-five dollars back."

I looked confused. "So if I keep twenty dollars for myself and offer Grace five dollars, and she rejects that, then both of us get nothing?"

"That's right," said Ray. "But if she accepts the deal, then you get to keep twenty dollars, and Grace gets to keep five dollars."

"Okay, I get it. And we're playing this for real?"

"Yes, you are. If Grace accepts your proposal, you both keep the money. But no communication between the two of you. And you have to promise you won't share the money afterwards."

I paused to consider. If I only offered five dollars to Grace, I knew she'd reject that because it wasn't fair. Even if I offered ten dollars to her and kept fifteen for myself, that wouldn't be acceptable either.

I realized I had to give her an even split. But twenty-five was an odd number, so I couldn't divide the one-dollar coins evenly. After some hesitation, I decided to keep thirteen dollars for myself, and moved twelve coins towards Grace.

Ray looked at Grace. "Bethany is offering you twelve dollars. What do you say?"

"I accept," replied Grace instantly.

"Great, that's the end of the game," said Ray. "Congratulations, you both win."

We gave each other a high-five, and put the money in our pockets. I was thirteen dollars richer.

Ray looked at me. "So was your strategy rational?"

"What do you mean?" I asked.

"No, she didn't play rationally," said Grace, answering her dad's question. "Bethany gave me way too much. She should have kept twenty-four dollars for herself and given me the minimum amount of one dollar."

"But you would have rejected that," I protested.

"No, I wouldn't have. We were complete strangers, so you had no reason to look out for me. You should have given me the minimum amount. And I would've been forced to accept it."

"Why?" I asked.

"Because one dollar is better than zero. Each player's goal is to maximize her money. That's the rational thing to do. So I would have accepted one dollar because it's better than nothing. And you, knowing that I would do this, should have offered me the minimum one dollar and kept the remaining twenty-four dollars for yourself. If we're both playing rationally, that's the correct strategy."

"That's right," said Ray. "My economics professor used this example in one of the classes I took, and he explained the solution pretty much the same way Grace just did. The rational and logical mathematical strategy is for Bethany to keep twenty-four dollars and offer Grace one dollar. Apparently there's a name for this. But I can't remember what it's called."

"The Nash Equilibrium," I said, recalling the famous bar scene from *A Beautiful Mind*.

"If you say so," said Ray. "So, Bethany, even in this simple game, there was a lot more involved than rational logical thought. In deciding how much money you would offer Grace, various thoughts and feelings ran through your head and heart: your honour, your sense of fairness, your conscience. Right?"

"Sure," I replied, a bit defensively.

"Even in a situation like this, you knew the difference between right and wrong, not because your logical mind told you, but because your inner conscience revealed itself to you. And so a natural question is *where* does this conscience come from? A Christian worldview says that this conscience comes from God. On the other hand, an atheist worldview says . . ."

"Come on Dad," interrupted Grace. "Just because we have a conscience, that doesn't imply conscience comes from God. And even if it does come from a God, that doesn't imply it comes from *your* God. There are other religions, you know."

"Fair enough," admitted Ray. "But what I'm trying to say is that for complex scenarios involving life's hardest questions, you can't just reason with your mind, and try to solve it like you would a math problem. That's why spiritual faith can't be based just on logic and science. It's so much deeper. Faith has to be examined with the heart and the soul. The most important things in life must always be examined with the heart and the soul."

"Everything?" I asked. "Even the Math Olympiad?"

"Especially the Math Olympiad," said Ray. "You two are trying to represent Canada at the Olympics. When I watch the sports Olympics, the people who succeed aren't always the athletes with the fastest legs or the best technique. The people who succeed have the strongest hearts, which gives them remarkable focus and the mental toughness to never quit. They succeed because they want it more. They succeed because their strength comes from inside of them."

He pointed to his heart to illustrate the point.

I was intrigued by the analogy to the sports Olympics, and was wondering if Grace told him about Mom and her story. Was Ray comparing math to figure skating?

"So you're saying that when we're stuck on some impossibly-hard Olympiad problem, we should try to solve it with our heart instead of our brain?"

"Exactly," said Ray. "Because that's where the solution will come from. To quote the Christian mathematician Blaise Pascal: 'We know the truth, not only by reason, but by the heart.'"

"Wow," I said, nodding in agreement. "That's deep."

Grace looked at Ray. "You're right, Dad. I agree with you."

"What was that, Grace?"

"I said you're right."

Ray smiled. "Sorry, Grace. I must have misheard you. Can you say that again?"

Grace folded her arms and scowled.

"Ha ha, Dad. Very funny."

# 32

“Yes! I got it.”

Grace looked up from her notepad, and listened as I explained the solution to Question #19 on Grace’s list of Geometry Challenges. It was a beautiful problem; the insight was using something called the Tangent-Chord Theorem to prove that two particular triangles were similar. It took me about half an hour to figure that out, but once I did, I got the rest pretty quickly.

“Cool,” said Grace. “I totally get it. Okay, want to try the last one?”

I looked up at the kitchen clock. It was 5:22 p.m. Grace’s dad had meetings and appointments all day, and would be home just after 6:00 p.m. We still had some time.

“Yeah, let’s do it.”

Unlike the day before, where we enjoyed a sunny afternoon hanging out in Granville Island and checking out an exhibit at Science World, today was rain, rain, and more rain. We stayed inside the whole day, with all of our papers and notes scattered on the dining room table. We took about an hour in the morning to create a “toolkit” of useful geometry theorems, and spent the rest of the day applying these results to work through a challenging twenty-question problem set that Grace received on the last day of the Waterloo Seminar. Other than a thirty-minute break for tuna sandwiches and potato chips, we spent the entire time uncovering properties of circles, triangles, quadrilaterals, and pentagons.

It was an awesome day.

Our toolkit consisted of fifteen useful geometrical theorems relating to properties of circles, with fancy names like “Power-of-a-Point” and “Tangent-Chord”, as well as key definitions like *circumcentre* and *centroid* that were commonly found in these geometry problems.

The first result in the toolkit was the “Star-Trek Theorem”; it was so cool that every other result in the toolkit could be derived from this one. We called it the Star-Trek Theorem because the figure looked like that thing worn by Captain Kirk.

In the Star-Trek Theorem, the central angle $\angle BOC$ had to be double $\angle BAC$, the angle at the top. Mathematically, $\angle BOC = 2\angle BAC$.

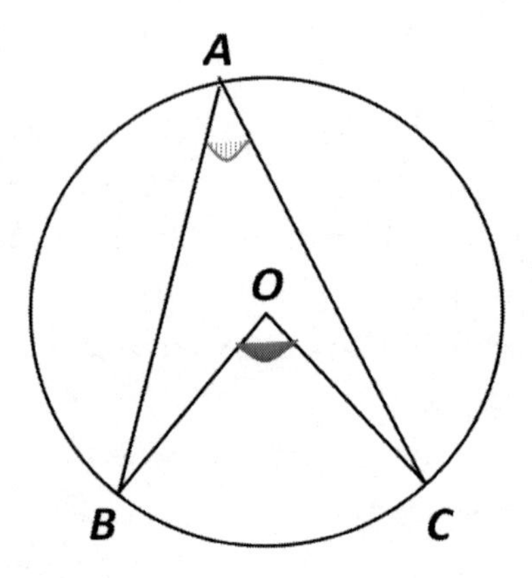


We had started the problem set just after 10:00 a.m. Seven hours later, we had solved every question on the problem set except for the final one. The problems got progressively harder: while the first few took us only a minute or two, Question #19 took us nearly forty minutes.

I looked at the statement of Question #20, and began drawing a diagram to capture all the information. I glanced over and saw Grace doing the same thing. It was weird; I felt like I had seen this problem before, years ago. But I couldn't remember where.

Question #20:

Let $P$ be a point inside triangle $ABC$.

The lines $AP$, $BP$ and $CP$ intersect the circumcircle of triangle $ABC$ again at the points $K$, $L$ and $M$, respectively.

The tangent to the circumcircle at $C$ intersects the line $AB$ at $S$.

Suppose that $SC = SP$.

Prove that $MK = ML$.

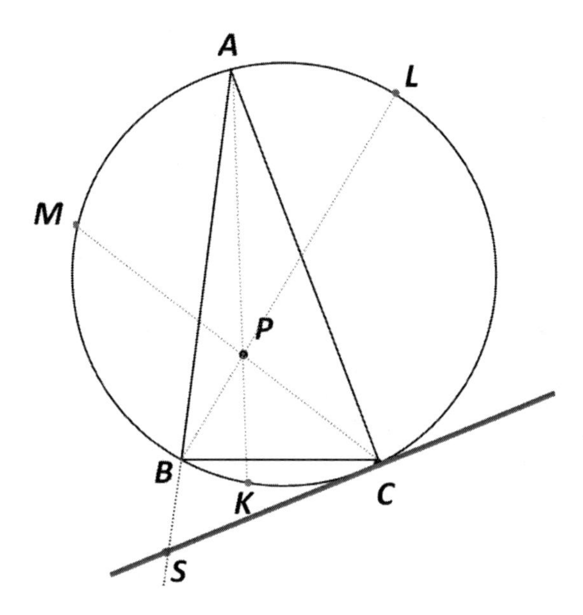


I drew a circle and selected three points *A*, *B*, and *C* on the circumference. This created my triangle $\Delta ABC$. By definition, I knew that this circle was the *circumcircle* of $\Delta ABC$.

I chose some random point *P* inside triangle $\Delta ABC$, and then marked *K*, the intersection point of line *AP* with the circle. I also marked *L*, the intersection point of line *BP* with the circle, and *M*, the intersection point of line *CP* with the circle.

I then used a thick pen to draw the *tangent* to the circle at point *C*, and I let *S* be the intersection point of this tangent line with the extended line *AB*.

Having completed all the steps in the diagram's construction, I re-read the problem statement:

Suppose that $SC = SP$. Prove that $MK = ML$.

I marked the relevant lines in my picture: my goal was to prove that if *SC* and *SP* were of equal length ($SC = SP$), then *MK* and *ML* also had to be of equal length ($MK = ML$).

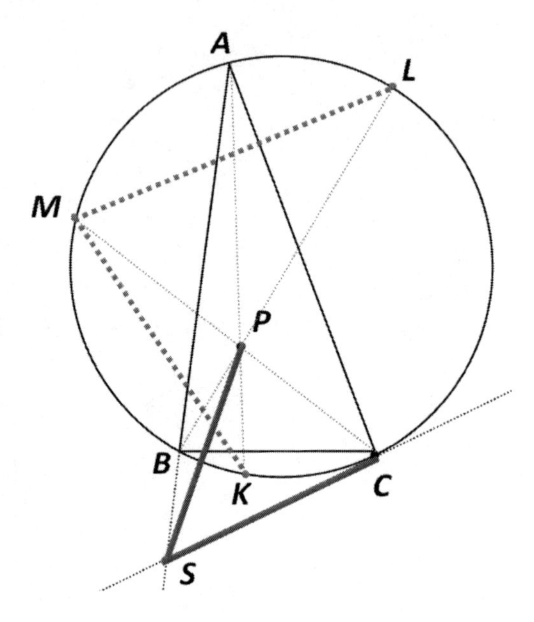


I looked at the picture, confused and unsure where to begin. There were eight different points marked in the diagram, and there was no obvious way to simplify this problem into smaller parts.

Grace looked at me. "Hey, I've got something."

She pointed to our toolkit of results that could be derived from the Star-Trek Theorem. "We can use the Butterfly."

I nodded. The Butterfly Theorem could be applied on any four points on a circle, where the two angles of the "butterfly wings" would be equal in measure. For example, by looking at the four points $M, A, C, K$ on the above circle, we could conclude $\angle KMC = \angle KAC$.

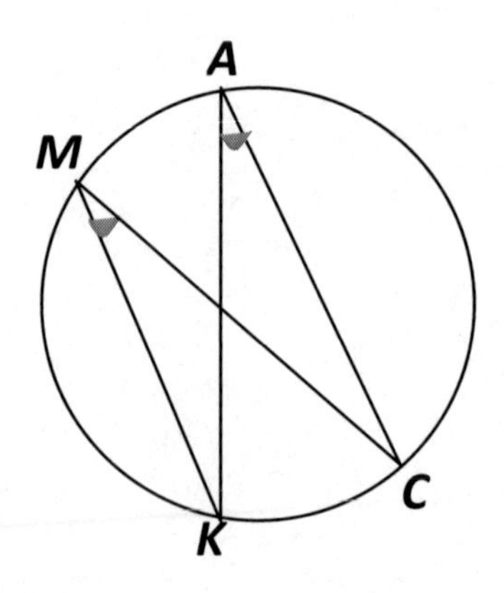

"Hey, I think I found some similar triangles," said Grace.

"Where?"

"Triangles $\Delta MKP$ and $\Delta ACP$ have to be similar. Also, triangles $\Delta MLP$ and $\Delta BCP$ are similar too."

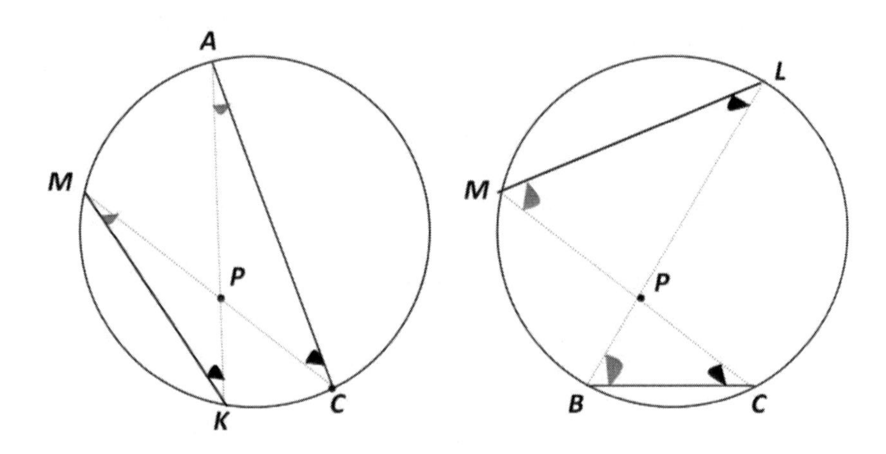


I looked at Grace's pictures, seeing how both sets of triangles were indeed similar. For the picture on the left, the two angles marked with my light highlighter ($\angle KMC$ and $\angle KAC$) were equal by the Butterfly Theorem, as were the two angles with dark shade ($\angle MKA$ and $\angle MCA$). This forced the remaining angle in each triangle ($\angle MPK$ and $\angle APC$) to be equal, which enabled Grace to conclude that the two triangles had the same three angles.

Since triangles $\Delta MKP$ and $\Delta ACP$ were *similar*, corresponding side lengths were in the same ratio. Therefore:

$$\frac{MK}{AC} = \frac{KP}{CP} = \frac{MP}{AP}$$

Similarly, triangles $\Delta MLP$ and $\Delta BCP$ were similar, and so:

$$\frac{ML}{BC} = \frac{LP}{CP} = \frac{MP}{BP}$$

Grace looked up from her notes. "We're trying to prove that if $SC = SP$, then $MK = ML$. So maybe these equations help."

She paused. "Or maybe I made it more complicated."

As I looked at her notepad, I saw that the length $MP$ appeared in both of her equations, and I had a clear understanding of what to do next.

"It's not more complicated. You just made it easier."

"How?" asked Grace.

I took the notepad from Grace and used some algebra to show that:

$$\frac{MK}{ML} = \frac{\frac{AC \times MP}{AP}}{\frac{BC \times MP}{BP}} = \frac{AC \times BP}{BC \times AP}$$

"If $SC = SP$, we need to prove that $MK = ML$. That's the question. But proving $MK = ML$ is equivalent to proving $AC \times BP = BC \times AP$. So we can rewrite the problem!"

I turned over a new page and wrote in big block letters:

THIS PROBLEM IS EQUIVALENT TO QUESTION #20:
If $SC = SP$, then prove that $AC \times BP = BC \times AP$

I ripped out the page from the notepad and made sure the new target problem was visible to both of us.

Grace smiled. "I get it. We just got rid of *K*, *L*, and *M*."

"Exactly. We just turned a crazy-hard problem into an equivalent easier problem."

We had used the Butterfly Theorem to find alternative expressions for *MK* and *ML*, and used the *ratio* of these alternative expressions to eliminate three points. So instead of a complicated diagram with eight points, we now had a much simpler diagram with just five points.

I paused to figure out what to do next. Re-reading the question, I noticed we hadn't yet used one important piece of information.

The point *S* was found by taking the tangent to the circle at point *C*, and finding its intersection point with the extended line *AB*. In other words, the line *SC* was *tangent* to the circle and *BC* was a *chord*.

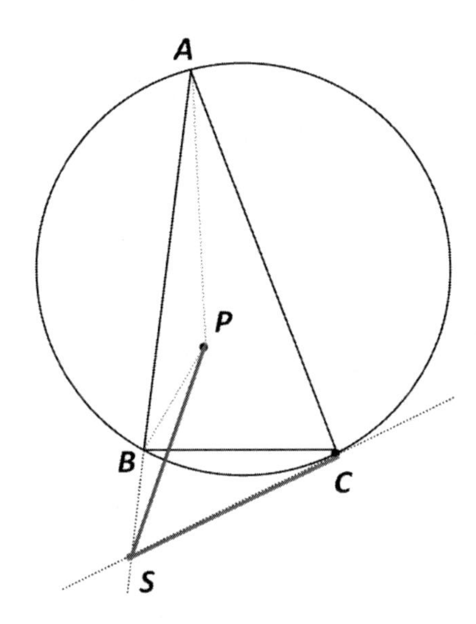


"Tangent-Chord," blurted Grace.

I laughed. "I was just about to say that!"

We needed the Tangent-Chord Theorem, a result that could also be proven from the Star-Trek Theorem. By applying this theorem, we could conclude that angles $\angle BCS$ and $\angle SAC$ were equal. And because $\angle BSC$ and $\angle ASC$ represented the exact same angle, they were equal as well.

Therefore we had found another pair of similar triangles: $\Delta ASC$ and $\Delta CSB$.

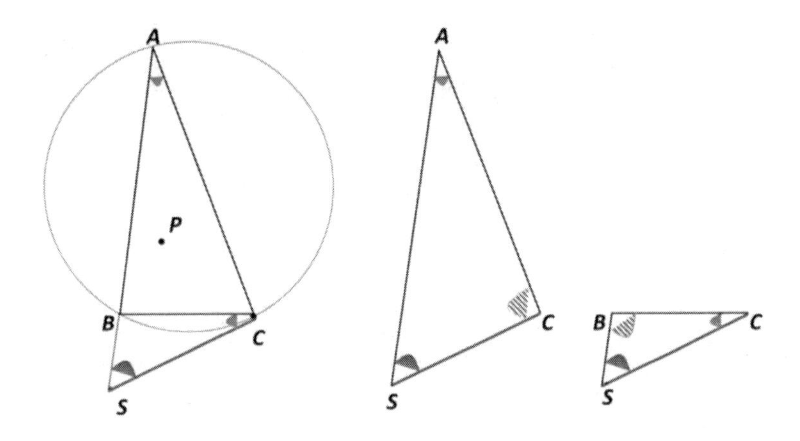


I could tell that we were close to solving this problem.

Since triangles $\Delta ASC$ and $\Delta CSB$ were similar, we had

$$\frac{AS}{CS} = \frac{SC}{SB} = \frac{AC}{CB}$$

"Check it out," Grace said. "From above, $AS \times SB = SC \times CS = SC^2$ and because we're given that $SC = SP$ that means $AS \times SB = SP^2$."

"Yeah, so?" I asked.

"That means triangles $\Delta SPB$ and $\Delta SAP$ are similar!"

It took me a moment to convince myself, but once I had, I completely understood.

"Okay, I get it," I said. "Since $\Delta SPB$ is similar to $\Delta SAP$, we know this."

$$\frac{SP}{SA} = \frac{SB}{SP} = \frac{BP}{PA}$$

"What's the new target question?" asked Grace. We reminded ourselves of what we wanted to prove, by looking at my equivalent re-formulation of the problem.

If $SC = SP$, then prove that $AC \times BP = BC \times AP$

"I have an idea," I said, seeing that each of the six side lengths $SC$, $SP$, $AC$, $BP$, $BC$, and $AP$ appeared in at least one of our earlier equations.

$$\frac{AS}{CS} = \frac{SC}{SB} = \frac{AC}{CB} \text{ implies that } SC = \frac{SB \times AC}{BC}$$

$$\frac{SP}{SA} = \frac{SB}{SP} = \frac{BP}{PA} \text{ implies that } SP = \frac{SB \times AP}{BP}$$

$$\text{Therefore, } \frac{SC}{SP} = \frac{\frac{SB \times AC}{BC}}{\frac{SB \times AP}{BP}} = \frac{AC \times BP}{BC \times AP}$$

Grace let out a yelp. "We got it!"

Indeed we had. We had proven that

$$\frac{SC}{SP} = \frac{AC \times BP}{BC \times AP}$$

Therefore if $SC = SP$, then the left side of the above equation had to equal 1, forcing the right side to equal 1 as well.

This implied $AC \times BP = BC \times AP$. Based on our earlier analysis, we showed that this statement was equivalent to $MK = ML$. In other words, if $SC = SP$, then we *had* to have $MK = ML$.

Our proof was complete. We found our solution, and completed the last of our twenty problems.

"Yes!" I yelled out, giving Grace a high-five.

"Oh man, I'm so tired," said Grace, throwing her pen on the table and leaning back against the chair.

"Me too," I said, exhausted from nearly eight hours of non-stop concentration. I looked up at the clock. It was 5:55 p.m. Ray would be home any minute.

I popped into the bathroom. As I walked back to the dining table, I saw Grace look into her binder and pull out a stapled document, which was folded in half.

"What's that?" I asked.

"The Waterloo prof gave us this. They're the solutions to the twenty problems we just solved."

"You had the solutions the whole time?" I asked in surprise.

"Yeah," she said. "As soon as I got this, I folded it up so that we wouldn't look at it. I wanted to know how many we could get on our own."

"I think we answered that!"

As we opened the five-page handout containing all the solutions, we noticed that many of our solutions were identical to the professor's – for a few of the problems, he had a simpler and more elegant approach, but for the most part, we had similar solutions for all twenty questions. In fact, our solutions to Questions #19 and #20 were essentially identical, step by step.

My jaw dropped when I got to the end of the document.

"Grace, look!"

Grace let out a shriek. We re-read the professor's closing paragraph.

> These twenty problems are representative of the types of geometry problems you will see on future contests. For the problems you did not solve, study and master the techniques involved in the solutions so that you can solve similar problems in the future. For your information, the last four problems were taken from the International Mathematical Olympiad (IMO). Specifically, the last four problems are 2007 IMO #4, 2008 IMO #1, 2009 IMO #2, and 2010 IMO #4.

In disbelief, I turned to Grace. "Can you get your laptop?"

We went online and downloaded the problems from the four IMOs between 2007 and 2010. Sure enough, the prof wasn't lying; those were the actual problems.

And we had solved them all.

Grace shook her head in amazement, and a huge smile lit her face.

I suddenly remembered where I had seen that final geometry problem: back in the summer before my Grade 7 year, the day I learned the sad news about Grandpa, the day I learned that Rachel had won a gold medal at the Olympiad.

And I had a thought – an exciting thought – that I believed for the first time in my life.

"I'm going to make the IMO team."

Grace looked at me coldly. "*We're* going to make the IMO team."

"Sorry," I said, turning red. "That's what I meant."

We heard the front door open, and poked our heads into the hallway. We saw Ray shaking off his umbrella and walking into the house. He waved at us.

"Hey, girls. Good evening."

As we waved back, Ray hung his rain coat and joined us in the kitchen. He was stunned at the sight of his dining room table, with nearly every square inch covered in paper. He leaned closer and saw the pages we had filled with various shapes and equations.

Ray was amazed. "You did all this today?"

"Yeah," I replied. "It was raining all day, so we just stayed inside."

"What did you work on?" he asked.

"The geometry set I got at Waterloo," said Grace. "The last four questions were all from the IMO."

"And we solved them all," I added.

"That's amazing," said Ray, with a proud smile on his face. "You solved four IMO problems today. I think that calls for a celebration."

Eager to take a break, Grace and I helped Ray get ready for dinner. I peeled carrots and potatoes while Ray prepared the chicken and Grace looked after the salad.

"First time in a kitchen?" asked Grace.

"Yeah," I replied, peeling off a bit too much of yet another carrot. "Can't you tell?"

Grace laughed and winked at her dad. "Bethany's mom does everything for her."

I stuck my tongue out at Grace, and we both started giggling.

Within thirty minutes, dinner was ready. We cleared our papers off the dining table and sat down together to eat. Grace and I began eating right away. I glanced over and saw Ray pause and bow his head for a few seconds before cutting into his chicken.

After Ray swallowed his first bite, I looked at him. "Was that a prayer?"

"Yes," said Ray, slightly embarrassed. "I just said a quick word of thanks."

"Thanks for what?" I asked, genuinely curious.

"I'm thankful for the day I had, where I was able to help someone going through a terrible crisis. I'm thankful for the two of you for everything you accomplished today. And I'm thankful for the food we're eating."

"That's cool," I said. "Do you always do that?"

"Always," said Grace, pointing to her dad. "I used to pray too, until I realized there wasn't any god out there who was actually listening."

I looked at Ray. "Is that a rule? Do you have to pray before every meal?"

Ray shook his head. "It's not a rule. It's just a small way I choose to remind myself that everything we have is an undeserved gift from God – which is the actual meaning of the word *grace*."

"That's cool," I said, nodding. "So you don't care that she doesn't pray?"

"Well, *don't care* isn't quite it, but I know what you mean. I respect Grace's choice to do as she wishes."

"Wow, look at the rain," said Grace.

Ray got the hint. "I brought home a couple of DVDs from the church library. Thought you would enjoy a break from math tonight."

"What did you get?"

Ray opened up his briefcase and put the two DVDs on the table.

"I loved this movie," I said, pointing to *Akeelah and the Bee*. "Mom and I saw it together."

"Yeah, it was awesome," added Grace. "Inner city girl, defies all the odds, wins the national spelling bee. It would have been even better if it was a true story."

"That one's a true story," said Ray, pointing to the other DVD.

Grace made a face. "*Soul Surfer*. Isn't that supposed to be preachy?"

"Absolutely not," replied Ray. "The main character is a Christian, but it's not at all preachy. It's a movie about never giving up on your dreams, no matter what you've gone through. It's a great message."

"Sounds preachy to me."

"What's the movie about?" I asked.

Ray looked at me in shock. "You haven't heard of it?"

I shrugged. "Maybe they didn't play it in Cape Breton."

Ray smiled. "Oh, Bethany, you have got to see this movie. Of all people, you need to see this."

"Why?" I asked, confused.

Grace glanced over at her dad. "This is about the one-armed surfer, right?"

"That's right," he nodded, and turned to me in excitement. "*Soul Surfer* is the true story of a teenager from Hawaii, who has the childhood dream of becoming a world champion pro surfer. But when she's thirteen, she loses her arm in a shark attack. It's the story of how she battles back and overcomes the odds to realize her dream of becoming a world . . ."

"Dad!" said Grace, horrified. "You always do this. Don't spoil it!"

"Sorry," said Ray. "I got a bit carried away there. Anyway, I think you two will enjoy this movie. I think you'll be inspired. Especially you, Bethany."

"The movie sounds really cool," I said. "But what's it got to do with me?"

"Well, you've got a lot in common with the main character. She's intelligent, she's determined, she's spunky, she never quits, and she's got this dream to achieve something big. And she's five-foot-eleven."

I looked at Ray, confused. Other than the bit about her height, wasn't all this true of Grace as well?

He handed me the DVD with the picture of a pretty blonde girl holding a surfboard.

I gulped when I saw the words below the title.

*The Incredible True Story of Bethany Hamilton.*

I didn't know anyone else named Bethany. I knew lots of girls named Stephanie, as well as two classmates at Sydney High named Melanie, but never another Bethany.

And yet, there was Bethany Hamilton, a young woman who had overcome unimaginable odds to achieve her dream, and we shared the same name. Amazing.

I waved the DVD at Grace.

"Let's watch this one."

# 33

"So where are we going?" I asked.

"Squamish. It's an hour from here."

"What are we doing there?"

"Not telling. It's a surprise."

I shook my head and followed Grace on to the Greyhound, using my jacket as an umbrella to avoid getting hammered by the rain. She walked to the back of the bus and took a window seat. I sat down next to her in the aisle and said nothing, silently watching the raindrop patterns forming on the window pane as the driver turned the bus and departed from Pacific Central Station.

My stomach was hurting, it was cold and wet outside, and I was feeling grumpy. I just wanted to stay at Grace's home and do math. A mystery road trip was the last thing on my mind. I told her I didn't want to go, but she insisted. I compromised and suggested we go the following day, but Grace said it had to be today.

Grace tapped me on the arm. "You're going to love this place."

"Whatever," I said, folding my arms and refusing to look in Grace's direction.

I reached into my backpack to get out my book, but it wasn't there. I must have left it on the kitchen table in our rush to leave Grace's house right after lunch.

Once the bus turned onto the highway, I turned to Grace.

"What have you got against your dad?"

"Nothing," replied Grace, taken aback. "My dad's great."

"But you're always criticizing him. Your dad is such a nice guy."

"I don't have a problem with my dad," snapped Grace. "I have a problem with his faith."

I sighed, annoyed we were back to this topic again. After watching *Soul Surfer* last night, Grace and I had our first major disagreement. I was so inspired by the movie, and was so moved by how Bethany Hamilton had the courage to get back on the board after she lost her arm, conquering obstacle after obstacle to achieve her dream of becoming a professional surfer.

But Grace hated it. She said the movie was patronizing and preachy, and said the shark attack was nothing more than a freak accident, instead of what the film director did – portraying the shark attack as an act of God to give the soul surfer a platform to share her Christian beliefs. She said it reminded her of how her dad described the car accident, not as a senseless tragedy caused by a speeding teenager running a red light, but as something God deliberately intended.

Grace went on this long rant last night about how her dad comforted her after the accident by saying that "God needed two more angels in heaven". What kind of loving god would intentionally take away a young girl's mother and brother, a man's wife and son, and rip apart a family?

After Grace finished her rant, I saw her cry for the first time. She was furious at how religion had deceived so many people, including her dad, and how the lies of Christianity blinded Grace until she finally "saw the light" less than a year ago.

As I glanced at the raindrops falling on the window behind Grace, I was reminded of Grace's reaction to the movie last night, and decided to make this as gentle as possible.

"It's so unfair what happened to your Mom and brother."

"Tell me about it."

"But the accident wasn't the fault of some higher power. You said so yourself – it was all because of that stupid teenager."

"Tell that to my dad. He thinks that the accident had a purpose."

I sighed. "Look, I know you don't agree with him, but do you have to argue all the time? Your dad respects your beliefs. So just let him believe whatever he wants to. It doesn't affect you."

"Yeah, it does."

"No, it doesn't," I said, a little too harshly. "Didn't you say that there is no such thing as absolute truth? For him, his faith is true. For you, it's not. For me, I'm not sure. It's all good."

"I was wrong about the 'absolute truth' thing. I take it back."

"What are you talking about?"

"That there's no such thing as absolute truth," she replied. "I changed my mind about that."

"Huh?"

Grace turned to face me. "No matter what, we have to assume some things – those were the *axioms* we talked about earlier. We have to make certain assumptions, such as the assumption that our senses aren't deceiving us. From those axioms you can figure out a bunch of things that must be true: it's raining outside now, and we're on a bus heading north. So there is absolute truth."

I shook my head, surprised by how Grace was contradicting herself. "But you and your dad have different axioms. He believes that the Bible was written by God, and believes in this Jesus guy. But you don't believe those things. That's why you end up with different conclusions, and that's perfectly fine."

"No, it's not! Because my dad believes in things that are scientifically false. So they're not axioms – they're delusions! God either exists or God doesn't exist. Exactly one of those two statements is true, and it's a universal absolute truth. God *is* or God *is not*. It can't be true for some and false for others. Ironically, this is the one point my dad and I actually agree on."

"So you're saying science can prove that your dad's axioms are delusional."

"Yes!" she shouted. "Everything the Bible says contradicts what we know to be true scientifically."

"Like what?" I asked, genuinely curious.

"Like everything! I mean, have you ever read the crazy stuff that's actually in the Bible? You know, that God formed the earth in seven days, that Moses parted the Red Sea in half, that a man got swallowed by a whale and came out perfectly fine three days later, that some invisible spirit made a virgin pregnant who then gave birth to Jesus, that this guy turned water into wine, walked on water, raised people from the dead, and even raised himself from the dead. Give me a break!"

"Grace, look . . ."

"I can't believe I used to think this nonsense actually happened!" she said, cutting me off. "I mean, everyone knows *Harry Potter* and *Lord of the Rings* are fantasy books – they're one hundred percent fiction. But when it comes to the Bible, people like my dad believe every word is true."

"Let's change the subject," I said. Talking about religion was making me feel uncomfortable, and my stomach cramps were getting worse. I could tell people on thc bus were staring at us.

"I don't want to change the subject," replied Grace, oblivious to the other people sitting nearby. "I can't live a lie anymore. We have to question everything. How else can we discover truth? We have to train our mind. How else are we going to make the IMO team?"

"You don't have to be a jerk to your dad, just to prove a point."

"That's not fair, and you know it."

I shook my head. "You have changed so much since Ottawa."

She turned away to stare at the rain beating hard against the window. I closed my eyes and reclined the chair as far back as it could go.

After a few minutes, Grace tapped me on the shoulder. "Bethany?"

I opened one eye and turned my head slightly. "What?"

"Can we talk?" Her voice had dropped to a gentle whisper.

"Fine," I sighed. I pulled my seat up to its normal position and turned to face Grace.

"Look, I'm sorry," she said. "I got carried away there."

"That's fine," I said. "But please forget about it."

"It's not easy to forget about it. Christianity was my whole life until ten months ago. And after I said goodbye to God, things haven't been the same with my dad."

"But he's such a nice guy," I said. "It's not like he's asking for a lot. Just go to church on Sundays, and do those five-second prayers with him at dinner. It seems like such a small sacrifice to make him happy."

"I can't live a lie."

I shook my head in resignation. This wasn't going anywhere.

"So where are we going?"

"Hey, did you ever read a book called *The God Delusion*?" asked Grace. "Raju recommended it to me."

I bit my lip, annoyed that we were back to where we started. "No, I haven't read it."

"Well, you should," she said, turning to face me. "This book opened up my mind so much – the first step towards me finally seeing the light. I discovered that I could be perfectly happy, balanced, and intellectually-fulfilled, without believing in some higher power who answers prayers, forgives sins, punishes people, performs miracles, and knows when we do bad things . . ."

"Grace, stop," I pleaded. "I don't want to talk about this anymore."

She kept going. "Christianity teaches that you have to believe in one specific God, or else you go to hell. The entire religion is based on fear. They do this by saying that faith, which is belief without evidence, is a virtue. And they scare you into . . ."

I couldn't take it anymore. I stood up from my seat.

"Hey, where are you going?" asked Grace.

"I'm tired. I'm taking a nap. Wake me up when we get there."

Without bothering to look back at Grace, I walked up to the front of the bus and took an empty seat in the second row. I plopped my bag on the aisle seat and leaned my head against the window. I closed my eyes, relieved to have some peace and quiet. The sound of the rain was great. Within minutes, I was feeling better, and my cramps were slowly subsiding.

Overall, I was having a good week. My confidence had never been so high, and I honestly felt that the dream was within reach – that Grace and I could make the IMO team in two years. Ray was so supportive of Grace and me having big dreams. And Grace was awesome, as long as we were training together and focussing on the stuff that mattered – getting to the Math Olympiad.

I was woken up by the PA system, hearing the words of the bus driver informing us that we'd arrived in Squamish, and he pulled up to a place called the Adventure Centre.

The rain was pounding on the pavement, so I put my jacket over my head and ran up the stairs into the Adventure Centre, not bothering to check whether Grace was behind me. I needed to use the washroom real bad, and was happy to find that one of the three stalls was empty.

After I was done, I walked towards the sink to wash my hands. Through the reflection in the mirror, I saw Grace's tiny pink shoes and knew that she was in one of the other stalls.

While waiting for Grace, I looked around the Adventure Centre and found a pamphlet on the Stawamus Chief, a 700-metre granite dome that was popular for hiking. That would have been cool to do if the weather was better. Too bad Grace was so stubborn, and insisted that we couldn't wait until tomorrow.

Out of the corner of my eye, I saw Grace at the far end of the Adventure Centre, talking to someone on her cell phone. I saw her approach me, and I pretended not to notice her.

Grace tapped me on the shoulder. "I'm sorry, Bethany."

"No problem," I said in resignation. "As long as you promise you'll stop talking about you-know-what."

"I promise."

"So what are we doing?" I asked. "I assume we're not going hiking."

"You'll find out in ten minutes. Someone's coming to pick us up."

I knew I wasn't going to get anything more from Grace, so I walked towards the Adventure Centre café, and found an empty table right by the window. Grace followed me and sat down too.

"Why are we here?" I asked. "I thought we were going to do math this week. You know, train together, so that we can make the IMO team."

"That's what we've been doing," she said.

"But we're doing all these other things. Pointless debates on religion, and now this mystery road trip. We're not focussing. We're wasting our time."

"It's not a waste of time," Grace replied, defensively. "Remember what you told me in Ottawa? That it's good to do things other than math, but still develop the problem-solving skills that open our minds? You called it cross-training."

"There's a big difference between doing puzzles and arguing with your dad. Sudoku puzzles and cryptic crosswords sharpen your mind."

"And so does arguing with my dad. Besides, I'm not arguing with him, I'm debating. I'm reading books on evolution and natural selection, and that opens up the way I think, and I've learned how to make clear and concise arguments against my dad whose job is to talk and defend religion. Believe it or not, becoming an atheist has helped me become a better mathematician."

"You promised you weren't going to talk about this."

"All I'm saying is we are training for the IMO – it's not just when we're doing the Olympiad problems. We're training our mind, we're expanding our perspectives, we're sharpening how we think."

She paused and looked at me. "I guess you could say that we're into the *liberal arts.*"

"The what?"

"The liberal arts," she replied. "That's what we love. Math, logic, music, rhetoric, philosophy . . ."

I looked out the window and fixed my eyes on a large puddle that was growing by the second. A dark green mini-van drove into the parking lot, and we saw a woman walk out.

"Hey," said Grace. "Our ride's here."

We watched her through the window, as she opened her umbrella and walked up the stairs into the Adventure Centre. The tall brunette looked around until her eyes spotted us. She smiled and started to head towards Grace and me. I noticed that her green polo shirt matched the colour of her car.

"Grace Wong and Bethany MacDonald?"

"Yes," replied Grace. She stood up to shake hands. "Thanks for picking us up. I'm Grace."

"It's a pleasure to meet you," she replied. "Welcome to Squamish."

The woman turned towards me and extended her right hand. I stood up and shook it.

"And you must be Bethany. It's nice to meet you. My name is Gillian."

I cringed. After all these years, I still hated that name.

We followed her to the mini-van, as she told us that "our quest" was just a short drive away.

"What the heck is this?" I whispered to Grace, just before we got in the car.

"It's the university I'm going to," she whispered back.

"Aren't you going to UBC?"

"I changed my mind once I heard about this place. It's Canada's only secular liberal arts university."

I was confused. "And what am I doing here?"

Grace hopped into the backseat and motioned for me to sit next to her. As I got in and shut the door, Grace tugged on my arm, and flashed a cheeky grin.

"It's the university *we're* going to."

# 34

The tall brunette drove us to the campus, winding around a tall mountain until we reached the very top.

As Grace stared through the window, this lady introduced herself as a member of the Admissions team at Quest University Canada, and explained how the sixty-acre campus bordered a provincial park with long trails for hiking and mountain biking.

Gillian said how honoured she was to welcome us to Quest's annual open house event, and was especially excited that I would fly all the way from Nova Scotia to be in attendance today.

I was about to protest when Grace cut me off.

"That's right, Gillian. Bethany came all the way from Cape Breton. We're really excited and can't wait to check out the campus."

"And we're excited that you chose to visit us," replied Gillian. "I'm just sorry we couldn't give you better weather. Grace, as we were exchanging e-mails last month, you mentioned how much you and Bethany enjoy math. Well, you two are going to love the presentation given by our math tutor, Robert Cooper."

I sat in the back of the car, seething. All along, I thought Grace wanted me to come to Vancouver so that we could train together for the Math Olympiad. But as we headed up the mountain with the rain pounding down on the pavement, I learned the real reason Grace used those air miles to fly me out west.

And what was the name Gillian just mentioned? *Robert Cooper*? It sounded too much like Cooper Robertson, and I didn't need any more reminders of Cooper.

I had a flashback to all those memories from last summer in Ottawa, of him wearing that dark suit and purple tie at the closing banquet, of how we touched fingers and held hands and kissed in the dark.

I missed him. And that was really annoying since I thought I was over Cooper.

My stomach cramps were coming back, and I felt like I was going to throw up.

Before I knew it, we had arrived at the campus. I stared through the window, shocked that the entire university consisted of just three buildings. Sydney High School was bigger than this.

"I'm going to drop you off here," said Gillian, parking the van right next to the largest of the three buildings. "This way you won't get soaked. Just head into the building in front of you. Walk around and enjoy the refreshments. I'll see you in a few minutes."

"Thank you, Gillian," replied Grace. "We'll see you soon."

As soon as we got inside, I turned to face Grace.

"You lied to me."

Grace looked crestfallen. "I didn't lie."

"Then what is this?"

"You're my best friend. I want us to go to university together."

"And you didn't bother asking me what I wanted? Where I might want to go?"

"But once I heard about Quest, I knew this was the right place for us."

"Don't we talk on Skype every Sunday? I thought friends don't hide stuff from each other."

Grace dropped her head.

I glared at her. "So you didn't invite me to Vancouver so we could train together. You brought me out here to recruit me to where you want to go to school. What kind of a friend are you?"

Grace looked away. When she turned back to face me, she had tears in her eyes.

"I wanted to surprise you," she whispered.

A few people clapped their hands to get our attention. We turned to face the front, and saw that there were about a hundred people assembled. It seemed that all the students had brought along their parents; Grace and I appeared to be the only ones who came by ourselves. I glanced over at Grace, who wiped away a tear with her arm.

A bearded white-haired man introduced himself as the university president, and welcomed everyone to the open house. He gave a speech about the importance of education, and the responsibility of each student to create their own education based on the questions that are meaningful to them.

Though the president spoke really well, I had a hard time concentrating because I kept staring at his shaggy white beard. I had a flashback to several Decembers in my childhood, when Mom would take me to Mayflower Mall in Sydney so that I could get a picture with Santa Claus.

My thoughts were interrupted by the university president, as he announced that all the guests now had the choice of four sample classes. He introduced four professors, whom he called "tutors".

Each tutor gave a short description of their sample class. The first one welcomed us to her class, on the *Evolution and Acquisition of Language*, followed by two men who invited us to *Political Thinking*, and the *Dynamic Geology of the Sea to Sky Corridor.*

I froze when the final tutor walked up to the microphone. He was tall and slim, had long brown hair that went down to his neck, and as he smiled, I saw a small dimple on his pale face. My face turned white, seeing all the resemblances to a boy about twenty years younger who I desperately wanted to forget.

"Good afternoon everyone. I'd like to invite you to my class, *Post-It Notes and Error-Correcting Codes.* I am a tutor in mathematics, and my name is Robert Cooper."

This was too much.

As the crowd filed into the four classrooms, I slowly walked in the direction of Room 315, where Professor Cooper's math class would be taking place. Grace walked up beside me.

"Look, I'm really sorry . . ."

I cut her off. "Grace, I don't want to talk to you."

Ignoring Grace, I walked into the classroom and saw that there were no rows of desks – just twenty seats around a massive oval-shaped table. I grabbed an empty seat, and Grace quickly grabbed the seat on my right.

"Is there anything that would make you forgive me?"

"No," I said. "This entire afternoon has been a waste of time."

"But maybe this prof will say something that will help us get to the IMO."

"Yeah, right," I said. "But if he does, I'll forgive you."

The professor came around and greeted all of us, asking all the high school students to sit around the oval-shaped table in the middle, while requesting that all the parents take the seats along the walls of the classroom. In the middle of the oval-shaped table were some nametags and black Sharpie markers, and we were asked to write down our names and stick the tag on our shirt.

Professor Cooper walked up to the front and greeted all of us, seeing that all of the nametags were in place. We were ready to begin.

"Okay, I need three volunteers. No parents, please. Just students."

Out of the corner of my eye, I saw Grace's hand shoot up. A few other hands went up too. I leaned back in my chair and decided to stay put.

The professor pointed to Grace and two girls. The three of them stood up and walked to the front of the classroom. Professor Cooper shook hands with each of them.

"Okay," said the professor, turning towards us. "I've got three eager volunteers – Deanna, Sophia, and Grace. We're going to play a game where all three of you are on the same team, and will share a small prize if you win. I've brought along a stack of pink and blue post-it notes. I'm going to randomly stick a pink note or blue note on each of your backs. The three of you can then turn around and look at each other's post-it notes, although you have no way of seeing what colour note you've got.

"Here's the object of the game. Once the three of you have had a chance to look at each other's backs, you must simultaneously guess the colour of your own post-it note. You have three options: you can guess 'pink', you can guess 'blue', or you can pass. The three of you win the game and get the prize if *at least one of you guesses correctly and none of you guess incorrectly*."

"What's the prize?" asked Grace.

Professor Cooper opened his backpack and took out three Kit-Kat bars, placing them on the table and grinning at his three volunteers.

"Here are the rules. Before I stick a post-it note on your back, you can have an initial strategy session so that you can determine how you're going to play. Once that's done, I'm going to stick the notes on your backs and after that, no communication of any kind is allowed. Any questions?"

"Do we all have to guess?" asked the girl named Sophia.

"No," said Professor Cooper. "Each of you can make a guess, or choose to pass. If all three of you pass, then your team doesn't win. At least one of you will need to guess. But everyone who guesses must guess correctly."

Deanna raised her hand. "So if all three of us guess, and even if one of us is wrong, then we all lose?"

"That's right," replied the professor. "You're playing as a team. Any other questions?"

After a few seconds of silence, it was clear that the team was ready.

"Okay, the three of you can begin your deliberation session."

They began to talk. Because the classroom was so small, I could hear every word they said. After about a minute, Grace had an idea.

"I think we should pick one person and have that person make the guess. Then we have a guaranteed fifty percent chance of winning."

"Great idea," said Deanna. "Let's do that."

"Yeah," agreed Sophia. "If two or three of us guess, then our chance of winning goes down."

"Exactly," said Grace. "And because no communication is allowed, we're not going to learn anything by looking at each other's backs."

"I agree," said Sophia. "The best we can do is have one person guess."

"Who wants to make the guess?" asked Grace.

"It was your idea, so you should," said Deanna, as Sophia nodded in agreement.

I noticed Professor Cooper walk to one side of the classroom, to give a coin to a freckle-faced boy.

"It seems like our team has their strategy. From this moment, the three of you are not allowed to communicate in any way. Now I'm going to ask Ian over here to flip this quarter three times, one for each of our volunteers. For each volunteer, if the coin lands heads, I'm going to stick a pink post-it note on your back. And if it lands tails, I'm going to stick a blue post-it note on your back. This way it's completely random. Ian, please begin."

As the coins were flipped, the professor asked each of the volunteers to turn around. He walked up and stuck a pink note on Grace's back, a pink note on Deanna's back, and a blue note on Sophia's back.

"Okay," said Professor Cooper, pointing to the clock on the wall. "As soon as the second hand reaches the top, you can guess. Everyone who guesses must speak at the same time."

We all watched the second hand move up, until it pointed directly north.

"Pink," said Grace, while the others remained silent.

"Congratulations!" said Professor Cooper, removing the pink post-it note from Grace's back and holding it up in the air. We all applauded as Grace exchanged high-fives with Deanna and Sophia before walking back to her seat holding the chocolate bar in her hand.

Grace walked back to her seat and broke the Kit-Kat in half. She smiled and offered me the bigger half.

"I'm not hungry."

Grace was about to say something when the professor spoke up.

"The team's strategy was to appoint Grace as the 'captain' and have her make a random guess on behalf of the entire group. This strategy guarantees the team a fifty percent chance of winning. It's much better than having two people randomly guess, which means the probability of winning is one-quarter, or twenty-five percent. Having all three guess is a terrible strategy, because that reduces the probability even more, to just one-eighth, or twelve and a half percent.

"Here's a challenge question for you. Believe it or not, there's a simple and elegant strategy that gives the team a seventy-five percent chance of winning the game. In other words, the comment that we heard just a few minutes ago, that you can't learn anything by looking at each other, isn't quite correct. Take some time on your own right now and see if you can figure it out. I'll give you a small hint: seventy-five percent is the same thing as six out of eight."

I thought about the professor's hint. A seventy-five percent chance of winning was equivalent to six wins out of eight. I realized something I could try: breaking the problem into smaller cases to find a pattern.

Grabbing my pen, I made a table listing the $2 \times 2 \times 2 = 8$ possibilities of how the colours could be assigned to the three people:

| Case | Player 1 | Player 2 | Player 3 |
|---|---|---|---|
| #1 | Pink | Pink | Pink |
| #2 | Pink | Pink | Blue |
| #3 | Pink | Blue | Pink |
| #4 | Pink | Blue | Blue |
| #5 | Blue | Pink | Pink |
| #6 | Blue | Pink | Blue |
| #7 | Blue | Blue | Pink |
| #8 | Blue | Blue | Blue |

I looked at the table, noting that we wanted a strategy that would win in six of these eight cases. That would give a winning probability of six out of eight, equivalent to seventy-five percent.

I added a new column to my sheet, listing the number of notes of each colour, for each of the eight possible cases. I wanted a strategy that would win in six of these eight cases. Staring at the sheet, I noted that the six cases in the middle all had a two-to-one split between pink and blue.

| Case | Player 1 | Player 2 | Player 3 | *Pink-Blue* |
|---|---|---|---|---|
| #1 | Pink | Pink | Pink | 3-0 |
| **#2** | **Pink** | **Pink** | **Blue** | **2-1** |
| **#3** | **Pink** | **Blue** | **Pink** | **2-1** |
| **#4** | **Pink** | **Blue** | **Blue** | **1-2** |
| **#5** | **Blue** | **Pink** | **Pink** | **2-1** |
| **#6** | **Blue** | **Pink** | **Blue** | **1-2** |
| **#7** | **Blue** | **Blue** | **Pink** | **1-2** |
| #8 | Blue | Blue | Blue | 0-3 |

I had a hunch that Professor Cooper's "simple and elegant" strategy would win in the six cases with the two-one split, and lose in the remaining two cases with the three-zero split.

All of a sudden, I saw it. The professor must have noticed my reaction because he came over beside me.

He looked at my nametag. "Bethany, did you solve it?"

"I think so," I whispered, feeling intimidated by the Cooper Robertson look-a-like.

"Can you tell me your strategy?"

I pointed to my sheet and stammered out my solution.

He nodded. "Perfect."

After a few minutes, Professor Cooper asked me to come up to the front to present my solution.

"Each person looks at the colours of the other two people," I said. "If both post-it notes are the *same* colour, she guesses her post-it note is the *opposite* colour. If the two post-it notes have different colours, she stays silent, and doesn't make a guess."

"That's right!" said Professor Cooper. "Can you walk us through each of the eight cases?"

I nodded and proceeded to draw out a large table so that I could clearly explain what happens in each of the eight cases: who guesses, who passes, and whether the team wins.

| Case | P1 | P2 | P3 | *Pink/Blue* | *What Happens?* | *Result* |
|---|---|---|---|---|---|---|
| #1 | Pink | Pink | Pink | 3-0 | All guess Blue | LOSE |
| #2 | Pink | Pink | Blue | 2-1 | Player 3 guesses Blue | WIN |
| #3 | Pink | Blue | Pink | 2-1 | Player 2 guesses Blue | WIN |
| #4 | Pink | Blue | Blue | 1-2 | Player 1 guesses Pink | WIN |
| #5 | Blue | Pink | Pink | 2-1 | Player 1 guesses Blue | WIN |
| #6 | Blue | Pink | Blue | 1-2 | Player 2 guesses Pink | WIN |
| #7 | Blue | Blue | Pink | 1-2 | Player 3 guesses Pink | WIN |
| #8 | Blue | Blue | Blue | 0-3 | All guess Pink | LOSE |

"Check out Case #1," I said. "Each person looks and sees pink post-it notes on the backs of the other two players. The strategy is you only make a guess if you see the same colour twice. And then you guess that your post-it note is of the *other* colour. So each person sees two pink post-it notes, and guesses their own post-it note is blue. And all three guess wrong. So the team loses.

"But check out Case #2," I continued. "Player 1 and Player 2 see a pink note and a blue note, so they both stay silent. But Player 3 sees two pink notes. So she guesses her post-it note is blue. Since she guesses right and no one guesses wrong, the team wins.

"And the same argument applies to all the other cases," I concluded. "The team loses in Case #1 and #8, and wins in the other six cases. So the answer is six out of eight, or seventy-five percent. That's the solution."

The class applauded, and I walked back to my seat.

"Good job," said Grace.

Deanna raised her hand. "I noticed something cool."

"What's that?"

"That in six cases, exactly one person guesses, and that person guesses right. And in the other two cases, all three people guess, and all three guess wrong. So the total number of right guesses is six, and the total number of wrong guesses is also six. You know, six times one equals two times three."

"That's a great observation," replied the professor. "As Deanna just said, if you look at the total number of guesses made, exactly half are right and half are wrong. You can increase the winning percentage if you can formulate a strategy where most of the time no one is wrong but occasionally everyone on the team is wrong."

A hand went up. "Is seventy-five percent the best you can do?"

"Yes," replied the professor. "It's possible to prove that seventy-five percent is the best you can do for three players. But for larger values of $n$, the probability goes up. If $n=3$, Bethany found a strategy that wins with probability 3/4. If $n=7$, there is a winning strategy with probability 7/8. If $n=15$, there's a winning strategy with probability 15/16. It's completely counterintuitive, but as $n$ gets bigger and bigger, you're *more* likely to win the game. That's remarkable, isn't it?"

"Why?" someone blurted out.

"We're going to investigate this for the $n=7$ case, and then you'll all understand how it works in general. For the seven-player case, we'll discover a strategy that guarantees victory seven-eighths of the time. To do this, I'm going to introduce ideas from *coding theory*."

"What's that?" asked Ian, the freckle-faced boy who flipped the coins.

"It's a branch of mathematics that has applications to all these things people do every day, like purchasing groceries, buying something on the internet, validating a social security card, checking out books at the library, playing a CD, and storing large data files on a computer. In fact, by the end of this session, you will understand how this post-it note problem connects to all these common activities."

Professor Cooper then gave a short mini-lecture on how to express numbers in *binary* (in 0s and 1s), a topic I knew already thanks to Mr. Collins. The professor then defined a strange concept called the "digital sum", or the "nim sum", which was a simple way to add binary numbers without carrying. He explained how this counter-intuitive concept, where $1+1=0$, was key to how logical circuits function on computer hardware. It was a bit weird, but I got the idea.

We spent the next thirty minutes solving the seven-player post-it note problem, and together, Professor Cooper guided us as we came up with a cool strategy based on this "digital sum" concept. We figured out that there were $2^7 = 2\times2\times2\times2\times2\times2\times2 = 128$ ways the post-it notes could be assigned to the seven players, and that in seven-eighths of the cases, or 112 out of 128, the team wins by having exactly one person guess correctly. And in the other sixteen cases, the same strategy forces all seven players to guess, and all will guess incorrectly.

Professor Cooper spoke up. "For the three-player problem, Deanna noted that the total number of right and wrong guesses is equal. As you can see, the same is true for seven players, since one hundred and twelve times one equals sixteen times seven. It's the same idea: most of the time no one is wrong, but occasionally everyone on the team is wrong."

The professor then wrote down a bunch of seven-digit binary numbers on the board. I quickly saw that there were sixteen of them.

| | | | |
|---|---|---|---|
| 0000000 | 0100101 | 1000011 | 1100110 |
| 0001111 | 0101010 | 1001100 | 1101001 |
| 0010110 | 0110011 | 1010101 | 1110000 |
| 0011001 | 0111100 | 1011010 | 1111111 |

"As we discovered, there are one hundred and twenty-eight ways you can stick the post-it notes on the seven players. For example, suppose the first three players get pink and the last four get blue. Say pink is one and blue is zero. Then this scenario can be represented by the seven-digit *binary code* 1110000.

"You can represent each of the one hundred and twenty-eight scenarios with a seven-digit code. I've just written sixteen of those codes on the board, namely the sixteen cases where the team *loses*. These are known as the seven-digit *Hamming codewords*, named after the person who discovered them. If you look carefully, you'll see that every pair of codewords differs in at least three positions."

To convince myself of what he just said, I randomly chose two of the sixteen codewords, to see how they matched up when compared bit by bit.

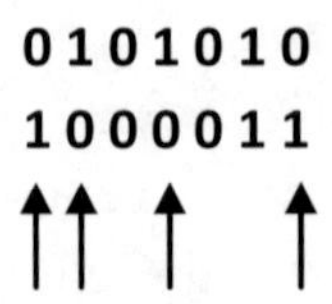


In my example, the two codewords differed in *four* positions: the first, second, fourth, and seventh.

"Let me show you something cool," continued Professor Cooper. "I'm going to associate each of the first sixteen letters in the alphabet (a four-digit binary string) to one of these seven-digit Hamming codewords. For example, A is 0000000, B is 0001111, C is 0010110, and so on.

"Say I write a message on my cell phone, and text it to you. My cell phone, just like any digital device, can only process 0s and 1s. So it has to take each letter in my message and convert it to binary form, in order to send my message over some wireless channel, bit by bit. Then your cell phone gets my string of 0s and 1s, converts it back to normal alphabet letters, so that you can read my message.

"But nearly all wireless channels are unreliable – that due to various conditions, data transmission isn't one hundred percent accurate, as sometimes the channel accidentally flips a 0 to a 1, or vice-versa. This is where the Hamming codes come in: the idea is that you're deliberately adding redundancy and complexity in order to eliminate ambiguity and unreliability. This extra redundancy allows the receiver to detect the errors that inevitably occur in various bits throughout the message, and automatically correct these errors without needing the message to be re-transmitted. Without adding this redundancy, you won't be able to read a message correctly if an error occurs somewhere during the transmission.

"Let me give you an example. Suppose I send you the letter B. So we convert this letter to its seven-digit Hamming codeword, which is 0001111. That's what I send you.

"We now have a seven-bit message. We send this message bit by bit over my unreliable channel. Now maybe an error occurs during transmission, and one of the bits is flipped and processed incorrectly. So at some position, a 0 becomes a 1, or vice-versa. Suppose the seventh bit is flipped.

$$000111\underline{0}$$

"Say you get this seven-digit binary number and see that it's not one of the sixteen Hamming codewords. But there's no need to panic. Because the set of Hamming codewords is an *error-correcting* code, you can determine where the error was made, fix it, and recover the intended message!

"Let's work through my example, and examine my messed-up code 0001110. Let's look at all seven possibilities for where the error could have occurred, and what the right message could have been:

| **Received Message** | **0001110** |
|---|---|
| Error in 1st Bit | 1001110 |
| Error in 2nd Bit | 0101110 |
| Error in 3rd Bit | 0011110 |
| Error in 4th Bit | 0000110 |
| Error in 5th Bit | 0001010 |
| Error in 6th Bit | 0001100 |
| Error in 7th Bit | 0001111 |

"It turns out that exactly *one* of these seven binary numbers is a Hamming codeword. See if you can find which of these seven numbers appears in our list of sixteen Hamming codewords:

| | | | |
|---|---|---|---|
| 0000000 | 0100101 | 1000011 | 1100110 |
| 0001111 | 0101010 | 1001100 | 1101001 |
| 0010110 | 0110011 | 1010101 | 1110000 |
| 0011001 | 0111100 | 1011010 | 1111111 |

I raised my hand, once I noticed that the number 0001111 appeared on both lists.

"Which one is it, Bethany?"

"It's 0001111. That's the only codeword."

"Correct," said Professor Cooper. "In the example we just did, the string 0001110 differs in just one position from the Hamming codeword 0001111, but in at least two positions from all the other Hamming codewords. You can show that for each non-codeword, there is exactly one closest Hamming codeword, which differs in precisely one position.

"Let me recap what we did. I sent you the letter B, which gets coded as the Hamming codeword 0001111. Unfortunately because of channel noise, the seventh bit gets flipped, and produces the non-codeword 0001110. And because this non-codeword differs in just one position from the Hamming codeword 0001111, we assume that the program meant to send this message instead, and so we correct it. I sent you the letter B, and you correctly received it, even though there was a transmission error. In summary, the losing positions of the Post-it Note Problem are, in disguise, an error-correcting code."

"That's really cool," I whispered.

Sophia raised her hand. "But what if there are two errors in a seven-digit string?"

Professor Cooper looked at her with a straight face. "Then we're completely screwed."

After some light laughter, he spoke again. "Sophia makes an excellent point. This Hamming codeword methodology doesn't work if there are two or more errors within a transmission of seven consecutive bits; fortunately, this is extremely unlikely to occur. That is why these error-correcting codes have important applications, such as playing CDs and DVDs."

"How does it work?" asked someone.

"Your DVD player comes with technology to detect and correct data transmission errors caused by scratches. So that's why your DVD player has no problem playing an entire movie, even if your disk contains scratches. By deliberately adding extra information, you can eliminate ambiguity and error."

"All this seems so complicated," said Sophia. "I mean, for every four bits, you have to turn that into seven bits in order to make this work. You're making it so much harder."

"Exactly," said Professor Cooper, with a huge grin. "By making it harder, we're making it easier!"

I stared at the professor. After a couple of seconds, I raised my hand.

"Yes, Bethany?"

"Could you repeat what you just said?"

"By making it harder, we're making it easier."

I put my pen down and closed my eyes.

*What an awesome problem-solving strategy.*

"Hey," I whispered, turning my chair to face Grace.

"Yeah?" replied Grace, in surprise.

I looked into the eyes of my best friend, and smiled.

"I forgive you."

# 35

"It's twelve o'clock. Time's up."

"Damn, I got stuck on the last problem," said Grace. "Did you get all five?"

"Yeah," I said. "But just barely."

We had an excellent morning, working for three hours straight right after breakfast. Grace and I downloaded a bunch of geometry problems from *artofproblemsolving.com*, one of the largest math forums on the internet. Grace told me about the site after she learned about it at the Waterloo Seminar, and we quickly signed up, joining over one hundred thousand other members from nearly a hundred countries, including thousands of IMO coaches, IMO contestants, and IMO wannabes.

After clearing our dishes at 9:00 a.m., Ray went to work and so did we. We found the Olympiad Section on the Art of Problem Solving (AoPS) website, and clicked on "Geometry". We downloaded problems that appeared on recent contests from all around the world, and started attacking them one at a time.

We first solved five hard geometry problems together, and then separately spent the remaining time writing up a full solution to each problem.

I stretched and walked over to the kitchen.

"Are you cooking for us, Margaret Atwood?" said Grace.

"Why are you calling me that?"

"Because she's the best writer in Canada," said Grace, reading through my five solutions. "When it comes to writing proofs, there's no one in Canada that can match you. Not even Albert."

"Thanks," I said, touched that Grace would say that. "And because you gave me such a nice compliment, I'll cook lunch for you."

"Awesome," said Grace. She walked over to the pantry and handed me a large package of Kraft Dinner.

As I filled up a pot with water, I noticed Grace open up her laptop to log on to her Facebook page.

"Hey, cool," said Grace. "Raju just posted a link to a puzzle game that Albert invented."

"What's the puzzle?" I asked, waiting for the pot to boil.

"It's hard to describe. I need to print it out. Be right back."

As Grace went off to connect her laptop to her dad's printer, I prepared lunch. Within ten minutes, everything was done, and I walked over to the table with two heaping bowls of yellow macaroni.

Grace saw me approaching and clapped her hands. "Thanks, Iron Chef."

I sat down and shoved a spoonful of macaroni into my mouth.

"What's the puzzle?" I asked.

Grace had a thin stack of printed pages in front of her. "It's some tiling game that Albert invented and uploaded on to his home page. Raju found it and posted the link on his Facebook wall."

"How does it work?" I asked, pointing to the first page of Grace's stack of paper.

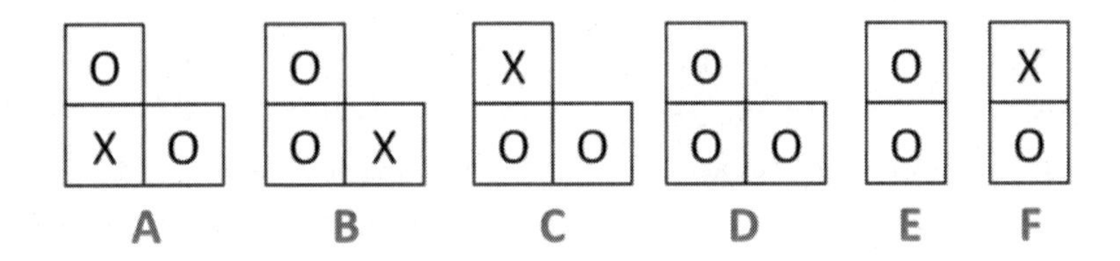


"Albert gives us six tiles. Four of the tiles are L-shaped, and the other two tiles are dominoes. He labels the pieces **A**, **B**, **C**, **D**, **E**, and **F**. So the six tiles have a total of sixteen unit squares."

"Yup," I said, doing a quick count to check there were sixteen squares in all. The four L-shaped pieces had three squares each, while the two dominoes had two squares each, so that 3+3+3+3+2+2 = 16.

Grace then showed me the second page in her stack, which was a square grid filled with four Xs and twelve Os.

| X | O | O | O |
|---|---|---|---|
| O | O | X | O |
| O | O | O | X |
| O | O | X | O |

She turned to the third page. "And what we need to do is *rotate* these six pieces and place them so they perfectly tile this 4×4 grid, with no overlap, so that all the Xs and Os are in the right positions. Like this."

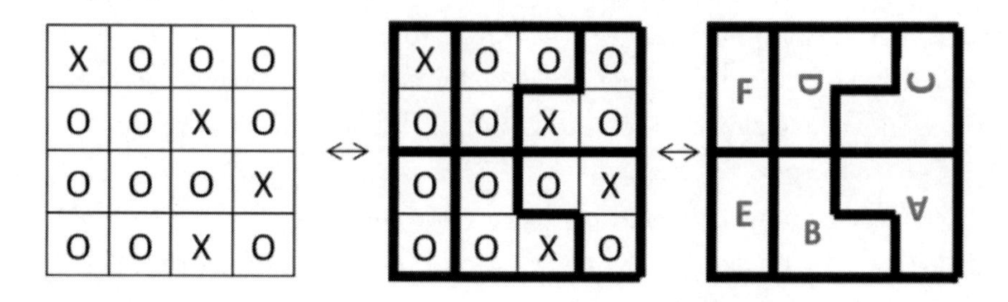


"Yeah, I get it," I said, seeing how the six pieces **A**, **B**, **C**, **D**, **E**, **F** could be rotated so they would fit exactly on the 4×4 grid, with all the Xs and Os lining up, and no pieces overlapping or placed outside the grid.

"Albert says there are exactly nineteen different grids, including the example he gave us, where there's only one correct tiling. He says it's like a Sudoku, because every Sudoku puzzle also has a unique solution. But he says that his puzzle is more interesting – the rules are simpler but the analysis is a lot deeper."

"Cool. Every piece has to be used once, and you're allowed to rotate it. What's the first grid?"

Grace turned over to the next page. "Albert says this one's the easiest – it's arranged from easiest to hardest. Every grid has exactly four Xs and twelve Os."

| O | X | O | O |
|---|---|---|---|
| O | X | O | O |
| O | X | X | O |
| O | O | O | O |

I looked at the six tiles, which Grace had cut out using her scissors.

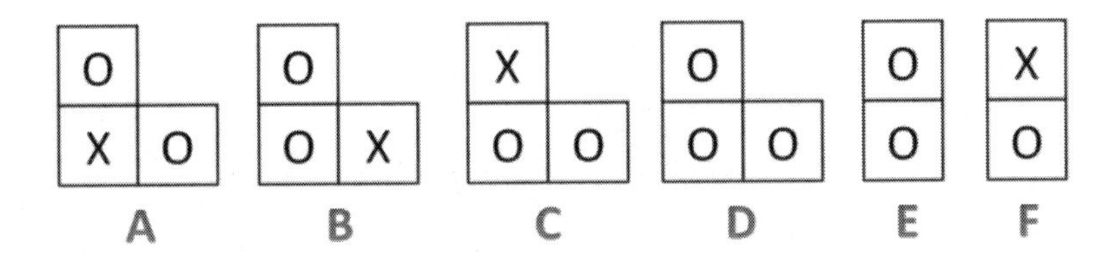


I thought of the different ways each tile could be positioned on the grid, trying to break the problem into smaller simpler cases. After some thought, I had an idea.

"Grace, check it out. The first tile, **A**, can only be put in one of three positions."

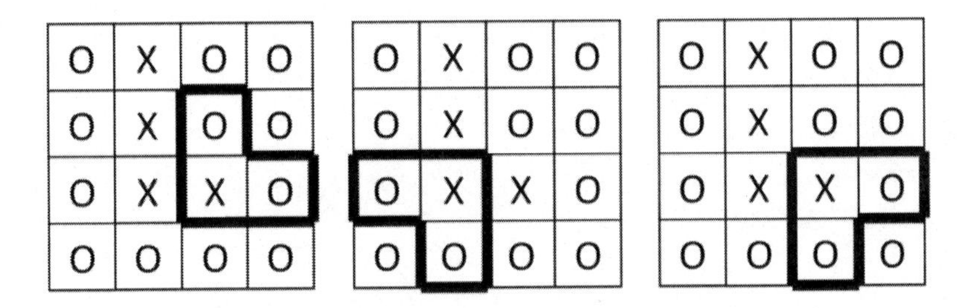


"Yeah, I see," said Grace, nodding her head to convince herself that the **A**-tile couldn't be placed anywhere else on the grid, since the tile had to be L-shaped, and cover exactly one X and two Os with the X in the 'centre'.

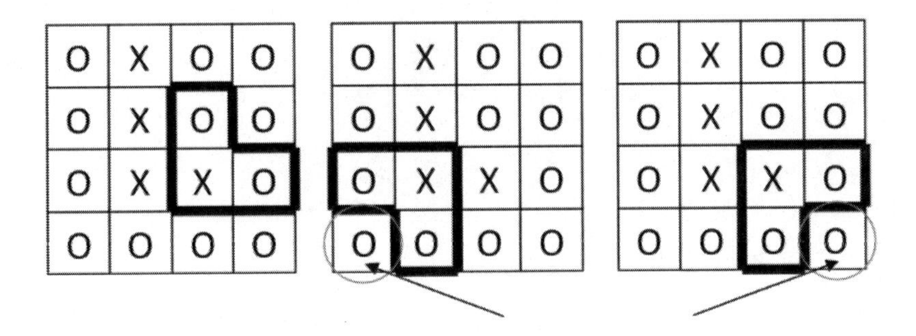


I continued: "And it can't be the either the second option or the third option, since in both cases, you'll get a single corner square on its own that can't be covered with a tile."

"Okay," said Grace. "So it has to be the first case."

I figured out the next two steps: the **E**-piece, the domino with two Os, had to go horizontally in the bottom right corner. And this forced the **D**-piece, the tile with three Os, to be rotated 180° into the top right corner.

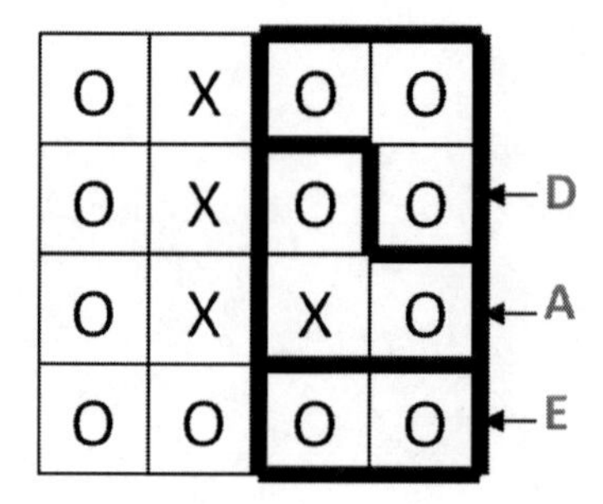


Within seconds I figured out the rest. The **F**-piece had to be placed horizontally in the top-left corner, forcing the **C**-piece to rotated 90° clockwise into the middle-left and the **B**-piece to rotated 90° counterclockwise into the bottom-left.

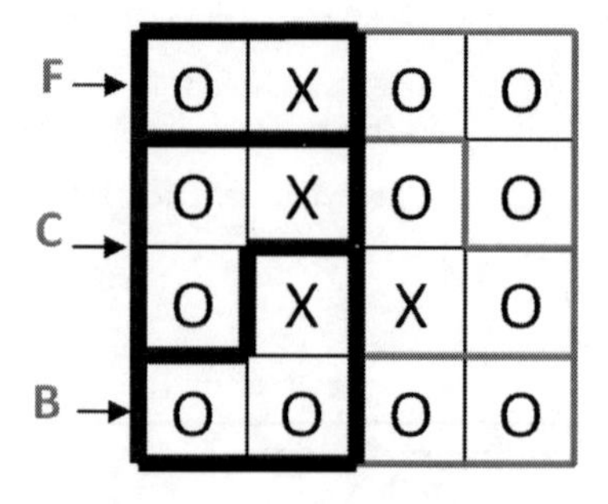


"Man, you're fast," said Grace, shaking her head in admiration.

I took another bite of my Kraft Dinner. "Let's do the next one."

"Let's finish our lunch first."

"I can multi-task," I said. "I'm not a boy, you know."

Grace insisted on finishing her macaroni. I happily grabbed the stack of papers from her and placed the top sheet in front of me.

Albert's second-easiest puzzle wasn't hard. Less than ninety seconds later, I figured out how to do it. I rotated the last of my six tiles so that they perfectly went on the grid. I took two more bites of my lunch.

"Hey, good job," said Grace.

She put her empty bowl aside and joined me as we did the next puzzle together.

This 4×4 grid was a little harder, since there were three separate positions for the **A**-tile and we had to look at each one. But within a few minutes, we had solved Albert's third grid tiling puzzle. I picked up my spoon and finished my lunch.

Grace shook her head. "Where does Albert get ideas like this? How is someone so young so creative?"

"Beats me," I replied. "In Ottawa, he was miles ahead of the rest of us, even Raju. His mind just works differently from the rest of us mortals."

"Yeah," nodded Grace. "And he made the IMO team this year, in Grade 9. So he might go to the Olympiad four times in total. That's crazy."

"I'd love to go just once," I replied. "That would be more than good enough for me."

"Yeah, me too," said Grace. "There's only one problem with that."

I turned to Grace, with a look of concern. "What's that?"

"Well, actually there are *five* problems," said Grace, fiddling with her spoon. "I'd been meaning to tell you this ever since you came to Vancouver, but didn't know how to bring it up."

"Bring what up?"

Grace took a deep breath. "Realistically, we're trying to make the IMO team two years from now, the year we graduate. Well, two years from now, five of the six spots are pretty much guaranteed. Albert, Raju, and three others I met at Waterloo."

"Says who?"

"These three Grade 10s were amazing. They were better than Raju – even Raju admitted that. I think these three will make the IMO team next year, so of course they'll all be going in two years' time. It sucks, but we graduate the same year they do."

"How come we didn't meet them in Ottawa?" I asked.

"Because all of them immigrated to Canada in the past eighteen months, so Marlene and Rachel didn't know about them. One guy's from Russia, and he moved to Alberta. The other two are from Hungary and Korea, and they now live in Ontario."

"How did they get so good?"

"They were brought up in countries where they do tons of math from a young age, and they were invited to special enrichment camps in junior high school. Unfortunately, that's five spots right there. That leaves just one spot for everyone else – including you and me."

"That's so unfair," I said, horrified by the thought of only one of us making the team. "They can be educated in some other country, learn all this math they don't teach in Canada, come over when they're in high school, and get on to the Canadian IMO team?"

"Yeah, it sucks," said Grace. "But there's nothing we can do. We've got to figure out a way to close the gap and become just as good as them."

"How are we supposed to do that in just two years?"

"Well, we went to Ottawa last summer. And I had the Waterloo Seminar, and you had the Math League. And now we've got this Art of Problem Solving website that we can use to train."

"Yeah, but everyone else has the website too," I protested. "You said Albert and Raju are on AoPS all the time, and I'm sure those other three guys are too."

Grace stared into my eyes. "What are you saying, Bethany? Are you giving up?"

"Giving up? On the IMO? What makes you think I want to give up?"

"You just sound like you want out."

"No, Grace, I want in! This is why we're working so hard. This is why I flew to Vancouver to train with you. But now you tell me this."

"There's still lots of time. Our shot at the Olympiad is two years from now."

"Yeah, but from the sound of it, there'll only be one spot for the two of us. And who knows, what if someone else comes from out of nowhere and takes that spot? All that hard work – for nothing."

"I know, it's so frustrating," said Grace. "But what if we work harder? You know, spend thousands of hours over the next two years, you and me, working together?"

"I don't know," I said, shaking my head. "You know Mom's story, right?"

"Of course. You've told me a bunch of times. The Olympic figure skater."

"The almost-Olympic figure skater," I said, correcting her. "Mom made it to the Olympic Trials, but missed the team by one spot. I thought you knew."

"I do," said Grace. "Sorry. You know what I meant to say."

"Mom said she trained *three thousand hours* to make the Olympic team. But she has nothing to show for it. She put everything into that one goal, and missed her dream by one spot. What if that happens to us?"

"But what if we put in three thousand hours? Others might be smarter, but I bet no one will work harder. If we do that, we'll be just as good as them. And you and I – we could both make it. Three thousand hours over the next two years – that's about thirty hours a week, around four hours a day. We can do this."

"Come on, Grace. Four hours a day? You know Grade 11 and 12 are the hardest years in school – there's not enough time in a week for us to do an extra thirty hours! We've got classes, exams, scholarship applications, and who knows what else. Are we going to wake up super-early to get in our training? That's what Mom did, you know. She woke up at five o'clock every day so that she could be at the rink by six-thirty and get in ninety minutes of training before her morning classes."

"Then that's what we should do. Everyone else is smarter, but we'll just get up earlier and work more."

"And we're going to drop everything else? You know, get A+ in math but get Cs and Ds in everything else? Then we won't even get into university!"

"No, we're not going to drop everything else," said Grace. "I'm still doing debating, and you said you were going to try out for the cross-country team. That's all part of our mental and physical training too. And we'll do just fine in our classes. But let's wake up two hours earlier – we'll start the day training, and do some more Olympiad prep after we finish our school homework. And then we'll talk about our solutions on Skype. Every day. We'll do more during breaks – you know, Christmas, March break, and of course, next summer. And let's promise that we'll cut out all the stuff we don't need – you know: TV, shopping, parties, and boys."

Grace grabbed on to my arm. Her face turned red. "Look, Bethany, I *need* this."

I saw a tear welling up in her eye.

"Are you okay?"

"No, I'm not okay," replied Grace, letting go of my arm. She stood up and put her arms on her hips. "I need the IMO, or my life doesn't work."

"Aren't you being overdramatic?" I asked, looking strangely at her. "Your life doesn't work without the IMO? Look at you. You've got an amazing life."

"No, I don't!" said Grace. Tears started gushing out of her eyes.

I handed her a tissue and waited until she was ready to speak.

"I'm so bored in school and you know I don't have any friends there. Me and my dad don't get along, so home life sucks too. I'm so lonely, Bethany!"

As she resumed crying, I put my hand on her shoulder. After a bit, she sniffled and looked up at me.

"Making the IMO team is more important than getting the highest marks at school, more than dating some boy, more than anything else in the world. I thought you felt the same way! You're my best friend. I thought we were in this together."

"We are," I said. "But come on, Grace. Three thousand hours? You can't be serious."

"I am serious!" yelled Grace.

She went to the washroom and slammed the door. Even though the door was shut, I could hear her sobbing.

I felt terrible. I closed my eyes and tried to process everything Grace had just said.

I tried to distract myself with Albert's tiling puzzle, but after a few minutes, I realized I couldn't concentrate. To clear my head, I stood up and walked around the kitchen, but found it difficult.

Not only was I feeling guilty, I was feeling angry at Grace for implying I wasn't willing to work for it.

Of course I was willing to work for it. I just wasn't willing to work three thousand hours for a goal that had no guarantee of success.

Shuffling around the kitchen, I came to the side wall where my eyes were level with the family portrait, the picture with three-year-old Grace, her dad, her Mom, and her brother.

I kept staring at the family portrait. As much as I wanted to gaze away, I couldn't.

As I looked into the eyes of Grace's dad, I was reminded of something he said on the Monday morning, telling me that he never regretted being married to Grace's Mom, that he was all-in, every day, committed to his marriage one hundred percent. I remembered Ray saying that with anything in life, you either do it or you don't do it, and that there's no point in doing something halfway, or three-quarters of the way. I recalled his story about the merchant who sells everything he owns so he can get the pearl that means more to him that anything else in this world.

My pearl was the International Mathematical Olympiad.

Ever since my twelfth birthday, that had been my one and only ambition.

If I wanted to get to the Olympiad, I had to do it all the way. It was all or nothing.

And if I didn't want it, I realized I might as well go home to Cape Breton right now, enjoy the rest of the summer with my Sydney High friends, and then go off to school in September, get high enough marks to enter university, graduate, and take a decent job that would promise me a stable and secure life.

But the prospect of that was so unsatisfying.

As I stared at the picture of Grace's mother, I thought of Mom and her merry-go-round life: the government job she hated but was afraid to walk away from; the same small group of friends she called from time to time; the predictable routine of our evenings at home, where she would cook the same seven meals each week, in the Monday to Sunday cycle that was in place for as long as I could remember.

Mom often talked about her regret of missing out on the Olympics, and how the pursuit of a foolish dream wrecked everything else in her life: that she was too focussed on her figure skating and denied herself an education; that she was now too old to go back to school and start over.

Mom didn't like me taking risks, because failing to achieve a goal could lead to a lifetime of disappointment and regret. She'd often remind me that taking risks made us vulnerable, and when we're vulnerable, we can get hurt.

But deep down, I knew that unless I took this risk, unless I made myself vulnerable, I would never know what I could achieve. I would never know what, and who, I could be.

As my thoughts returned to the portrait of Grace's family, I realized I had to make a decision about the IMO, whether I would settle for a life of comfort or pursue a life of courage.

The decision was obvious.

I heard a flush, followed by a running faucet. Grace came out and looked at me sheepishly.

"I'm sorry, Bethany. I shouldn't have yelled at you."

I turned to my best friend. "No, Grace. I am sorry."

She walked up to me and gave me a hug, her arms wrapping tightly around my waist. I returned the hug and lightly pressed on her shoulders, not wanting to squeeze too tightly and crush her.

"I want to do it," I whispered. "All in, three thousand hours, for the IMO."

Grace looked up at me. "Look, Bethany, you don't have to."

"I know I don't have to," I replied, putting my arms on her shoulders and staring into her eyes. "I want to. I want this just as much as you. And if it takes three thousand hours, then that's what it takes."

"And what if you're not chosen? What if you come seventh in Canada and just miss out on the IMO?"

I paused, thinking about what happened to Mom. I knew the right answer, but it took me a full minute to convince myself that I really meant it, before I could actually say the words out loud.

"Yeah, even if I miss the team by one spot, it would still be worth it."

"And you wouldn't regret it?" she asked.

"I'd only regret it if I knew that I didn't give it my absolute best. I want to do this, Grace. I'm all in."

"Yay!" said Grace, squeezing me. "I'm so happy!"

After a long pause, she looked up.

"Three thousand hours. How do we do it?"

I sat down and took out a fresh sheet of paper and made some calculations. I quickly figured out a way to hit the target, and motioned for Grace to sit down next to me.

"When we've got school, we do twenty-five hours a week: three hours each weekday, five hours on Saturday, and five hours on Sunday. There are thirty-six weeks of school in a year, so that's thirty-six times twenty-five, which is nine hundred.

"In the sixteen weeks we don't have school, we do forty hours a week. Let's train for fifteen weeks, and then take one week vacation, say at Christmas time, when we don't do any math at all. Then that's fifteen times forty, which equals six hundred. Nine hundred plus six hundred is fifteen hundred. That's fifteen hundred hours in one year, and we multiply that by two. That's three thousand hours over the next two years."

"It's crazy but we can do it," said Grace. "Oh yeah, and we should count anything that helps us in our training. You know, the debate team for me, the math league for you."

"I can't believe we're doing this."

Grace grabbed my arm. "I just had the best idea. Before we head to the airport tomorrow night, let's stop by the PNE Playland."

"PNE?"

"Pacific National Exhibition. They have a famous ride in the Playland, and we have to go on it together before you head home."

"What kind of ride?" I asked, before stopping. I knew the answer.

Of course I knew the answer.

# 36

The gentleman behind the Air Canada counter smiled as I walked up to him.

"Hello, young lady. What's your destination?"

"Halifax, via Toronto," I sighed.

"Can I see your booking confirmation?"

I wordlessly handed over a printed page containing all of my flight reservation details, and waited for the man from Air Canada to give me my boarding passes, and allow me to check in my suitcase.

I couldn't believe it was already Friday night. I wasn't ready to go home.

After I got my tickets, I turned around and saw Grace and Ray waiting for me. We slowly walked towards the food court in the domestic terminal of Vancouver Airport, and found an empty table next to the Tim Hortons coffee stand.

It was now 9:24 p.m. I still had a bit of time before I needed to clear security and catch my 10:30 p.m. flight to Toronto.

"How long's your layover?" asked Ray.

"Two hours," I said. "I get to Halifax at eleven o'clock tomorrow morning, local time. Mom's picking me up at the airport."

Grace wiped a tear from her eye, and I could tell she was thinking the same thing I was. I had wonderful friends at Sydney High, but no one who understood me the way Grace did; no one who encouraged and challenged and supported me like Grace could; no one who shared the same crazy dream and wanted to go all out and reach it together.

And I was that person for Grace. It sucked that we lived on opposite ends of the country.

I looked at my best friend. "We'll see each other in Waterloo next summer."

"For sure," she replied. "And we'll see each other at the IMO the summer after. Well, we'll try."

"Yup," I said. "We'll try our best."

Ray raised an eyebrow. "We'll try our best? Is that what you just said?"

Grace glared at her father. "What's wrong with that?"

Ray cleared his throat. "I'm sorry to bring this up right now, but I need to. Is '*we'll try our best*' your approach to the Math Olympiad?"

"Yes," I replied defensively. "Grace and I will do everything we can to try to get on to the IMO team. Of course, there aren't any guarantees."

"But you two have committed to doing three thousand hours of training over the next two years. And both of you have this relentless tenacity that will keep you focussed on your goal. So why not just go out and say what I *know* is true – that in two years, both of you *will* be Math Olympians for Canada!"

"Come on Dad," said Grace. "There aren't any certainties in life. Even in math."

"Really?"

"Yes, Dad. Really. Remember our chat on Monday about axioms and incorrect assumptions? Even 'one plus one equals two' is based on five axioms we take for granted, none of which we can prove with absolute certainty. So if that's not certain, how can you say it's certain we're going to make the IMO team?"

"Because you have more determination and passion than anyone else I've ever met – more than even this Albert boy you've told me about, more than anybody else that's also trying to reach this goal. That matters."

"It's not that easy," I protested. "Only six people get to represent Canada. It's a big country."

Ray paused and glanced over at the clock, checking to make sure we had enough time left.

"Bethany, you saw the *Soul Surfer* movie a few nights ago. I know Grace wasn't impressed with the film, but you said you loved it. What was your favourite scene?"

"The climax of the national surfing championship, where she rides that monster wave in the final seconds."

"That was amazing," said Ray, nodding in agreement. "For me, though, it was actually a scene after the movie ended and they rolled the credits showing clips of the real Bethany Hamilton. Right after the shark attack, a reporter asks thirteen-year-old Bethany, 'Do you think you'll ever surf again?' She stares through the reporter. She clenches her jaw and tells him: 'I think? I *know* – I'm gonna surf again.'"

"Yeah, I remember," I replied, bracing for what was coming next.

"After the shark attack, Bethany Hamilton could have given up her dream of becoming a pro surfer. But she didn't. She accomplished her dream only because she had absolute certainty that she'd make it. And sure her journey was brutal, competing against people far stronger with twice as many arms, and it took her six long years to get onto the professional surfing circuit. But she had the faith that she'd eventually succeed. And because of that, it was easier for her to wake up at the crack of dawn every day, knowing each day of training was leading her one step closer to her target.

"I believe it's the same with the two of you, and the bold decision you made this week to sacrifice everything else in your life so that you can focus on achieving this one dream. So don't say I hope I can make the Olympiad team, or I think I can make the Olympiad team – say I know I will make the Olympiad team! Because if you're not one hundred percent convinced that you'll make it, then it's probably not going to happen. If you spend even one percent of your time doubting, then that's one percent of precious training time you've lost, focussing on something other than your goal."

"But Dad," protested Grace, "what if, hypothetically, seven people have that exact same attitude? Then one person gets their heart crushed. There's no way around it."

"You're right," said Ray. "But over the next two years, you don't have to deal with hypothetical situations. It's just the two of you, putting in the three thousand hours, to get two of the six spots on the IMO team. I know that you'll make it. Stop saying *think* and *might*, and replace those words in your vocabulary with *know* and *will*. Stop doubting!"

"What's the point of your sermon, Dad? Stop doubting, have faith, turn to God?"

"No, Grace, I'm talking about you and Bethany. This isn't a Sunday morning sermon. This is about your quest for the Math Olympiad. The more faith you have that something will happen, the more likely it is that it will happen. Questioning yourself does you no good."

"I disagree," I replied. "It's only when we question ourselves that we find out who we really are."

"But you and Grace have already done that!" said Ray. "And you've stripped off all the layers, and discovered the core of who you are, and you've found your answer. So there's no point going back, checking and re-checking, making sure that you've got it right."

Grace looked at her dad. "Without doubt, how can you find truth?"

"After a certain point, you need to stop doubting and *decide* what is truth! There are five ways in which humans receive knowledge, and in each way, you can produce an infinite cycle of doubt. Grace, you said so yourself that even 'one plus one equals two' is based on axioms that are impossible to verify. So our powers of rational thought can be doubted. Also, our senses can be doubted – a colour-blind person like me can't tell the difference between red and green, so I might see one colour but it might be the other. Also, we can doubt our memory, because no one has perfect recall of every situation in our past. And we can doubt the testimony of others – what we read about and what others tell us is true. So you can doubt and doubt forever, and refuse to come to the conclusion that anything is true."

"That's four sources of knowledge," I said, counting with my fingers. "Rational thought, our senses, our memory, and what we learn from others. Didn't you say there were five?"

"Yes," said Ray, "there's a fifth source of knowledge, where we discover truth by having our conscience moved by a spirit that's working inside our hearts – it's called *revelation*."

Grace rolled her eyes. "Here we go again."

"Can you give me an example?" I asked Ray, ignoring Grace.

He paused. "Did Grace tell you about her Uncle Roy?"

"I don't think so."

Ray turned to Grace. "Do you want to tell Bethany the story about the air miles?"

"Go ahead, Dad."

Ray glanced up at the clock before turning to me.

"About two months ago, Grace was crying in her room because she learned that her best friend from Nova Scotia wasn't going to the Waterloo Seminar with her. I tried to encourage Grace, saying that the two of you would make it together next year, but she was inconsolable. And I felt so terrible. The following day, and I'm not joking – the following day – I'm having lunch with my brother Roy, when he tells me about these bonus air miles he got from Air Canada. And these were his exact words: 'I can't really explain this, but I feel like I'm supposed to give this to Grace.'

"Roy didn't know anything about Grace's sadness from the night before, or that her best friend lived in Nova Scotia, but he felt something inside him compelling him to donate these air miles to Grace. For me, that was an answer to prayer. Grace and I decided that instead of her flying out to see you in Cape Breton that we'd use the air miles to bring you to Vancouver. She said she had a secret motive for wanting you to visit this province."

"Yes, she did," I said, laughing at the memory of our mystery road trip.

"That's an example of knowledge by revelation," said Ray. "Grace's uncle felt his conscience moving inside him, directing him to do something. Of course, he could have been wrong, or he could have doubted his conscience, ignored his heart, and kept those air miles for himself. But I'm glad he decided that an important piece of truth was being revealed to him. Clearly it worked out for the best."

The intercom went on and a deep male voice announced that flight AC156 to Toronto was delayed by fifteen minutes. I checked my boarding pass, and confirmed that AC156 was my flight.

"Yes!" said Grace, standing up. "We have more time. Okay, I gotta pee – I can't hold it anymore."

As Grace sprinted to the washroom, I looked at Ray.

"Thank you. For everything."

I was deeply inspired by Grace's dad, and appreciated all the wisdom he had shared with me over the past week, especially about the importance of going all out to obtain the pearl dearest to my heart, and having the belief that I would succeed in getting it. But as my mind flashed to the family portrait in their kitchen, my heart broke for this gentle man sitting across from me.

"I really respect you. Especially after what happened thirteen years ago, I'm amazed that you can still believe in a higher power."

Ray told me that as he read more about the life of Jesus, his heart began to change. It took many years, but eventually he was able to release the anger he had towards the seventeen-year-old who ran through the red light and killed his wife and son; and that over the past decade, a large part of his professional ministry has been leading grief recovery workshops throughout

Vancouver, offering support to others who have endured the sudden loss of loved ones.

He mentioned how his work honoured the memories of Carmen and Jesse, and gave him so much purpose and meaning. Instead of having the car accident turn him away from faith, Ray said that it miraculously led to a deeper and more intimate connection with God.

Ray looked at me. "I'm absolutely certain that God did not cause the car accident, that this was not deliberately pre-determined as part of his plan. Instead, God found a way to reconcile and redeem this horrible tragedy to serve a higher purpose. Now I'm fully restored. I'm no longer bitter or resentful about the accident. And because God has also experienced intolerable suffering, he understands me, and I can relate to him. And through his strength, I can help provide restoration to others."

"Jeez, Dad, I can't believe you're talking about the accident."

Grace was standing behind her father and must have overheard the last bit. She sat down and faced Ray, her face flushed with anger.

"What's your problem, Dad? Bethany has to leave in a few minutes, and you've got to spend that time preaching at her?"

"He wasn't preaching," I quickly replied. "He was just telling me his story."

"I heard what you just said to Bethany, that God has also experienced suffering. If that were actually true, Dad, then maybe you'd be right about God wanting a 'personal relationship' with us. But that's a bunch of crap. God doesn't know what intolerable suffering is, and you know it!"

"He does know," said Ray, softly.

She clenched her teeth. "Does God know what it's like to lose his wife? Can God relate to a three-year-old who had her mother taken away from her? Give me a break, Dad! How can you respect a God, let alone believe in a God, who has no friggin' clue what that feels like?"

Grace was shouting at her father. I felt too stunned to say anything. I couldn't believe our amazing week was ending like this.

Ray put his head down, and a tear formed in one eye. He spoke gently, bringing his voice down to a level that was barely audible.

"He knows what that feels like."

"Bullshit!" she screamed.

Getting even angrier, Grace pressed on.

"And what about baby Jesse? How can you say that God can relate to your pain? Your son died! How the hell would God know what it feels like to lose his only son?"

There was an uncomfortable silence throughout the food court. I could tell a bunch of eyes were staring at us. All of a sudden, Grace started to sob.

Grace cupped hands to her mouth and began shaking.

"Oh my God."

I jumped up. "Are you okay?"

Tears continued falling. As I stood next to her and put my hand on her shoulder, I could hear incoherent blabbing.

"It makes sense . . . it makes sense . . ."

"What makes sense?" I asked.

Grace mumbled something but I couldn't figure it out.

I turned to Ray, in complete confusion. "What's she saying?"

Ray tried to explain but couldn't form the words because tears started flowing down his cheeks too.

*What was happening?*

As I helplessly watched Grace and Ray, I looked up at the clock and gulped. I stood up and put my backpack around my shoulder.

I had to leave right now.

Grace stood up and threw her arms around me. Ray came over and did the same. They were still sobbing, and couldn't get anything out, so the last words fell to me.

"I love you guys."

## The Canadian Mathematical Olympiad, Problem #4

Let $\Delta ABC$ be an isosceles triangle with $AB = AC$. Suppose that the angle bisector of $\angle B$ meets $AC$ at $D$ and that $BC = BD + AD$. Determine $\angle A$.

## Solution to Problem #4

I look at my diagram yet again, desperately searching for something.

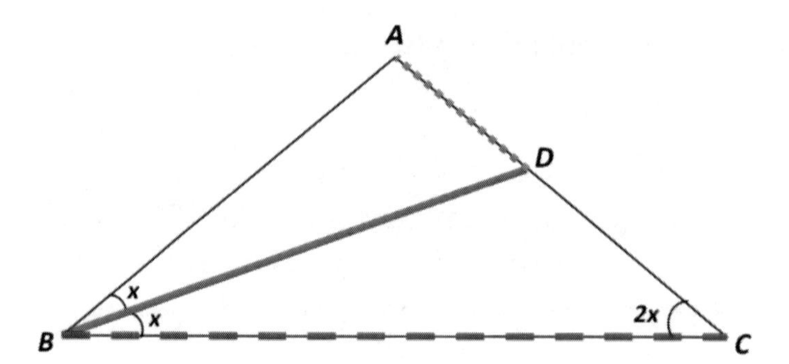


I'm not seeing anything.

Triangle geometry is messy, but circle geometry is beautiful. I can solve circle geometry problems and it's too bad I can't change this triangle problem into a circle problem.

*Draw a circle.*

I sit up straight with my back against the chair, and quickly look around the room to make sure I'm not hallucinating.

*Draw a circle.*

I'm not hearing a voice; I *feel* it.

And yet, the idea doesn't make any sense. Drawing a circle, for no apparent reason, is dumb. But I'm desperate – and I need to try something.

Given any triangle $XYZ$, I know that there is exactly one circle (called the "circumcircle") that passes through the three vertices of the triangle.

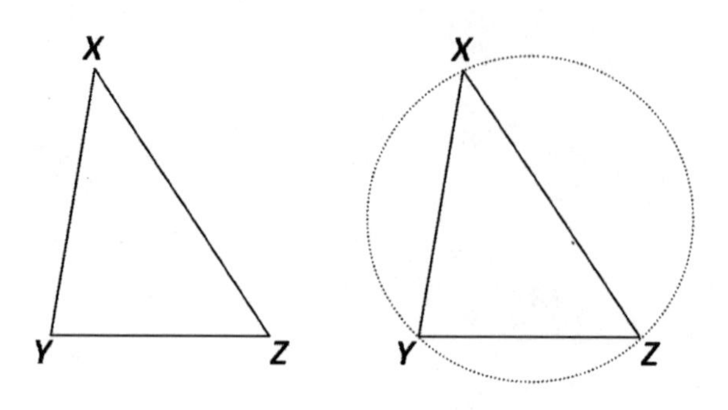

There are three triangles in the diagram: $\Delta DBC$, $\Delta ABC$, and $\Delta ADB$. If I'm to draw a circle, then surely it would be the circumcircle of one of these three triangles.

I create a fresh diagram, and draw the circle passing through the vertices of $\Delta DBC$. It looks terrible – I can't even fit the picture on a single page.

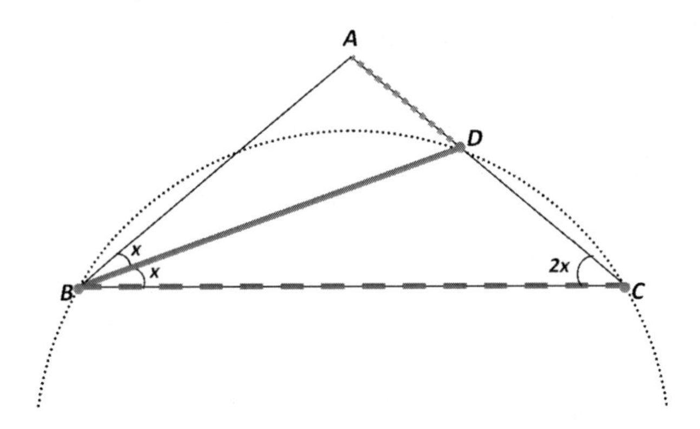


Surely this isn't it. To solve my problem, I need to use a property of line *AD*, and find the measure of angle $\angle A$. My circle has to pass through point *A*.

I try my second triangle. I take out a new sheet of paper, and re-draw the diagram, and construct the circle passing through the vertices of $\Delta ABC$.

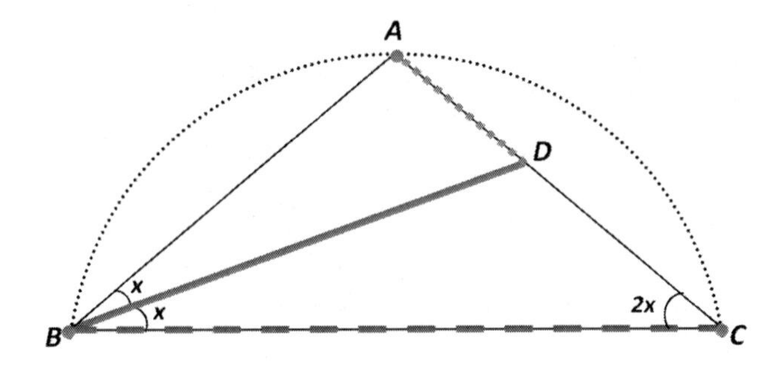


Once again, I draw the diagram too big and can't draw the entire circle.

But it doesn't matter because I realize that this can't be it either. My circle doesn't contain the point *D*, which is definitely important because I need to use the fact that *BD* is an angle bisector.

I've now exhausted every possibility but one. If this case doesn't work, I have no idea what else to do.

I create my last diagram, and draw the circle passing through the vertices of $\Delta ADB$.

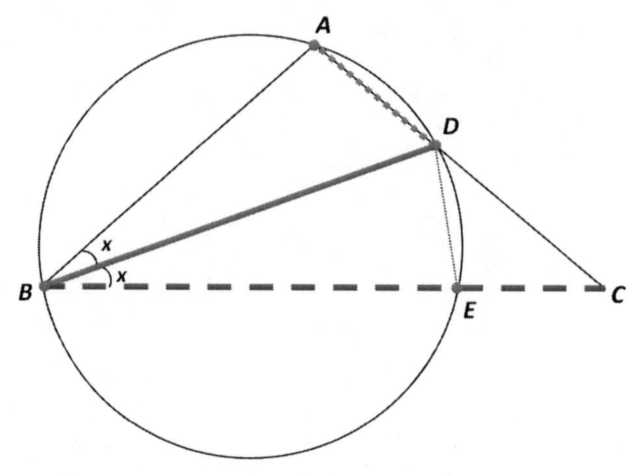


I stare at this new diagram. Noticing that the circle hits the dashed line *BC* at some point *E*, I mark that information down in my diagram.

So instead of a triangle and four points (*A*, *B*, *C*, *D*), my diagram now contains an extra circle, as well as an additional point (*E*).

I shake my head, questioning the wisdom of what I just did. What the heck is the point of drawing a circle?

This must have been a mistake. All I've done is make the problem harder.

*By making it harder, we're making it easier.*

My eyes focus on the diagram in front of me. I know that I have to solve for angle $\angle A$, which is equivalent to angle $\angle BAD$. But because the four points *BADE* form a *cyclic quadrilateral*, I can exploit various properties.

I know that in any cyclic quadrilateral, opposite angles sum to 180°. If I can determine the measure of angle $\angle BED$, I can use the equation $\angle BAD + \angle BED = 180°$ to solve my problem.

I know this will work. I don't think – I know.

I draw in the lines *AE* and *DE*, and use the Butterfly Theorem to show that angles $\angle EAD = \angle EBD = x$ and $\angle DEA = \angle DBA = x$, which implies that $AD = DE$.

From these two equal angles, I see that angle $\angle EDC$ is $2x$, which happens to be the measure of $\angle DCE$. This means that $DE = EC$.

My heart starts to pound, knowing I'm close.

I've now determined that this special point $E$, formed by constructing a seemingly-useless circle and finding its intersection point with the line $BC$, has the special property that $AD = DE = EC$.

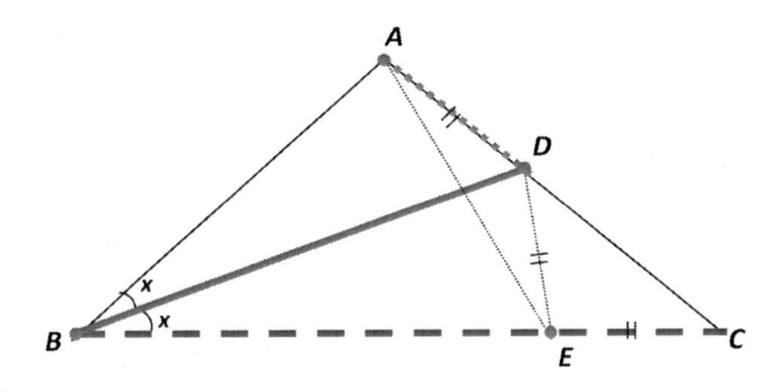


I haven't yet used a critical piece of information given in the question: $AD + BD = BC$. Since I've just determined that $AD = EC$, I develop a chain of logical implications.

$$\begin{aligned} & AD + BD = BC \\ \rightarrow\ & BD = BC - AD \\ \rightarrow\ & BD = BC - EC \\ \rightarrow\ & BD = (BE + EC) - EC \\ \rightarrow\ & BD = BE \end{aligned}$$

Out of nowhere, I've discovered that $BD$ must equal $BE$.

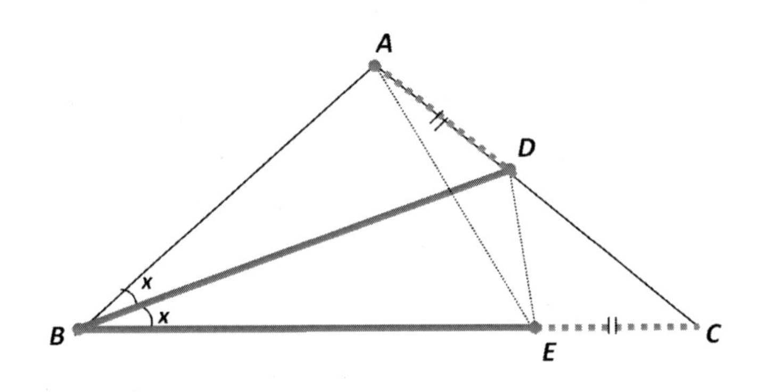

This implies that triangle $\Delta BDE$ must be isosceles, with sides $BD$ and $BE$ of equal length. Thus, $\angle BDE = \angle BED$.

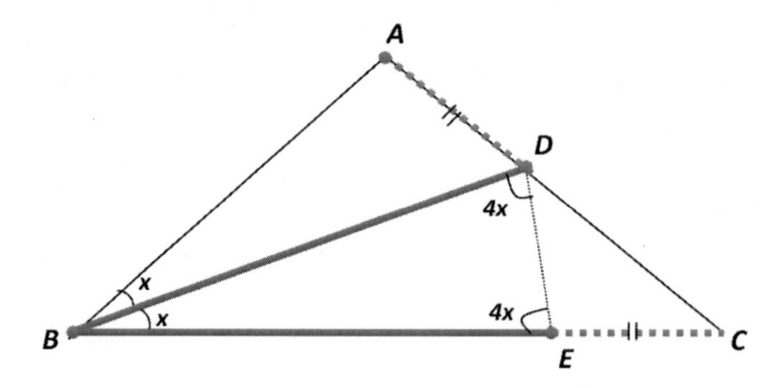


From earlier, I know that $\angle EDC = \angle DCE = 2x$. This forces $\angle DEC = 180° - 4x$, and so this implies that $\angle BED$ has to equal $4x$.

*Yes!*

I know that $\angle BED$ and $\angle BDE$ have to be equal. Since $\angle BED = 4x$, that means $\angle BDE = 4x$ too. Combined with the final angle $\angle DBE = x$, a simple equation enables me to solve for the unknown angle $x$.

$$180° = \angle BED + \angle BDE + \angle EBD = 4x + 4x + x = 9x$$

Therefore, $x = 20°$. Finally, because opposite angles in a cyclic quadrilateral add up to 180°, $\angle BED = 4x = 80°$ forces the measure of $\angle BAD$ to be 100°. Therefore, $\angle A = 100°$.

Alternatively, since $\angle ABC = 2x = 40°$ and $\angle ACB = 2x = 40°$, that implies the third angle of the big triangle, $\angle A$, must be $180° - 40° - 40° = 100°$.

My goal is to find the measure of $\angle A$.

I've done it. The answer is 100°.

I let out a little scream. I'm amazed at how beautiful the solution is: challenging my incorrect assumption that the problem can only be solved by using methods in triangle geometry; drawing a circumcircle to transform it to a problem in circle geometry; making the problem harder in order to make it easier.

As I'm nearing the end of my solution, all those memories of Vancouver begin flooding back. I try hard to block out the thoughts and just concentrate on writing the last few sentences, but find myself fighting a lost cause.

I remember Ray telling me that for life's hardest and most important questions, we can't reason only with our minds. That's why complex questions relating to faith can't just be based on logic and science – it has to go much deeper, and needs to be examined with the heart and the soul.

As I finish writing the solution to the fourth problem of the Canadian Mathematical Olympiad, I finally make the connection to what Ray told me back in Vancouver.

That my journey to the IMO isn't about my mind. It isn't about my ability to answer ridiculously-hard contest questions like this. It's so much deeper than that.

It's about reaching to the core of my being, down to the depths of my soul.

That evening at Vancouver Airport, Grace finally understood the message of Christianity in her heart and was moved to tears because of it. She still has questions about her faith, but that evening, Grace learned that her life journey wasn't just about the knowledge in her head.

Something, or someone, had touched her heart.

And just like Grace, it's at the heart that I will discover my true self.

Not my head, but my heart. Not my mind, but my soul.

And after three thousand hours of training, my heart and my soul are strong.

I've now solved the first four problems of the Canadian Mathematical Olympiad. One more gets me to the IMO.

*11:28 a.m.*

I've got just over thirty minutes. That's enough time – I think.

That's the wrong thought, and I shake my head.

*I have enough time. I don't think, I know – I'm gonna get this next problem.*

I take a deep breath, and open the envelope containing Problem #5.

**Problem Number:** 4
**Contestant Name:** Bethany MacDonald

We will prove that $\angle A = 100°$.

Let $D$ be the point on $AC$ so that $BD$ is the angle bisector of $\angle B$. We are given that $BC = BD + AD$.

There is a unique circle passing through $A$, $B$, and $D$. Let this circumcircle intersect line $BC$ at point $E$. Then $ABED$ is a cyclic quadrilateral. Then construct line segments $AE$ and $DE$, as illustrated:

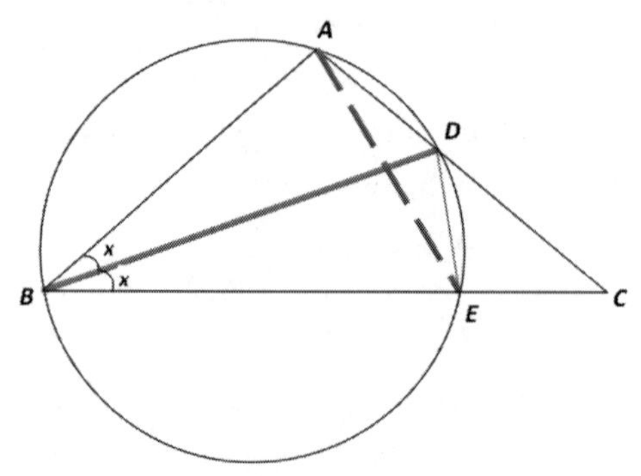


Let $\angle B = \angle C = 2x$. Since $BD$ is the angle bisector, $\angle ABD = \angle EBD = x$.

Since $ABED$ is a cyclic quadrilateral, $\angle EAD = \angle EBD = x$ and $\angle AED = \angle ABD = x$. Therefore, $\Delta AED$ is isosceles with $ED = AD$ and $\angle AED = \angle EAD = x$. Thus, $\angle EDC = 180° - \angle ADE = 180° - (180° - 2x) = 2x$.

Since $\angle EDC = 2x$ and $\angle DCE = \angle C = 2x$, $\Delta EDC$ is isosceles with $ED = EC$, and $\angle BED = 180° - \angle DEC = 4x$. In the previous paragraph we showed that $ED = AD$, and so we must have $AD = EC$.

We are given that $BC = BD + AD$. Since $BC = BE + EC$ by definition, this implies that $BD + AD = BE + EC$. Since we just established that $AD = EC$, we have shown that $BD = BE$, i.e., $\angle BED = \angle BDE$.

From above, $\angle BED = 4x$, and so $\angle BDE = 4x$ as well. Since the angles of $\Delta BED$ sum to 180°, we have $180° = \angle BED + \angle BDE + \angle EBD = 4x + 4x + x = 9x$, implying that $x = 20°$ and $\angle BED = 4x = 80°$.

Since $ABED$ is a cyclic quadrilateral, we conclude that $\angle A = \angle BAD = 180° - \angle BED = 180° - 80° = 100°$.

Our proof is complete.

## The Canadian Mathematical Olympiad, Problem #5

Let $m$ be a positive integer.

Define the sequence $x_0, x_1, x_2, \ldots$ by $x_0 = 0$, $x_1 = m$, and $x_{n+1} = m^2 x_n - x_{n-1}$ for $n = 1, 2, 3 \ldots$.

Prove that an ordered pair $(a, b)$ of non-negative integers, with $a \leq b$, gives a solution to the equation

$$\frac{a^2 + b^2}{ab + 1} = m^2$$

if and only if $(a, b)$ is of the form $(x_n, x_{n+1})$ for some $n \geq 0$.

# Problem #5: Sequence of Integers

I re-read the question, and read it yet again, sentence by sentence.

After the third time, it clicks. I know what's being asked.

*11:30 a.m.*

To get a feel for the problem, I decide to try the simple case $m = 2$, to see if I can find any interesting patterns.

Letting $m = 2$, I have the sequence of integers $x_0, x_1, x_2, \ldots$ whose initial values are $x_0 = 0$ and $x_1 = 2$. All other terms are defined by the recurrence relation $x_{n+1} = 4x_n - x_{n-1}$, where each term is a "linear combination" of the two previous terms.

Substituting $n = 1$, I get $x_2 = 4x_1 - x_0 = 4 \times 2 - 0 = 8$, so $x_2 = 8$.

Substituting $n = 2$, I get $x_3 = 4x_2 - x_1 = 4 \times 8 - 2 = 30$, so $x_3 = 30$.

Substituting $n = 3$, I get $x_4 = 4x_3 - x_2 = 4 \times 30 - 8 = 112$, so $x_4 = 112$.

So my sequence $x_0, x_1, x_2, x_3, x_4 \ldots$ starts with 0, 2, 8, 30, 112.

I check the problem statement, and see that each pair $(a, b) = (x_n, x_{n+1})$ of consecutive terms in my sequence has to satisfy the equation $\frac{a^2+b^2}{ab+1} = 4$.

That means (0, 2), (2, 8), (8, 30), and (30, 112) all need to satisfy this equation. And indeed they do:

$$\frac{0^2+2^2}{0\times2+1} = 4\,, \quad \frac{2^2+8^2}{2\times8+1} = 4\,, \quad \frac{8^2+30^2}{8\times30+1} = 4\,, \quad \frac{30^2+112^2}{30\times112+1} = 4$$

So there are many solutions $(a, b)$ to the equation $\frac{a^2+b^2}{ab+1} = 4$, including (0,2), (2,8), (8,30), and (30, 112). All have the form $(a, b) = (x_n, x_{n+1})$ for some non-negative integer $n$.

I see the expression "if and only if" in the problem statement, and realize I have to prove two directions:

Direction 1: If $(a, b) = (x_n, x_{n+1})$ for some integer $n \geq 0$, then we must have $\frac{a^2+b^2}{ab+1} = 4$.

Direction 2: If $(a, b)$ satisfies $\frac{a^2+b^2}{ab+1} = 4$, then $(a, b)$ must be of the form $(x_n, x_{n+1})$ for some integer $n \geq 0$.

It's not enough to show that each pair in the infinite sequence $(x_0, x_1), (x_1, x_2), (x_2, x_3), \ldots$ satisfies the equation $\frac{a^2+b^2}{ab+1} = 4$. I also need to prove that *no* other solutions exist. I have to first justify that the sequence of pairs beginning with (0, 2), (2, 8), (8, 30), (30, 112) are all valid solutions to the equation. But then I need to prove that these are the only solutions that will satisfy this equation.

I know that the statement "if and only if" requires proving two directions: sufficiency and necessity. Being born and raised in Nova Scotia is a *sufficient* condition for being a Maritimer, but not a *necessary* condition as New Brunswick and PEI are also part of the Maritimes. Having a license is a *necessary* condition to drive around in a Ferrari, but it's not *sufficient* – since most people can't afford a luxury car.

I've seen lots of examples of "if and only if" statements in math, such as:

$$p \text{ is an even prime number if and only if } p = 2$$

$p = 2$ is definitely an even prime (sufficiency); furthermore, $p = 2$ is the only even prime (necessity). So for $p$ to be an even prime number, it is both necessary and sufficient that $p = 2$.

In this problem, I have to prove that $(a, b) = (x_n, x_{n+1})$ is a necessary and sufficient condition for $(a, b)$ to be a solution to the equation $\frac{a^2+b^2}{ab+1} = 4$.

I start by trying to prove that the condition is *sufficient*. For this direction, I need to show that all pairs $(a, b) = (x_0, x_1), (x_1, x_2), (x_2, x_3), (x_3, x_4), \ldots$ satisfy the equation. I have a hunch that "mathematical induction" will work, a technique Mr. Collins taught me a long time ago.

Induction starts by proving the first pair $(a, b) = (x_0, x_1)$ satisfies the equation. And then we show that for any $k \geq 1$, *if* the pair $(a, b) = (x_{k-1}, x_k)$ satisfies the equation, *then* the next pair $(a, b) = (x_k, x_{k+1})$ must too.

Mr. Collins compared induction to the domino effect: if the dominoes are positioned close to each other so that if the $k^{\text{th}}$ domino falls then the $(k+1)^{\text{th}}$ domino must fall, then the only thing you need to do is to push the first domino. Then this guarantees that *all* the dominoes must fall.

Using the recurrence relation $x_{k+1} = 4x_k - x_{k-1}$ and a bit of algebra, I quickly prove this logical statement:

$$\frac{x_{k-1}{}^2+x_k{}^2}{x_{k-1}x_k+1} = 4 \text{ implies that } \frac{x_k{}^2+x_{k+1}{}^2}{x_k x_{k+1}+1} = 4$$

So if $(x_{k-1}, x_k)$ satisfies the equation, then $(x_k, x_{k+1})$ must too. This logical implication is true for all values of $k$. Then all I need to do is push the first domino, representing the base case $k = 1$.

I see this is clear, since $\frac{x_0{}^2+x_1{}^2}{x_0x_1+1} = \frac{0^2+2^2}{0\cdot 2+1} = 4$. So $(x_0, x_1)$ satisfies the equation, implying that $(x_1, x_2)$ does too, implying that$(x_2, x_3)$ does too, and so on. All the dominoes have to fall.

Using induction, I have a proof for the *sufficiency* direction:

If $(a, b) = (x_n, x_{n+1})$ for some $n \geq 0$, then we must have $\frac{a^2+b^2}{ab+1} = 4$.

I'm half done! But now I need to prove the other direction, that this condition is also *necessary*.

If $(a, b)$ satisfies $\frac{a^2+b^2}{ab+1} = 4$, then $(a, b)$ must be of the form $(x_n, x_{n+1})$ for some value of $n \geq 0$.

And this looks really hard. I take a deep breath and close my eyes, hoping for some insight.

I simplify the equation $\frac{a^2+b^2}{ab+1} = 4$, re-writing it as $a^2 + b^2 = 4ab + 4$, and use some algebra to show that this is equivalent to $(b - 2a)^2 - 3a^2 = 4$.

This looks familiar – it's a certain type of equation in Number Theory whose name I can't recall. I don't remember the technique for how to solve the equation – it has something to do with combining square roots, but the details are all fuzzy.

During my three thousand hours of preparation for this day, Grace and I did hundreds of old contests and then, over Skype, we studied the solutions to all the problems we couldn't get. So I know there's an elegant technique for solving equations of the form $x^2 - 3y^2 = 4$. I know it.

I close my eyes and realize it's called "Pell's Equation". But I can't recall the method.

*Think, Bethany!*

After a couple of minutes, I realize it's no use.

*11:35 a.m.*

Okay, forget Pell's Equation. There's got to be another way to solve this. I only have twenty-five minutes left.

What else can I try? I tap my pen on the table until another idea comes to mind.

I re-write the equation $a^2 + b^2 = 4ab + 4$ as a function of a single variable, $b$.

$$b^2 - 4ab + (a^2 - 4) = 0$$

This is good. This is a *quadratic equation.* I can solve for $b$, using a formula that I know by heart:

$$b = 2a \pm \sqrt{3a^2 + 4}$$

Ugh. This goes back to ugly square roots and Pell's Equation, and I don't want that. I decide not to solve the quadratic equation.

I have another idea. If I pretend that $a$ is some fixed number that doesn't change, then I can consider a simple quadratic equation in terms of $x$:

$$x^2 - 4ax + (a^2 - 4) = 0$$

And this quadratic equation *must* have $b$ as a root, as well as some other value $c$.

For example, if $a = 8$, the quadratic equation is $x^2 - 32x + 60 = 0$, and I can quickly see that there are two solutions: $x = 30$ and $x = 2$. These numbers are familiar: both (2, 8) and (8, 30) are solutions to my original equation, found when I toppled the dominoes in my induction proof.

I see what I need to prove. I have to show that if $a$ is *not* one of the numbers appearing in the sequence 0, 2, 8, 30, 112 . . . then the quadratic equation's roots are *not* integers.

For example, if $a = 3$, then $x^2 - 4ax + (a^2 - 4) = 0$ simplifies to $x^2 - 12x + 5 = 0$, and I can see that the roots are ugly expressions involving square roots.

From my induction proof with the dominoes, I know that if $a$ is any number in {0, 2, 8, 30, 112 . . .}, then my quadratic $x^2 - 4ax + (a^2 - 4)$ has integer roots.

If $a = 0$, then my quadratic is $x^2 - 0x - 4 = (x + 2)(x - 2)$, with roots -2 and 2.

If $a = 2$, then my quadratic is $x^2 - 8x + 0 = (x - 0)(x - 8)$, with roots 0 and 8.

If $a = 8$, then my quadratic is $x^2 - 32x + 60 = (x - 2)(x - 30)$, with roots 2 and 30.

If $a = 30$, then my quadratic is $x^2 - 120x + 896 = (x - 8)(x - 112)$, with roots 8 and 112.

I see that this pattern continues: for each number $a$ in my sequence, the two roots of the quadratic are the number before $a$ and the number after $a$. For example, if $a = 30$, then my two roots are 8 and 112, the numbers before and after 30 in the sequence 0, 2, **8**, 30, **112** . . .

So my challenge is to show that if $a$ is not one of these numbers, then the roots are not integers. But I have no idea how to do this.

I think of every technique I know, but nothing comes to mind.

*11:41 a.m.*

With less than twenty minutes remaining in the Canadian Math Olympiad, I'm stuck.

I'm so close, yet so far. After three thousand hours, training and committing my entire life to this one goal, I fall just one question short – half a question short – of my Olympic dream.

And to be stumped on a Number Theory problem involving a sequence of integers. Unbelievable.

*Come on, Bethany – it's a Number Theory problem involving a sequence of integers.*

If anyone can solve this, it's me.

But how?

# 37

"Time's up."

I tossed my pen down on to the table, and leaned my head back, disgusted at myself.

The young student-teacher came by my desk and picked up my exam booklet, containing my solutions to the Canadian Open Math Challenge (COMC), including the blank pages to all the problems I couldn't answer. He walked towards Gillian and collected her booklet, and told us that we were free to go off to lunch.

I was in no mood to eat. I just sat there, stone-faced, wondering how I could have messed up so badly.

"How'd you do?" asked Gillian.

I pretended not to hear her, and instead, closed my eyes and tried to process what had just happened.

All the rooms at Sydney High were being used, so Gillian and I had to write the COMC exam in the library, cramped together in a tucked-away corner, sharing one long table. Patrick was supposed to write the contest as well, but he had caught a bad cold and was resting at home this morning. Maybe Gillian had caught some of the germs going around school, since she was sniffling throughout the contest, hundreds of times, during the past two and a half hours.

Gillian's sniffling was breaking my train of thinking, and I found myself unable to concentrate on the hard questions at the end of the contest. I wanted to tell Gillian to go grab a tissue, but I knew that no talking was allowed, and besides, if Gillian knew that her sniffling was getting on my nerves, she would have been even more annoying.

But I couldn't use Gillian's sniffling as an excuse. I had trained so hard to prepare myself for this, my fourth and final Canadian Open Math Challenge. And I choked.

Fifteen months ago, during that amazing week in Vancouver, Grace and I made our 3000-hour pact. And we'd stuck to our promise: every weekday, we woke up at 6:00 a.m. to devote two full hours to Olympiad training before school, and we ended each day by doing an hour of IMO prep after finishing our homework. On weekends we did even more, and shared our solutions and ideas via Skype. We tracked our progress and realized that we had hit the 2000-hour mark in early September.

We were now in Grade 12, and so this was our last chance to make the IMO team. The Canadian Open Math Challenge, written by nearly five thousand students across the country, was the qualifying exam for the Canadian Math Olympiad (CMO). Only the top fifty students were invited to write the CMO.

Of course, not all five thousand students were going for the IMO, but for those of us who had a realistic shot at the team, the COMC was hugely important. And I blew it. If I didn't make the top fifty, I wouldn't be invited to write the CMO. If I didn't qualify for the CMO, then of course they wouldn't select me for the IMO.

I always did well on practice contests, but maybe that was because I always did the practice contests alone, in a quiet home, at the kitchen table after breakfast while Mom was still sleeping. But still, the types of problems were the same. I would get perfect, or close-to-perfect, on all the practice contests. But when it came to the actual contest, why did I always choke?

I opened my eyes and saw Gillian standing a few feet away. She stared at me.

"How many did you get?"

I leaned my head back and closed my eyes again.

The same thing happened last year, in Grade 11. In every contest I wrote, the same story unfolded each time. I would always get stuck on a few problems during the exam, only to figure all of them out right after the contest ended.

Last year, I was one question away from qualifying for the CMO, just missing the top fifty. And I was also one question away from becoming the provincial champion of the Grade 11 Waterloo contest, which meant that I missed out on the Waterloo Seminar yet again.

Grace was the British Columbia champion of last year's Grade 11 contest, so she made the Seminar. And not only did Grace qualify for the CMO, she had an amazing performance, with one of the top scores in the country. Grace later found out that she was the seventh ranked mathlete in Canada, meaning she missed out on the IMO by just one spot.

When we trained together, we were at the same level. We were mathematical equals, in every topic. We loved the joy of discovery, struggling with hard problems and having a massive rush whenever we saw a key insight or found a key pattern, and watched a seemingly-insurmountable problem crumble before our eyes. Grace and I loved the process of Olympiad training, knowing what we enjoyed doing the most was helping us prepare for the goal that both of us longed for.

But I hated contests, which gave me so much stress and anxiety. Grace could just ignore all of that and perform above her expectations. But contests had the opposite effect on me. Sure, I got perfect on my very first math contest, way back in Grade 7, and I did awesome in the Math League, but that was a team event. On individual contests, I just couldn't do it. The pressure was just too much.

I opened my eyes and saw Gillian looking at my copy of the COMC contest, lying face-up on the table.

The Canadian Open Math Challenge consisted of twelve problems, with four "easy", four "medium", and four "hard". Gillian saw my blue check-marks next to the first nine problems. I had made partial progress on the last three problems, and would get part marks for sure, but couldn't finish any of them.

It was embarrassing that I could only solve nine of the twelve questions – the same number as I got in the fall of my Grade 10 year, right when I started at Sydney High: even before I heard about the Nova Scotia Math League. I had trained for two thousand hours, and couldn't beat a result from two years ago.

As she looked at my copy of the COMC contest, Gillian's face registered both surprise and delight. Her look of shock was quickly replaced by a knowing smile.

"I got the same number of questions as you."

I ignored her and started putting my stuff back in my backpack, ready to leave. I didn't need anyone, especially Gillian, rubbing it in any further. She pressed.

"Hey, aren't you going to the Math Olympics this year?"

My face flushed, and I stood up. I knew that there were other people milling around in the library so I didn't want to make a scene. I kept my voice to a whisper.

"What's your problem, Gillian?"

"Oh, nothing," she replied, coolly. "It's just that you announced to the whole world that you were going to be an Olympian in math."

"Yeah, so?" I said, cringing at the memory of that front-page article in the Cape Breton Post two summers ago, where I did indeed 'guarantee' that Grace Wong and I would become Math Olympians in our Grade 12 year.

"Well, if you're so smart, why didn't you get more than nine questions? I mean, I got nine questions, and I haven't been wasting thousands of hours 'training' with my imaginary best friend in Vancouver. You're a total loser, Bethany."

"Look who's talking."

"Is that your best comeback?" asked Gillian, looking smug. "In case you need to be reminded, I have the highest grade-point-average at Sydney High, not you. When we graduate from here, they're going to choose me as the class valedictorian, not you. And I will be getting the President's Scholarship to St. FX University, not you."

"Yeah, so?" I retorted, racking my brain for something better to say.

Gillian continued: "And also, maybe you remember that in Grade 10, I was the provincial champion for the Nova Scotia High School Math League."

Ironically, that was the one day when Gillian and I actually got along.

I snorted. "You were our team's replacement for Logan, and you know it."

"Maybe," said Gillian, stone-faced. "But the St. FX scholarship committee doesn't. All they see is that Gillian Lowell was on the provincial championship team in Grade 10. And in Grade 11, when Sydney High chose to send you, Tommy, Patrick, and Logan to the provincial finals, the team came in second."

I bit my lip, remembering the provincial math league championship last year, when Halifax Prep finished the final relay five seconds before we did, and won the provincials by one point.

"And that's why I'm going to get the $32,000 President's Scholarship to St. FX University. On the other hand, you risked everything on a stupid dream, trying to be an Olympian in *math*."

I glared at her.

She laughed and shook her head. "Bethany, get a life."

"At least I have dreams, Gillian."

"So do I," she replied, but with a slight hesitation to her voice.

I suddenly remembered that evening in the Halifax hotel room the day before the Math League Championship, when I overheard Gillian on the phone, yelling at her mother for pushing her about some scholarship, and that she didn't care what her older sisters had accomplished.

"And is the President's Scholarship your dream, or your Mommy's dream?"

Now it was Gillian's face that flushed.

"It's mine," replied Gillian.

"Is it really?" I asked. "It's not because you want to please Mommy and finally earn her love."

"Go to hell, Bethany."

The librarian walked over to us and asked us to leave. As we turned around, we saw twenty people staring at us.

Gillian marched towards the exit, while muttering something under her breath, and ignoring everyone who stared at her as she left the library and slammed the door behind her.

It was just after midday, so I was sure Bonnie and Breanna were eating lunch in the cafeteria; I had asked them to save me a seat. Just as I turned the corner towards the cafeteria, I ran into Mr. Marshall walking in the other direction. He stopped me and asked how I was doing.

"Not good. I absolutely bombed the COMC."

Mr. Marshall had a look of sympathy on his face, the same sense of concern and care I saw after Sydney High missed a second consecutive provincial Math League Championship by five seconds.

"I am sorry, Bethany. I know how hard you trained."

"I can't do this anymore," I said. "It's too much."

Mr. Marshall silently motioned for me to walk with him towards the Math Department office. We went around the crowd of people heading towards the cafeteria and walked into his office, pulling up two chairs.

"It's too hard, Mr. Marshall. I just feel so much pressure."

"Pressure from who?"

"From everyone! Ever since that stupid newspaper article came out, I've screwed up contest after contest. But people in Cape Breton don't know that – they think of me as some math genius destined for the Olympics – and you have no idea how much pressure that puts on me!"

"Please, Bethany."

"Just last weekend, I was hanging out with Bonnie and Breanna, and some stranger approached me and told me he was rooting for me this year. Rooting for me! And what happens if I don't make the team? What happens when I don't make the team? I'll be letting down all of Cape Breton!"

"Stop, Bethany. Please, stop. You have extraordinarily high expectations of yourself, and I applaud that, but no seventeen-year-old should feel the weight of an entire community on her shoulders. I don't need to tell Lucy MacDonald's daughter how unfair it is for anyone to feel that kind of pressure."

I nodded, and took a deep breath before answering.

"What do I do?"

"Take a break from math contests, and come back only when it becomes fun again. And the same advice holds for the Math League. Perhaps you should take a break from that too, and come back only when you no longer feel pressure or stress."

"I don't feel pressure with the Math League. I love running the practice sessions. I'm good at it."

"Indeed you are, Bethany. You are a natural teacher, and I appreciate how you have taken those Grade 10 students under your wing. Thank you for taking the initiative with them."

"You're welcome," I replied, feeling a bit better about myself.

Mr. Marshall handed over a thick stack of pages marked "Grade 12 Pre-Calculus".

"By the way, here are today's notes."

Mr. Marshall handed over the notes from the class that Gillian and I missed this morning because of the contest. It was the only class I had without Bonnie and Breanna, and I was looking forward to catching up with them over lunch. Just as I was leaving, I looked back at Mr. Marshall.

"What's the President's Scholarship?"

Mr. Marshall seemed surprised by the question.

"It's the big entrance scholarship to St. Francis Xavier University in Antigonish. Why do you ask?"

"How do I apply for it?"

Mr. Marshall raised an eyebrow. "Patrick says you have already decided to move to the west coast. He said you were applying only to one university in British Columbia, and that nothing would change your mind."

"I'm just keeping my options open," I said, shrugging.

"Well, I should tell you that Sydney High School has a special application process, where the school formally nominates one person for the President's Scholarship. Of course, anyone can apply to St. FX University, but the school recommends just one person for the top scholarship."

"Has anyone from Sydney High ever won?" I asked.

Mr. Marshall hesitated. "Yes, a student named Caroline won the President's Scholarship five years ago. And two years before that, her older sister Victoria did as well."

"Sounds like a smart family."

"Indeed," said Mr. Marshall, eyeing me suspiciously. "Both left a tremendous legacy. They each had the highest grade-point average in their graduating years, were chosen as the valedictorian, won one of the most prestigious undergraduate entrance scholarships in all of Canada, and then four years later, graduated from St. FX with full honours – with both of them winning the Rhodes Scholarship to Oxford."

I nodded in understanding.

"Thanks for the chat, Mr. Marshall. I appreciate it."

"You're welcome, Bethany."

I left Mr. Marshall's office and leaned against one of the lockers.

For the first time, I realized that Gillian faced just as much pressure as me, but in a totally different way.

# 38

We drove into the parking lot of the baseball field, right behind the church on the corner of Kings Road and Peacekeepers Way. It was bright and sunny, fifteen degrees, with no rain and light wind.

A perfect day for running.

Mom and I stepped out of the car, and walked towards a group of one hundred people who had gathered for the annual Terry Fox Run in Sydney.

Bonnie and Breanna came over and said hello. They had met Mom a few times before, and were looking forward to going to lunch with us after the five-kilometre run. Mom had already booked a table-for-four at the Olive Tree Pizzeria.

Bonnie turned to Mom. "Are you going to wait for us at the finish line, Ms. MacDonald?"

"Actually, you'll be waiting for me, since I'll be finishing a few minutes after all of you."

Breanna raised an eyebrow. "You're running?"

Mom smiled. "I haven't run in many years, but as Bethany might have told you, I was a pretty good athlete when I was your age."

I glanced over at Mom, so proud of her for doing the Terry Fox Run with me today. Much had changed for her in recent months, especially after she started seeing a therapist recommended by her friend. Ever since the first appointment with her new counsellor, Mom surprised me by doing things she had never done before, like opening up about some events in her past. And I was amazed that she wanted to be involved in the things I was now into, like long-distance running.

I enjoyed the long jogs with Mom, and both of us wondered why we had never done this before. Other than her job, which Mom still hated, she was as happy as I had ever seen her.

While I was a bit faster than Mom, that was only because I had a full year of training from being on the Sydney High Girls' Cross-Country team. But Mom was the more natural athlete, still slim even though she was nearly forty years old. I was sure that after we ran together some more, we'd be running at the same pace.

Running with Mom wasn't a contest, and never would be. Running was just something I enjoyed doing, as a personal challenge. I was so grateful that I didn't have to compete with anyone on the track, since I had enough of that in my life – whether it was competing for a prestigious university scholarship or competing for a spot on Canada's IMO team.

On second thought, I wasn't competing for the Math Olympiad anymore – I was realistic, and I knew that the IMO dream was pretty much over, especially after I bombed the Canadian Open Math Challenge last week.

I sighed, thinking about my Skype call with Grace later this evening. I was planning to tell her I was stopping the three thousand hours of training, leaving Grace to pursue the Olympiad on her own. After all, the IMO was out of my reach, and so there was no longer any point in me pursuing this dream. Grace and I had done two thousand hours already – but the final one-third would have to be done by Grace alone.

I knew Grace would be disappointed to hear my news, but I knew she'd get over it – especially because she was going to be an Olympian; I had no doubt about that. She didn't need me anymore.

Last night, I told Mom that I'd be giving up my quest for the IMO. She was supportive of my decision, but she was also sad for me because she knew how badly I wanted it. But math contests were too stressful and no longer fun, and I had to move on and pursue a new challenge in my life.

I looked towards the main stage, and saw our Phys. Ed. teacher Mr. Campbell talking to someone. Mr. Campbell was the coach of our Girls Cross-Country team, and also one of the main organizers of today's Terry Fox Run. He had insisted that each member of the team participate in today's five-kilometre run, and as I looked around the crowd of one hundred or so people, I saw all my teammates there.

I was happy that Bonnie and Breanna decided to do Cross-Country with me this year, and it didn't matter that we were the three slowest runners on the team. Thankfully, we had Jennifer MacNeil, last year's provincial champion, on our team. No girl in Nova Scotia could keep up with her.

Once the projector was connected, Mr. Campbell took the microphone to welcome all of us. He talked about the significance of today's run in continuing the legacy of Terry Fox, who ran five thousand kilometres across Canada, on one leg, after having his other leg amputated due to cancer. Mr. Campbell mentioned that the Terry Fox Run had raised over $600 million dollars since its inception in 1981, and thanked the Cape Breton community for raising money among our family and friends to contribute to cancer research. He gave a touching tribute to his parents, who both died of cancer, and shared his hope that his children and future grandchildren would live in a cancer-free world.

He turned it over to two ladies, who led us in some fun stretching exercises while hip-hop music played in the background.

Before starting the run, Mr. Campbell motioned towards the big screen and began playing a music video of *Never Give Up on a Dream*, telling us that Rod Stewart wrote the song as a tribute to Terry Fox. He told us to continue stretching our arms and legs, because the run would begin as soon as the song ended.

The music video began with the twenty-one-year-old Canadian dipping his prosthetic leg into the Atlantic Ocean to begin his Marathon of Hope.

I stared at the screen, seeing Terry Fox running by himself, taking two hops on his left leg before taking a step with his artificial right leg, and repeating that three-step sequence again and again and again.

Just like every other high school student in Canada, I knew that Terry Fox ran from Newfoundland to Ontario on one leg, raising awareness for cancer research and pursuing the goal of raising one dollar for every Canadian – a goal that everyone said was crazy and impossible.

He ran thousands of hours – tens of thousands of hours – stopping only when the cancer returned and spread to his lungs. And just before he died in 1982, more than twenty-five million dollars had been donated for cancer research: one dollar for every Canadian.

While I knew Terry Fox's story, I had never actually seen him run. Listening to the inspiring lyrics of the Rod Stewart song, I stared at the screen, amazed at our national hero who never quit running – in the cold, in a snowstorm, in the dark, uphill, on rain-soaked highways, with cars honking and drivers screaming at him to get off the road.

Breanna looked over at me. "Hey, are you okay?"

"I'm fine," I replied, using my arm to wipe a single tear trickling down my right cheek.

Mom looked concerned. "What's wrong?"

"Nothing," I whispered, clenching my fists. "I'm fine."

I blocked out the concerns of Mom and my two friends, and stood mesmerized at the image of Terry Fox on the screen, soaking in the words of the raspy-voiced singer repeating, "no, you never, never gave up on a dream".

Another tear fell out of my eye.

I realized I couldn't stop now. I had come so far, and sacrificed so much to quit now.

The first two thousand hours were hard, and the final one thousand would be even harder, especially if I finished outside the top fifty and failed to qualify for the Canadian Math Olympiad. But no matter what, I needed to keep running. I couldn't stop until the IMO team was officially announced.

No, I'd never give up on my dream.

I closed my eyes, and felt a strange sense of peace and calm.

The music ended and Mr. Campbell started the race.

"Go!"

I was jolted back into reality, as a bunch of people ran past me. I hit my stopwatch and began jogging, catching up to Mom, Breanna, and Bonnie, a few seconds later.

"Hey, are you okay?" asked Bonnie. "What happened there?"

"Yeah, I'm good," I replied, taking a breath between strides. "I'm really good."

The four of us ran, side by side, at a nice and comfortable pace. The race was five laps around a flat one-kilometre course around the baseball field which adjoined a small park.

I glanced at Mom on my left, and over at my two closest friends from Sydney High on my right, and realized how happy I was. For everything.

We finished the first lap. I glanced at my watch.

*6:12.*

Six minutes and twelve seconds. Multiplying by five, I saw that we'd finish the run in exactly thirty-one minutes if we continued this pace. My personal best was just over twenty-eight minutes, and I knew we could pick it up.

"Hey guys, let's run a bit harder," I said.

We picked up the pace. My heart started beating a bit faster, and I was feeling good.

Halfway through the second lap, I glanced over at Mom. "How are you doing?"

"I'm fine," said Mom, taking deep breaths. "Not sure how long I can hold this."

"Let's finish this lap," I said. "Five hundred metres to go. You can do it, Mom."

Less than three minutes later, we were running up to complete the second kilometre. Just as I crossed the start-finish line, I glanced at my watch.

*11:52.*

We did that lap in five minutes and forty seconds. I did a quick calculation: if we continued at that pace for the remaining three laps, we'd finish the five kilometres in 28:52.

I glanced over at the other three. "How are you doing?"

"Sorry," said Mom, heaving gulps of air between sentences. "It's too fast for me. You girls go on ahead. I'll see you at the finish."

"Okay, Mom," I replied, glancing back at her.

I picked up the pace, with Bonnie and Breanna following along.

We were all sweating quite a bit, and running a lot faster than any of us had before. I didn't know exactly how fast we were running, but I could calculate that in just a few moments.

We approached the start-finish line again.

*16:50.*

"We did that lap in five minutes!" I shouted, my voice hoarse.

"Can we slow down?" asked Breanna. "I can't hold this pace."

"Me neither," added Bonnie.

I needed to make a split-second decision. I was feeling too good to slow down.

"I'm gonna run up ahead."

"Fine," said Bonnie, heaving gulps of air. "See you later."

Without looking back, I kept going, advancing my stride until I found a pace that I felt I could hold for the next ten minutes.

I was feeling better by the minute, and was shocked to find myself passing runners, including two of my teammates on the Sydney High Cross-Country team. When they glanced over, I could see how surprised they were to see me – the six-foot broad, who outweighed them by forty pounds – surging past them.

The end of lap four was approaching. I passed the start-finish line.

*21:28.*

I couldn't believe I just ran that lap in four minutes thirty-eight seconds. I'd never run that fast before.

Taking long even breaths and looking straight ahead, I tried to block the pain in my knees and the voice in my head urging me to slow down.

*One more kilometre, I can do it!*

If I could sustain this pace, and run the final lap in four minutes thirty-eight seconds, my final time would be 26:06, destroying my personal best by two minutes.

Ignoring my right calf muscle, which was tightening with every step, I kept pressing ahead.

As I turned the corner and headed towards the finish line, I realized I had only two hundred metres to go. I began pumping my arms as hard as I could.

I saw one of my Sydney High Cross-Country teammates just up ahead, as well as two slim men running just in front of her. Every second, the gap between us got smaller and smaller. With five metres to go, I passed all three of them.

As I crossed the finish line, I hit the button on my stopwatch, let momentum carry me another thirty metres or so, before I collapsed on the ground by the side of the road, my stomach heaving.

I turned over on to my back, stretched my arms over my head, and closed my eyes.

*That was awesome.*

I needed a full minute before I could sit up again, and rose up slowly. I turned my head towards my stopwatch.

*26:00.*

My eyes widened. That was impossible. I couldn't have run five kilometres in twenty-six minutes.

That meant I ran my last lap in 4:32. There was no way I could run that fast.

But I did.

I turned towards the finish line and saw Breanna and Bonnie cross the line together. I gingerly stood up and after stretching my calf muscles, I slowly walked towards them, giving them a high-five.

"Great job," I said.

"Oh man, I am so sore," said Breanna, lying face up on a grassy patch.

Bonnie joined her on the grass. She looked up at me and shook her head. "Holy cow, you were fast today."

"Yeah," I said, "I was really feeling it."

"Hey! Here comes your mom!" said Bonnie, pointing towards the finish line.

I turned around and saw Mom crossing the finish line with a big smile on her face. I ran towards her and gave her a hug.

"Way to go, Mom!"

Mom held on to me, too exhausted to speak. After a minute, she walked over and got high-fives from Bonnie and Breanna.

We shuffled over to the refreshments station and helped ourselves to water, an orange, and a banana.

"I am so tired," said Mom, with her hands on her hips. She took a few deep breaths. "But I feel great. You know, I haven't run five kilometres in almost twenty years – before you were born."

"We're going to enjoy the pizza," I said. "We deserve it. Especially you, Mom."

The four of us sat down on an empty patch of grass, and stretched out. I was still in shock that I had run so fast. Of course, twenty-six minutes was a walk-in-the-park for Jennifer MacNeil, who broke the Nova Scotia record at the provincial championships last year. But for me, it was miraculous.

Tonight I was going to call Grace and tell her about the run, and let her know that the IMO would work out the same way for us – that we'd complete the three thousand hours of training and finish strong.

I got up from my stretching and walked towards the refreshments section to get another orange.

"Bethany."

I looked up and saw Mr. Campbell standing in front of me. He shook my hand.

"Great run, and an outstanding finish."

"Thanks, Mr. Campbell. Today was awesome."

"Nobody on the team finishes as strongly as you do. How fast was your final kilometre?"

"I ran the last lap in 4:32. My overall time was twenty-six minutes on the dot."

"That's over two minutes better than your previous personal best. Way to go."

"Thank you."

Mr. Campbell looked at me. "Bethany, there's something I've been meaning to do. And after what I saw out there today, I'm now absolutely certain."

"What's that?" I asked, a bit nervously.

"I'm naming you the captain of the Girls' Cross-Country team."

I burst out laughing. It took me a few seconds to realize that Mr. Campbell wasn't joking.

"You can't be serious."

"I am completely serious," he replied.

"Come on, Mr. Campbell," I protested. "Look at me. I'm not a runner! I ran as hard as I could and it still took me twenty-six minutes. Jennifer could run twenty-six minutes on one leg."

"That's exactly right, Bethany," he replied. "You ran as hard as you could. And you ran that final lap in 4:32, which means you can run a one-kilometre lap faster than the majority of your teammates. You get more out of your body than anyone else on the team, because you have so much grit. That's why you succeed. I want that grit to rub off on your teammates."

"Thank you, sir. But what about Jennifer? Shouldn't she be the captain?"

"Actually, Jennifer was the one who recommended you be our team captain. She recognizes that she's a natural runner, but not a natural leader – which is the opposite of you."

"She thinks I'm a leader? Why?"

"Because her little brother does the Math League. He says you coordinate the training sessions for all the Sydney High teams, and organize all the practices. He says you're an excellent teacher."

"Thank you," I said, making the connection that Bobby MacNeil in Grade 10 was Jennifer's younger brother.

"So will you be our team captain?"

"Sure," I responded. "What do I do?"

"There are many responsibilities for being the team captain, and I'll explain all that at school. But here's one thing you can start thinking about right away. I need you to meet with each of your teammates before the regional finals, to figure out the right strategy for each runner to beat their personal best time. Each runner has to know how they need to pace themselves throughout the race, and I need you to calculate their ideal split times for each lap, so they can finish strong – just like you always do."

He paused. "There's some math involved, and I hear you're pretty good at that stuff."

I laughed out loud.

"Yes. I'm pretty good at that stuff."

# 39

I shuffled into Grade 12 Pre-Calculus, and took my usual seat in the second row, next to Patrick.

Patrick turned towards me. "How's it going?"

"Fine," I replied, a bit too strongly.

I was feeling tense, since Sydney High would make its decision at the end of this week, formally nominating one Grade 12 student for the $32,000 President's Scholarship to St. Francis Xavier University. Many people from Sydney High were applying to St. FX for their undergraduate studies, but we all knew that only two of us would be considered for the big scholarship.

At $8,000 a year, the scholarship would cover tuition, meals, and housing, and would enable me to go to university without having to take out a massive student loan. The stakes were high.

Out of the corner of my eye, I noticed Gillian walk into the classroom and take her usual seat in the first row. She fiddled with her hair and I could tell she was both nervous and distracted.

Over the past few weeks, I had spent a long time thinking about whether I should consider St. FX. After chatting several times with Mr. Marshall and Mom, as well as with Bonnie and Breanna and Patrick, I realized that I absolutely should apply for the President's Scholarship. Recently, I'd begun having second thoughts about moving all the way to British Columbia for university, and wondered whether staying in Nova Scotia was maybe the right decision for me and my friends.

Grace was angry when we chatted on Skype last night, accusing me of betraying our promise to go to Quest together. I told her that St. FX was just a backup option in case I didn't get a scholarship to Quest, and felt bad lying to my best friend. In an ideal world, Grace and I would both get big scholarships to the same university. But Grace wouldn't entertain the possibility of applying to St. FX too; for her, all of her eggs were in one basket, and she was going to Quest for sure.

Thankfully there was at least a month before the Canadian university application deadline, so I had lots of time to convince Grace to apply to St. FX as well, and then we could decide where to go together. No matter what, we had made a pact to go to the same university and be roommates for four years, and neither of us was breaking that promise.

I heard loud footsteps behind me, and turned around. In walked Mr. Marshall, followed by our principal Mr. Davis, as well as a lady in her fifties who dwarfed both of the men standing in front of her.

The three adults stood at the back of the classroom, and I gazed at the white-haired woman who was at least six-foot-four. She put her hands on her hips so that her thumbs were facing outwards and her elbows were at the same level as the principal's eyes.

I was shocked by the woman's stature and wondered if she had experienced the same kind of bullying I did when she was growing up. As she surveyed the room, our eyes met. She smiled at me, revealing a large gap between her two front teeth.

I liked her right away.

"Hey," whispered Patrick. "It's the guest speaker."

I nodded, remembering Mr. Marshall's comment that we'd have a mathematics professor speaking to our class today. I was annoyed at myself for assuming the speaker would be a man.

I noticed Gillian stand up tall in her chair, looking particularly attentive, especially in the presence of our scary principal Mr. Davis, who had the final say in which one of us got nominated for the President's Scholarship.

Mr. Marshall walked up to the front and got our attention.

"Good morning, everyone. It is my tremendous pleasure to welcome Dr. Anna Weber to class today. Dr. Weber is originally from Germany, and moved to Canada after completing her Ph.D. in mathematics at the University of Berlin. For the past twenty years, Dr. Weber has been a professor in the mathematics department at St. Francis Xavier University in Antigonish, and we are honoured to have her teach us today. Please join me in welcoming Dr. Weber to Sydney High School."

I glanced over at Gillian, who was clapping vigorously. As soon as Mr. Marshall mentioned the speaker's connection to St. FX, I tensed up. I wondered if that was the reason our principal was visiting the class.

"Hello, everyone," said the professor, with a faint European accent. "Thank you for welcoming me here today. Our math department organizes an outreach program where several of us visit high schools each year. And I'm very happy to be here at Sydney High."

Dr. Weber opened up her laptop and projected a Powerpoint presentation onto the main classroom wall.

"You're taking Pre-Calculus this semester, and I'm sure many of you will take Calculus next semester. From the brief chat I had with your teacher Mr. Marshall this morning, you've been learning how to model real-life problems in the language of calculus, and explore its connections to physics and biology, computer science and statistics, engineering and economics."

From her pocket, Dr. Weber took out a little remote control, and used it to advance to her first slide, showing us two pictures: some chemical equations and graphs on the left, and a young woman in a lab coat pointing at a computer screen and explaining something to an older man.

"I have a colleague who specializes in mathematical medicine, where she works with pharmaceutical companies to figure out how drugs decay inside a person's body. She uses calculus to calculate the correct dosage for each drug."

Dr. Weber moved to the next slide, a screenshot of a computer game. She clicked a button to play a fifteen-second clip of players running up and down a soccer field, ending with one player running past two defenders to score a goal.

"I have another colleague who's a consultant for a well-known Canadian video game company, using calculus to model the behaviour and movement of three-dimensional models under rapidly-changing conditions. As a result, his company's games feature characters that are more realistic than any competitor."

Dr. Weber then flipped through five more slides, briefly describing the research of her St. FX colleagues, and how each professor's work involved a real-life application of calculus. Her stories were interesting but I was quickly losing my focus. My thoughts were on the scholarship.

Her next slide was a simple picture, of a familiar pattern.

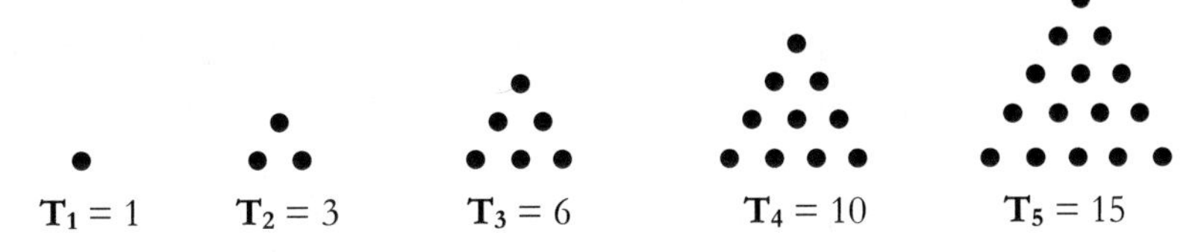

"These are known as *triangular numbers*. The first five triangular numbers are 1, 3, 6, 10, and 15. What's the pattern? Can anyone tell me the next number in this list?"

A bunch of hands went up, including mine. Dr. Weber pointed to Gillian, who raised her hand first.

"The answer is twenty-one, Dr. Weber."

"Correct," said the professor, looking at Gillian. "And what is the pattern?"

"The sixth triangular number is the sum of the first six positive integers. That's 1+2+3+4+5+6, which equals 21."

"Thank you," said Dr. Weber. "Now this particular sequence of triangular numbers has nothing to do with calculus. But this is what I do. More generally, I'm known as a *number theorist*, and in my research, I study the properties of integers.

"Before I tell you about my research, and how it connects to applications such as internet cryptography, I want you to do some number theory. Look at this sequence of triangular numbers and find one interesting property. In fact, take a few minutes on your own, and tell me something surprising."

I tapped my pen a few times, looking at this sequence of numbers to see whether I could discover anything that Dr. Weber would deem "surprising".

My eyes locked on to the last picture in her slideshow.

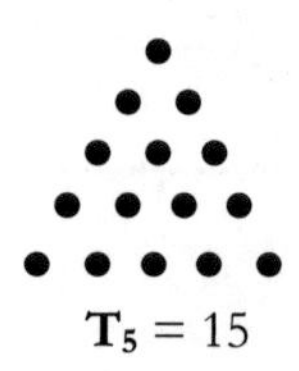

$\mathbf{T_5} = 15$

There were 15 dots in the picture, with 1 being the first digit and 5 being the second digit. And it just so happened that there was 1 dot in the top row and 5 dots in the bottom row.

In other words, $\mathbf{1} + 2 + 3 + 4 + \mathbf{5} = \mathbf{15}$. That was cool.

I wondered if there were any other triangular numbers that had that property, where the sum of the first *n* positive integers equalled the two-digit number 1*n*.

I quickly saw that the answer to my question was no, since the next triangular number was 21, which meant that there were no other two-digit triangular numbers starting with the digit 1.

Gillian just told us the sixth triangular number. In my notepad, I wrote down the seventh:

$$1 + 2 + 3 + 4 + 5 + 6 + 7 = 28$$

After staring at the equation for a few seconds, I saw something.

Subtracting 1 from both sides, we get $2 + 3 + 4 + 5 + 6 + 7 = 27$, which means that the sum of the integers from 2 to 7 adds up to the two-digit number 27.

That was definitely surprising. It's the same property as before: there were 2 dots in the top row and 7 dots in the bottom row, and the total number of dots was 27. I scribbled the idea in my notepad.

$$\mathbf{1} + 2 + 3 + 4 + \mathbf{5} = \mathbf{15}$$

$$\mathbf{2} + 3 + 4 + 5 + 6 + \mathbf{7} = \mathbf{27}$$

When does $x + \cdots + y$ equal $x$ stitched together with $y$ ?

I noticed Dr. Weber pacing around the room, passing by me before she walked back to the front of the classroom.

"Okay, let me have your attention. I'd like to invite you to share any ideas if you have them."

She pointed to Patrick. "Let's start with the young man over here."

Patrick nodded and slowly walked up to the front. He wrote down the first five triangular numbers on the board: 1, 3, 6, 10, 15.

He turned to the class, nervously looking at the back towards his father and the principal.

"If you add up any pair of consecutive triangular numbers, it looks like you always get a perfect square."

$1+3 = 4 = \mathbf{2^2}$     $3+6 = 9 = \mathbf{3^2}$     $6+10 = 16 = \mathbf{4^2}$     $10+15 = 25 = \mathbf{5^2}$

"That's excellent," said Dr. Weber. "Can you prove that for us?"

"Yes, I can," he replied. "I found a formula for the $n^{\text{th}}$ triangular number, $T_n = \frac{n(n+1)}{2}$. You can use this formula to prove the result."

I smiled, knowing that Patrick and I solved a similar problem during last year's Math League. He wrote down a short equation, demonstrating that the sum of any two consecutive triangular numbers must be a perfect square.

$$T_{n-1} + T_n = \frac{(n-1)n}{2} + \frac{n(n+1)}{2} = \frac{n^2 - n + n^2 + n}{2} = \frac{2n^2}{2} = n^2$$

"I can prove that differently," said Gillian, not bothering to raise her hand.

"Okay," said Dr. Weber enthusiastically. "Come up and show us."

With pomp and flourish, Gillian grabbed the chalk from Patrick and began drawing *staircases*, showing how consecutive triangular numbers could be joined together to form a perfect square, producing a visual proof of Patrick's result.

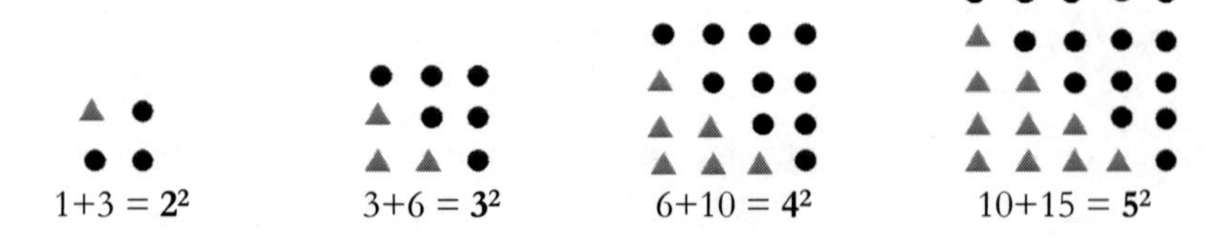


Gillian pointed to the picture on the very right, representing the picture $10+15 = 5^2$. She explained that the first staircase, made up of triangles, represented the sum $1+2+3+4 = 10$, and the second staircase, made up of circles, can be flipped upside down, producing the sum $1+2+3+4+5 = 15$. By joining the two staircases, Gillian formed a $5 \times 5$ square, showing that $(1+2+3+4) + (1+2+3+4+5) = 5^2$.

"That's a beautiful proof," said the professor, flashing her gap-toothed grin and nodding in approval. "What's your name?"

"Gillian Lowell," she replied, beaming.

"And where did you come up with the idea of using a staircase?"

"It's not that hard," said Gillian, with a dismissive wave of her hand. She moved three steps to the right, putting herself in a spot right between the professor and me.

"Hey, didn't she get that from you?" whispered Patrick, leaning towards me.

I ignored Patrick's comment, and glared at Gillian. She calmly returned to her seat, refusing to look in my direction, though my eyes followed Gillian all the way.

"Young lady," said Dr. Weber.

I was surprised to see the professor looking down at me.

"Yes?" I asked.

"Is there something you would like to show us – perhaps an idea or pattern that would build upon what Gillian just did?"

I hesitated and looked over at Gillian, who stared back at me defiantly.

Her eyes and face said it all: *you think you're better than me, Bethany – but you're not.*

"Yes, Dr. Weber. I've got something," I said, standing up.

As I walked up to the front of the classroom, I could see Mr. Marshall lean over towards Mr. Davis, and whisper something in his ear. The principal nodded in understanding, and his blank expression turned into a broad grin.

Before I could process the meaning of that exchange, Dr. Weber handed me a piece of chalk, and I was taken aback by how much she towered over me.

I wrote down an equation and turned my body towards her.

$$\mathbf{1} + 2 + 3 + 4 + \mathbf{5} = \mathbf{15}$$

"The fifth triangular number, fifteen, has a cool property."

"And what is that?" asked Dr. Weber.

"That if you add up the numbers from one to five, the answer matches those two digits: fifteen."

I wrote another equation on the board: **2** + 3 + 4 + 5 + 6 + **7** = **27**.

"And the same thing happens if you add up the numbers from two to seven, even though it's not really a triangular number. The answer is twenty-seven. I'm curious if there are any other pairs of numbers when this happens – that if you add up all the numbers between $x$ and $y$, the answer is the number produced by stitching $x$ and $y$ together."

"What an interesting question," said Dr. Weber. "What is your name?"

"Bethany MacDonald," I said, standing up straight.

"So, Bethany," said the professor, staring at the board. "Maybe you can answer your own question. Suppose $x$ and $y$ are one-digit numbers. What are all the possible values for $x$ and $y$ that have this special property?"

I raised my eyebrow. "You want me to tell you right now?"

"Oh, I'm sorry," said Dr. Weber. "I shouldn't put you on the spot like this. Perhaps we could . . ."

"I can do it," I said, grabbing a piece of chalk from the board.

$$x + (x + 1) + (x + 2) + \cdots + (y - 1) + y = 10x + y$$

Ignoring the class, I wrote down an equation and looked at Dr. Weber.

"Suppose $x$ and $y$ are one-digit numbers that have this special property. Then the sum of all the integers between $x$ and $y$ must be the number formed by stitching $x$ and $y$ together. So if $x$=2 and $y$=7, then the stitched two-digit number is $10x + y$, which works out to 20+7 = 27."

With the professor nodding in encouragement, I simplified the left side of my equation.

$$\begin{aligned} & x + (x + 1) + (x + 2) + \cdots + (y - 1) + y \\ &= [1 + 2 + \cdots + (x - 1) + x + \cdots + y] - [1 + 2 + \cdots + (x - 1)\,] \\ &= [1 + 2 + \cdots + y] - [1 + 2 + \cdots + (x - 1)] \\ &= \frac{y(y+1)}{2} - \frac{(x-1)x}{2} \end{aligned}$$

"The left side is the sum of the numbers from 1 to $y$, minus the sum of the numbers from 1 to $x$-1. This has to equal the right side, which is $10x+y$."

"Correct," said Dr. Weber, pointing to the left side of my equation. "The same formula that Patrick just showed us, which you can also derive visually using Gillian's Staircase."

"Yes," I said, biting my lip. "By Gillian's Staircase."

Dr. Weber pointed to the equation I had written down.

$$\frac{y(y+1)}{2} - \frac{(x-1)x}{2} = 10x + y$$

"Can you solve this? Or should we turn it over to one of your classmates?"

"I've got this."

I grabbed the chalk and began simplifying this equation, getting into a zone and becoming completely unaware of anyone else in the class. It was just the professor and me. Within a couple of minutes, I produced an equivalent equation that I knew how to solve, and looked up at her triumphantly.

$$(9 + x + y)(10 + x - y) = 90$$

"Excellent. Let me stop you there. Do all of you follow how Bethany got to this point?"

I looked around, and saw a few people, including Patrick, nod their heads in agreement. Gillian was scowling with her arms folded across her chest. I glanced at the two teachers in the back. Mr. Davis had a proud look on his face, while Mr. Marshall nodded and gave me a silent thumbs-up.

Dr. Weber invited me to continue my solution.

"I've shown that my original equation is equivalent to this one, where we have two terms, $9 + x + y$, and $10 + x - y$, which multiply to give ninety. Then you can list out all the possible ways that these two whole numbers can multiply to ninety."

| | | | | | |
|---|---|---|---|---|---|
| 1×90 | 2×45 | 3×30 | 5×18 | 6×15 | 9×10 |
| 10×9 | 15×6 | 18×5 | 30×3 | 45×2 | 90×1 |

"But $x$ and $y$ are both one-digit positive integers, so the first number, $9 + x + y$, has to be at least 9+1+1 = 11, and be no more than 9+9+9 = 27. That reduces all of our options down to just two cases."

| | | | | | |
|---|---|---|---|---|---|
| 1×90 | 2×45 | 3×30 | 5×18 | 6×15 | 9×10 |
| 10×9 | (15×6) | (18×5) | 30×3 | 45×2 | 90×1 |

"So we have only two cases: in the first case, $9 + x + y = 15$ and $10 + x - y = 6$. In the second case, $9 + x + y = 18$ and $10 + x - y = 5$."

I summarized this information into a simple table, which I wrote on the board.

| | Case #1 | Case #2 |
|---|---|---|
| $9 + x + y$ | 15 | 18 |
| $10 + x - y$ | 6 | 5 |

I took a deep breath to calm myself. I normally had no problems presenting in front of a class, but with Mr. Davis in the back and the St. FX math professor standing next to me, I realized that a lot was at stake. I was so close to finishing the solution, and didn't want to stumble right at the end.

"And then, for each of my two possible cases, we can just solve for the variables $x$ and $y$."

I then added another row to my table, indicating the solution $(x, y)$ for each of my systems of equations.

| | Case #1 | Case #2 |
|---|---|---|
| $9 + x + y$ | 15 | 18 |
| $10 + x - y$ | 6 | 5 |
| $(x, y)$ | (1, 5) | (2, 7) |

"This gives us the two solutions we had earlier. Not only have we generated my two earlier solutions, we've just proven that no other solutions are possible."

**1** + 2 + 3 + 4 + **5** = **15**

**2** + 3 + 4 + 5 + 6 + **7** = **27**

Dr. Weber invited the class to applaud as I walked back to my seat.

"Great job," said Patrick, giving me a high-five. I sat down and breathed a sigh of relief, and out of the corner of my eye, could see Gillian staring at her desk with a frown on her face.

The professor beamed at me. "I came to class today prepared to give a short lecture on the theory of Triangular Numbers and use that to introduce my work in computational number theory. But I'm going to switch directions and do something better, thanks to your classmate Bethany who just presented us with an astonishing idea."

She walked over to her laptop, and closed her Powerpoint presentation. From the desktop, she clicked on a red icon marked Maplesoft.

"I'm opening up a program called *Maple*, a computer algebra system developed in Waterloo, Ontario. I use this program all the time in my research, using Maple's high-powered computational software to discover numerical patterns, which I then prove mathematically. Let's try that with Bethany's idea.

"As Bethany just explained, say that positive integers $x$ and $y$ form an *astonishing pair* if the sum of the integers from $x$ to $y$ is equal to the number formed by stitching together $x$ and $y$. So $(x, y) = (1, 5)$ and $(x, y) = (2, 7)$ are both astonishing pairs, and Bethany just proved these are the only astonishing pairs when $x$ and $y$ are one-digit numbers. But what if $x$ and $y$ are two-digit numbers? Can anyone tell me the right equation?"

Patrick and Gillian raised their hands. Dr. Weber invited Patrick to come up.

He walked up to the board, explaining that if $x$ and $y$ are two-digit numbers, then the right side would have to be $100x + y$ to produce a four-digit number. For example, if $x = 12$ and $y = 34$, then the stitched number would be $100x + y = 1200 + 34 = 1234$. So the correct equation had to be:

$$x + (x + 1) + (x + 2) + \cdots + (y - 1) + y = 100x + y$$

Patrick used my earlier formula to show that this equation could be re-written as:

$$\frac{y(y+1)}{2} - \frac{(x-1)x}{2} = 100x + y$$

"Thank you," said Dr. Weber, as she walked over to her laptop and wrote down a few lines of computer code, which she explained to us step by step:

```
for x from 10 to 99 do
 for y from 10 to 99 do
  if y·(y+1)/2 - (x-1) ·x/2 = 100·x + y then
   print([x, y]):
  end if:
 end do:
end do:
```

"What Maple will do now is to look through all possible two-digit numbers $x$ and all possible two-digit numbers $y$, and list out all the astonishing pairs."

She clicked the Enter key, and a split-second later, four solutions were returned.

[13, 53]
[18, 63]
[33, 88]
[35, 91]

Dr. Weber smiled. "So what Maple has done is produce all the two-digit astonishing pairs. There are only four of them."

**13** + 14 + 15 + … + 52 + **53** = **1353**
**18** + 19 + 20 + … + 62 + **63** = **1863**
**33** + 34 + 35 + … + 87 + **88** = **3388**
**35** + 36 + 37 + … + 90 + **91** = **3591**

"Interesting," she said, looking at the screen. "If you add up all the integers between 13 and 53, you get 1353."

She then repeated the same process with three-digit numbers, making a tiny fix to her computer code. She hit the Enter key, and a half-second later, her fancy computer program produced the following:

[133, 533]
[178, 623]

In other words, the only three-digit astonishing pairs are:

**133** + 134 + 135 + … + 532 + **533** = **133533**
**178** + 179 + 180 + … + 622 + **623** = **178623**

Dr. Weber looked at us. "This is what I do as a number theorist. I take ideas and concepts relating to the positive integers, explore their properties using Maple, find surprising patterns, investigate their characteristics using various techniques in my field, and discover new results and theories. And through this process, I'm discovering just how rich and mysterious these integers are. We discover brand-new insights all the time, and who knows, maybe we even discovered a new theorem – an astonishing new theorem – through our work today."

I raised my hand. "I notice something interesting."

"What is that, Bethany?"

I walked up to the board and wrote down three equations: the first "astonishing" pair that I found, followed by two more that Dr. Weber just found on her computer.

**1** + 2 + 3 + 4 + **5** = **15**
**13** + 14 + 15 + … + 52 + **53** = **1353**
**133** + 134 + 135 + … + 532 + **533** = **133533**

I looked at Dr. Weber and pointed to the board. "I think this pattern continues. Could you check something on your Maple program?"

"Sure," she said, looking surprised by the pattern that was surely apparent to her as well.

"What would you like me to do, Bethany?"

"Add up all the integers from 1333 to 5333," I said. "Please."

"I sure can," she said, with a smile. "Before I do, why don't you tell me what the answer will be?"

"I'm pretty sure it will be the eight-digit number 13335333."

Dr. Weber took the formula $x + \cdots + y = \frac{y(y+1)}{2} - \frac{(x-1)x}{2}$, and substituted $x = 1333$ and $y = 5333$. Instantly, Maple outputted the correct answer; and sure enough, it was 13335333.

As I sat down, the professor then substituted $x = 13333$ and $y = 53333$, and got Maple to compute the sum of the integers between 13333 and 53333. And the answer was no surprise: 1333353333.

The professor was speechless. The pattern was now obvious to everyone in the class.

**1** + 2 + 3 + 4 + **5** = **15**
**13** + 14 + 15 + … + 52 + **53** = **1353**
**133** + 134 + 135 + … + 532 + **533** = **133533**
**1333** + 1334 + 1335 + … + 5332 + **5333** = **13335333**
**13333** + 13334 + 13335 + … + 53332 + **53333** = **1333353333**
etc.

Patrick leaned over towards me. "Awesome."

"We have just a few minutes left," said Dr. Weber, looking at the clock. "So in closing, I'd like to thank all of you for your active participation today. I find that every time I do mathematics, I get a *sense of wonder* – and the more I do mathematics, the more I understand the structures and patterns that form the foundation of this world, and the more I appreciate the beauty that emerges from these structures. Today, through the process of discovering and creating your own mathematics, I hope you too experienced this sense of wonder. I most certainly did."

She opened up her Powerpoint presentation and scrolled to the very last slide. "Here is a quote by a famous mathematician named Bertrand Russell. Let me end with this."

> Mathematics, rightly viewed, possesses not only truth, but supreme beauty — a beauty cold and austere, like that of sculpture, without

> appeal to any part of our weaker nature, without the gorgeous trappings of painting or music, yet sublimely pure, and capable of a stern perfection such as only the greatest art can show.

We all applauded, and Mr. Marshall walked up to give a wrapped gift to Dr. Weber on behalf of the school. As the bell rang, and we started to file out of the classroom, the principal stopped me.

"Well done, Bethany," he said, shaking my hand.

"Thank you, Mr. Davis," I replied. Mr. Marshall also came over to congratulate me.

I felt a tap on my shoulder. I turned around and saw Dr. Weber looking down at me, smiling.

"Bethany, thank you for today," she said, handing over a card. "Here's my e-mail address and phone number. Let's stay in touch."

As I took the professor's business card, I heard a loud crash and turned to see Gillian running out of the classroom, bawling, with tears streaming down her face.

# 40

It was 8:00 p.m. in Sydney, so it was 4:00 p.m. in Vancouver.

Grace would be coming home anytime now. I saved my Word file marked "St. FX Scholarship Application", and got out my phone to text Grace.

*Skype me when you get in . . . got exciting news (x2) to share.*

I couldn't wait to share with Grace the two pieces of news I learned today, and was still unsure which one was more exciting.

The Canadian university application deadline was now less than a week away. After a lot of discussion and debate, Grace and I decided to put "all our eggs" in two baskets, applying for Quest in the west coast, and St. FX in the east coast. I finished my Quest application last night, and was putting the finishing touches on my St. FX application when I realized it was already eight o'clock.

Shortly after Dr. Weber's visit, Sydney High School decided to nominate me for the St. FX President's Scholarship. Mr. Davis said that the decision was very hard, since Gillian had a higher grade point average, but the selection committee felt I better represented the qualities required of the scholarship recipients: "leadership and dedication in service to others as well as academic achievement".

I knew that Gillian took the news really hard, and in the cafeteria that afternoon, I heard her yelling at Vanessa after she said something to try to cheer up Gillian. Part of me felt really guilty, especially because I knew Gillian was going to St. FX for sure, while the probability of me going there was only fifty percent.

Grace was still insistent on pursuing the "liberal arts" option at Quest while I was equally insistent on pursuing the "math specialization" option at St. FX. Grace wanted to take weird-sounding courses like *Cornerstone* and *Rhetoric*, while I wanted to pursue *Modern Geometries* and *Differential Equations.* Hopefully the decision would make itself clear to both of us, especially if we both got scholarships to the same university.

I heard the familiar ring coming from my laptop, and clicked on the green "Accept Call" icon.

"Hey, what's up?" I said.

"I'm good," said Grace. "So what's *x2*? Some new secret code word?"

"Thought you'd be able to figure it out," I teased. "Didn't you learn multiplication?"

"Yeah, yeah," replied Grace. "Exciting news *times two*. I get it. So tell me, what's the first one?"

"I qualified for the CMO!"

"Yay!" said Grace, with a big smile on her face. "Congrats. That's so awesome."

Earlier today, I learned that I had finished fiftieth on the Canadian Open Math Challenge (COMC). Since only the top fifty qualify for the Canadian Math Olympiad, my IMO dream was still alive. But I knew that it was close, since only one point separated me from a half-dozen students who tied for fifty-first.

Fortunately, I got full marks for every problem I solved on the COMC, and despite all the possible pitfalls involving careless errors, missing cases, incorrect logic, and poor explanation, I didn't lose a single point on any solution. On every question I got, my score was ten out of ten, pushing me into the top fifty, passing others who solved an additional question but did a poor job justifying their solutions.

"Like I always say, Bethany. You're the best solution-writer in Canada."

Grace was just saying that to be nice, but deep down, I knew she had a point. The problem was I couldn't solve every question. If only I had Albert's problem-solving ability on top of my solution-writing ability, I'd become an Olympian for sure.

On the hard questions that separated the Olympians from everyone else, it didn't matter whether I had great solution-writing skills. Zero times ten was still zero.

"How did you do?" I asked, already knowing that she'd rank amazingly high since she solved the first eleven problems on the COMC and got halfway through the final problem before time ran out.

Grace excitedly told me that she finished third in Canada, just one point behind Raju, and five points behind Albert who obtained a perfect score on the COMC. Grace texted a bunch of people earlier in the day and found out the scores of everyone she knew, including the three "star students" she met at Waterloo two summers ago who Grace felt were shoo-ins for the

IMO team. Those three students, who had now all become permanent residents of Canada, finished fourth, fifth and sixth.

"That's awesome, Grace," I said. "You're going to make the team for sure."

"Nothing's set in stone, Bethany. You know that."

"Yeah, yeah, I know."

While I was thrilled for Grace, I knew that my prospects were bleak. There was a long climb from top fifty to top six, especially given the big gap between me and the six students most likely to be chosen.

My only hope was The Rule: that the winner of the Canadian Math Olympiad would automatically get selected for the IMO team.

We weren't sure whether The Rule actually existed, but we knew that the CMO winner was always chosen for the IMO team, ever since Canada started sending a team back in 1981. And in the one year when two people tied for first place, both winners represented Canada.

I needed history to repeat itself. While Grace needed a strong performance on the CMO to become a Math Olympian, I needed to win it. Because Albert was going to get a perfect score for sure, my only hope was to get a perfect score too. I needed fifty out of fifty, and tie Albert for first.

Just thinking about that prospect gave me a headache, and I decided to change the subject.

"Let me tell you my second bit of news."

"Go for it," said Grace.

"I told you about those cool sums, right? You know, like 13+ . . . +53 = 1353. Remember?"

"Yeah, so?"

"And remember that math professor I told you about? You know, the one from St. FX?"

"St. FX? Where's that?" asked Grace.

"Very funny," I said. "Anyway, Dr. Weber invited me to the university on Friday afternoon, to do research with her. She's invited me to come every Friday, from now till the end of the semester."

Grace sounded surprised. "Wow. Every Friday?"

"Every Friday," I said. "Mom's going to drive me to Antigonish. We leave Sydney at eleven o'clock, right after I finish Period 2."

I was shocked at Dr. Weber's weekly invitation. She called me shortly after her visit to Sydney High, saying she loved my "astonishing pairs" idea and felt it would be worthwhile to pursue it further and see if we could write a research paper together. Mr. Marshall and Mr. Davis were "astonished" by the request of a professor inviting a high school student to collaborate on a research project, and gave me the permission as long as I prioritized my school work.

The timing of Dr. Weber's invitation was perfect, since my only Friday class after Period 2 was Biology, which was a joke since Mrs. Finley just taught out of the textbook, and her tests were always easy to ace as long as we memorized the summary page at the end of each chapter. Lucky for me, I was only cutting Mrs. Finley's class and not Mr. Marshall's.

"Your Mom is going to drive you to Antigonish every Friday? What about her job?"

"Mom has this thing called 'compressed hours' which means she can leave the office as long as she puts in her 37.5 hours before 11:00 a.m. on Friday. She says it's a perk of working for the government."

"But what if something urgent comes up?" asked Grace.

I laughed. "Nothing about Mom's job is urgent. Especially on a Friday."

"That's awesome," said Grace. "Your Mom's really cool."

I nodded, realizing how close Mom and I had become, especially in the last couple of years. Mom was now running four times a week, feeling a lot happier at home, and eager to drive six hours from Sydney to Antigonish and back, every single Friday for the rest of the semester, just for me.

Yeah, Mom was definitely cool.

Grace cleared her throat and spoke with some hesitation. "I've been learning about Cape Breton and the famous MacDonald clan."

"Really?"

"My English teacher started a new unit. We're now reading *No Great Mischief.*"

"Seriously?"

I was pleasantly surprised that Grace's class in Vancouver was reading Alistair MacLeod's famous novel, which looked at the themes of loyalty, family, and discovering your identity, and was recently selected as the greatest Atlantic Canadian book of all time.

"Yeah, the book is amazing," said Grace. "I can tell why you love Nova Scotia so much. Maybe this is a sign I'm supposed to move there."

"Maybe this is a revelation from your God."

"Maybe."

I smiled. "My hope is constant in thee, Clan Donald."

"Huh?"

"Don't worry," I said, laughing. "You'll get to that part of the book soon."

"Congrats on qualifying for the CMO," said Grace, changing the subject. "We both did it. And if we do well on the CMO, we'll both make the team."

"Easier for you than me," I said, suddenly feeling anxious about the big contest in March. I shook my head in resignation.

"When it comes to contests, you know how much I tense up. The pressure just gets to me."

Pressure didn't affect Grace the same way, since she could magically eliminate all negative thoughts and just focus on the math problems in front of her. But I was constantly distracted by the fear of failure, or distracted by people sitting and sniffling around me.

"Bethany, remind me of your score on the CMO last year."

"Zero," I snapped, annoyed at Grace for bringing this up. "I didn't qualify, remember?"

"That's not what I meant," said Grace, sighing. "You wrote the CMO a few days after the contest, working for three hours straight at your kitchen table. How many did you solve?"

"The first four."

"That's right," said Grace. "You nailed each one. And because of how you write solutions, you would have gotten forty out of fifty. Remember I came seventh in Canada, with thirty-seven points."

"Yeah, yeah," I said, reminded of Grace's pep talk last year that a score of forty would have placed me sixth in Canada, behind Albert, Raju, and three Grade 12 students who were also chosen for the IMO team and had gone off to university.

"Grace, you know that doing a practice contest in my kitchen is different from the real thing, where there's all this pressure. It's not the same!"

"You should just do what I do before a big contest."

"Like watch an inspirational movie? The underdog overcomes all the odds and becomes the champion? That would stress me out even more!"

"It depends on what the movie is. I've already picked out mine for the day before the CMO."

"Really?" I said, curious to know. "What is it?"

"It's a movie my Dad recommended. It's about a British politician named William Wilberforce who spends forty-five years fighting to abolish slavery, and comes a little bit closer each year. Right at the end of his life, the parliament passes the final law that abolishes slavery forever, and then he dies, just three days later, knowing that he fulfilled his life's mission."

"That sounds cool," I said. "But it doesn't sound like a Grace movie."

"It is," she said, laughing. "The politician was inspired by his Sunday school teacher who wrote a famous song, and the title of the movie is this song. It's very much a Grace movie."

"What's the title?" I asked.

"*Amazing Grace*."

I laughed and shook my head. "You're lucky. It's not like there's a movie of an Amazing Bethany."

"Sure there is," replied Grace. "*Soul Surfer*."

# 41

I finished the last line of my proof and put down the whiteboard marker.

Dr. Weber gave me a thumbs-up. "Well done."

I smiled, and glanced at the clock in Dr. Weber's office, surprised that it was already 3:42 p.m. I couldn't believe that nearly two hours had already passed. I wasn't tired at all.

"Bethany, your mother's meeting you here in about fifteen minutes. Before she arrives, let's recap what we did. Please start from the beginning, and summarize what we found."

I nodded, and pointed to the first line of the big whiteboard in the professor's office.

"We say that $(x, y)$ is an *astonishing* pair of integers if the sum of the integers from $x$ to $y$, inclusive, is equal to the digits of $x$ followed by the digits of $y$. To give an example, (1, 5), (13, 53), (133, 533), (1333, 5333), (13333, 53333) . . . is an infinite sequence of astonishing pairs."

$$\mathbf{1} + 2 + 3 + 4 + \mathbf{5} = \mathbf{15}$$
$$\mathbf{13} + 14 + 15 + \ldots + 52 + \mathbf{53} = \mathbf{1353}$$
$$\mathbf{133} + 134 + 135 + \ldots + 532 + \mathbf{533} = \mathbf{133533}$$
$$\mathbf{1333} + 1334 + 1335 + \ldots + 5332 + \mathbf{5333} = \mathbf{13335333}$$
$$\mathbf{13333} + 13334 + 13335 + \ldots + 53332 + \mathbf{53333} = \mathbf{1333353333}$$

etc.

"If $y$ has exactly $n$ digits, we showed that astonishing pairs satisfy

$$\frac{y(y+1)}{2} - \frac{(x-1)x}{2} = x \cdot 10^n + y.$$

We re-wrote this as a quadratic equation, and solved it to find a *generating formula* for all astonishing pairs."

$$x = 2^{n-1}q - 10^n + \frac{p+1}{2} \quad \text{and} \quad y = \frac{1+|2^n q - p|}{2}, \text{ where } pq = 5^n(10^n - 1)$$

"Very good, Bethany. Please continue."

"For $n = 2$, we have $pq = 5^n(10^n - 1) = 25 \times 99 = 2475$. And we went through all the possible cases to show that there are only five astonishing pairs $(x, y)$ if $y$ is a 2-digit number."

| | Case #1 | Case #2 | Case #3 | Case #4 | Case #5 |
|---|---|---|---|---|---|
| $p$ | 75 | 165 | 55 | 45 | 225 |
| $q$ | 33 | 15 | 45 | 55 | 11 |
| $x$ | **4** | **13** | **18** | **33** | **35** |
| $y$ | **29** | **53** | **63** | **88** | **91** |

"For example, if $p = 75$ and $q = 33$, then $pq = 2475$, and we can show that this gives $(x, y) = (4,29)$ from the above generating formula."

I then pointed to a section of the whiteboard where we had listed the sums corresponding to each of the five astonishing pairs for the case $n = 2$.

**4** + 5 + 6 + … + 28 + **29** = **429**
**13** + 14 + 15 + … + 52 + **53** = **1353**
**18** + 19 + 20 + … + 62 + **63** = **1863**
**33** + 34 + 35 + … + 87 + **88** = **3388**
**35** + 36 + 37 + … + 90 + **91** = **3591**

Dr. Weber looked at me intently, with her thumb tucked in under her chin.

"When I came to your high school, I programmed this on Maple in front of all of you. I got four solutions for the two-digit case. But we just came up with five. Did we make an error?"

"No error," I said, seeing the answer immediately. "When you came to Sydney High, you programmed it so that $x$ and $y$ were *both* two-digit numbers, but here, we just specified $y$ to be two digits. So that's how we came up with this extra solution 4 + 5 + 6 + … + 28 + 29 = 429, where $x$ has just one digit."

"Excellent, Bethany," said Dr. Weber. "Mathematics requires careful language – defining terms in a very precise way to avoid ambiguity. When it comes to creativity and imagination, we're similar to artists; but when it comes to ambiguity, we differ."

"What do you mean?" I asked.

"In art, ambiguity leads to complexity and confusion and multiple interpretations, and this is a good thing. But in math, it's not. We need to define our concepts exactly, and insist upon rigorous logical justifications for every single statement, so that we can build a rock-solid foundation. And because of this insistence on absolute precision, mathematics gives us the freedom to imagine the unimaginable, concepts like infinity and infinitesimals, which are free of ambiguity and misinterpretation."

"That's cool."

"You noticed the tiny but subtle difference between how we defined these astonishing pairs today, compared to how we defined them at Sydney High last week. Well done. Please continue."

I pointed my marker to our generating formula.

$$x = 2^{n-1}q - 10^n + \tfrac{p+1}{2} \quad \text{and} \quad y = \tfrac{1+|2^n q - p|}{2}, \text{ where } pq = 5^n(10^n - 1)$$

"If we let $p = \frac{5}{3}(10^n - 1)$ and $q = 3 \cdot 5^{n-1}$, then we have $pq = 5^n(10^n - 1)$. Now substituting these values for $p$ and $q$ into our formulas for $x$ and $y$, we get a simple formula for $x$ and $y$ in terms of $n$.

$$x = \frac{2 \cdot 10^n - 5}{15} \quad \text{and} \quad y = \frac{8 \cdot 10^n - 5}{15}$$

If we substitute $n = 1$, we get $(x, y) = (1, 5)$.
If we substitute $n = 2$, we get $(x, y) = (13, 53)$.
If we substitute $n = 3$, we get $(x, y) = (133, 533)$.
If we substitute $n = 4$, we get $(x, y) = (1333, 5333)$.
If we substitute $n = 5$, we get $(x, y) = (13333, 53333)$.

"Isn't that beautiful?" asked Dr. Weber. "We get the same sequence as before."

"Definitely," I said. "It's cool that there's a simple formula to generate these astonishing pairs, and we can actually prove that this pattern continues forever."

"That's right," she said, smiling. "This is an infinite sequence of astonishing pairs. Now recap for me what you just discovered – a completely different infinite sequence of astonishing pairs."

"We figured out that if we let $p = \frac{5}{9}(10^n - 1)$ and $q = 9 \cdot 5^{n-1}$, then we also have $pq = 5^n(10^n - 1)$. And if we substitute these values for $p$ and $q$ into our formulas for $x$ and $y$, we get a formula for $x$ and $y$ in terms of $n$ that generates a new infinite sequence of astonishing pairs."

$$x = \frac{8 \cdot 10^n + 10}{45} \quad \text{and} \quad y = \frac{28 \cdot 10^n + 35}{45}$$

If we substitute $n = 1$, we get $(x, y) = (2, 7)$.
If we substitute $n = 2$, we get $(x, y) = (18, 63)$.
If we substitute $n = 3$, we get $(x, y) = (178, 623)$.
If we substitute $n = 4$, we get $(x, y) = (1778, 6223)$.
If we substitute $n = 5$, we get $(x, y) = (17778, 62223)$.

I loved the simplicity and beauty of this pattern, that we could generate infinitely many astonishing pairs by repeatedly sticking the digit 7 into each $x$, and repeatedly sticking the digit 2 inside each $y$. The first line, $(x, y) = (2, 7)$, didn't fit the pattern, but everything after that did. Ironically, this first pair that didn't fit the pattern had the same two digits: 7 and 2.

"I just noticed something," said Dr. Weber, pointing to my simple formulas for $x$ and $y$ above. "The fraction $y/x$ equals $7/2$ for every single value of $n$."

$$\frac{7}{2} = \frac{7}{2} \qquad \frac{63}{18} = \frac{7}{2} \qquad \frac{623}{178} = \frac{7}{2} \qquad \frac{6223}{1778} = \frac{7}{2} \qquad \frac{62223}{17778} = \frac{7}{2}$$

I shook my head in wonder. "The numbers 7 and 2 keep popping up everywhere! That is so weird."

"Actually, it's not weird at all," said Dr. Weber, with a gleam in her eye. "When it comes to mathematics, the word 'amazing' is a better adjective to use."

"Or astonishing," I added.

"Yes," said Dr. Weber. "It's astonishing how the simplest of ideas can sometimes lead to results of surprising depth. There is something significant with the (2, 7) pair, but we haven't figured that out yet; all we did was scratch the surface. Let's keep working every Friday afternoon this semester, and I'm confident we'll unlock the secret behind this pattern, and be able to explain why this works. You know, your astonishing pairs idea reminds me of the Heegner numbers."

"The what?" I asked.

Dr. Weber glanced at the clock and saw that we still had a few minutes remaining. She opened her laptop and loaded up Maple, and invited me to sit down next to her.

"Before I explain the Heegner numbers, tell me what the special numbers $\pi$ and $e$ represent."

"$\pi = 3.14159\ldots$ and it's the ratio of the circumference of a circle to its diameter. And $e = 2.71828\ldots$ and it's the base of the natural logarithms. We see the number $e$ all the time in Mr. Marshall's class."

"Good," said Dr. Weber. "Both $\pi$ and $e$ are *irrational* numbers. Also $\sqrt{163}$ is an irrational number. So now I'm going to take these three irrational numbers, $\pi$, $e$, and $\sqrt{163}$, and make a little calculation to fifteen decimal places."

$$e^{\pi\cdot\sqrt{163}} = 262{,}537{,}412{,}640{,}768{,}743.999\,999\,999\,999\,250\ldots$$

I stared at the result, shocked that the first twelve digits after the decimal place were .999999999999.

"That's so weird . . . I mean, that's really amazing."

Dr. Weber smiled. "At first, this looks like a complete coincidence, that three irrational numbers, when combined in this way, produces a number that is less than $10^{-12}$ away from an integer. But there's a sophisticated result in number theory that explains this phenomenon, dealing with the 'class number' of various 'quadratic imaginary number fields'. It turns out that

163 is the largest of the 'Heegner numbers', which are all the integers with class number 1. And *because of this* special class number property, when you calculate $e^{\pi \cdot \sqrt{163}}$, the result *has* to be close to an integer."

"So it's not a coincidence."

"Exactly. In mathematics, if something looks like a coincidence, then there's almost certainly a reason why it's *not* a coincidence. Another Heegner number is 67. And we get:

$$e^{\pi \cdot \sqrt{67}} = 147,197,952,743.999\,998\,662\ldots$$

"Neat," I said. "Another almost-integer. But why do they call them Heegner numbers?"

"Heegner was the German high school teacher who solved the Class Number Problem, one of the hardest unsolved problems in number theory, originally proposed by Gauss."

"Another German," I said, remembering a conversation with Mr. Collins just before the Grade 7 Gauss Contest. "My math coach said that Gauss was one of the greatest mathematicians in modern history."

"Yes," said Dr. Weber, "he most certainly was. When Gauss was young, he did his research in number theory, and the ideas he generated during that time inspired other ideas – like his method of 'least squares' in surveying land, which produced more accurate maps. Then Gauss tackled hard problems in surveying, triggering his breakthrough insights in differential geometry, which was exactly what Einstein needed eighty years later for his theory of general relativity. Mathematics is structured like a chain, where one theory builds upon another, stimulating curiosity, and giving birth to new ideas."

"Einstein's relativity comes from number theory?"

"Indirectly," said Dr. Weber. "But that's always the case, because every generation builds upon the work of the previous generation. As a result, Grade 7 students in Nova Scotia can solve algebraic problems that would have stumped the greatest minds in Euclid's day. And my first-year calculus students learn basic techniques that go far beyond anything that was even conceivable four hundred years ago by Leibniz and Newton, the inventors of calculus. Because we're 'standing on the shoulders of giants', we can take the results of our predecessors and build upon them, enabling us to see farther and farther ahead. That's how mathematical research always works."

"Will anyone build upon this research?" I asked, pointing at the astonishing pairs.

"You never know," said Dr. Weber. "Maybe someone, hundreds of years from now, will need precisely this idea of the MacDonald-Weber pair to discover a new breakthrough."

"The MacDonald-Weber pair," I repeated, with a smile. "It has a nice ring to it . . . even if it's not practical or useful in the real world."

Dr. Weber reached her bookshelf with her massively-long arm, and pulled out a thin book titled *A Mathematician's Apology*.

"G. H. Hardy was a British number theorist. He said that mathematics was the queen of the sciences, and that number theory was the queen of mathematics. Hardy pursued his research in number theory because it was 'pure' mathematics, free of applications, and he insisted that nothing he worked on would have any practical use. Here's a famous quote of his."

She flipped through the book and found the page she was looking for.

> I have never done anything "useful". No discovery of mine has made, or is likely to make, directly or indirectly, for good or ill, the least difference to the amenity of the world.

"Of course, Hardy was wrong about his own research, whether it was the Hardy-Weinberg principle that helps biologists understand and model population genetics, or the Hardy-Ramanujan formula that has everyday applications to quantum mechanics and thermodynamics. And Hardy was wrong about his field of number theory, which enables physicists to calculate atomic energy levels, help sound technicians design diffusers for optimal acoustics, among other applications."

"Like cryptography?" I asked.

"Exactly. Number theory is the foundation of modern-day cryptography, without which the internet couldn't function. Hardy died in 1947 – so naturally, he never witnessed how number theory enables us to send and receive a private e-mail, or buy a book securely on Amazon."

We heard a loud knock.

"Come on in," said Dr. Weber, in a booming voice.

Mom tentatively opened the door and peeked her head in.

Dr. Weber stood up. "Come on in, Ms. MacDonald. We've just finished."

Mom took a few steps inside Dr. Weber's office, holding a thick stack of brochures and pamphlets marked with the university logo. She folded the brochures under her arm and shook Dr. Weber's hand.

"I really appreciate you taking the time to work with Bethany today. Thank you so much for this, Professor."

"You're welcome, Ms. Macdonald," said Dr. Weber. "Please call me Anna."

"Then you must call me Lucy," said Mom.

Dr. Weber pointed to the corner of the whiteboard. "When you dropped Bethany off less than two hours ago, this whiteboard was completely empty. Look at what she discovered."

$$\mathbf{2} + 3 + 4 + \ldots + 6 + \mathbf{7} = \mathbf{27}$$

$$\mathbf{18} + 19 + 20 + \ldots + 62 + \mathbf{63} = \mathbf{1863}$$

$$\mathbf{178} + 179 + 180 + \ldots + 622 + \mathbf{623} = \mathbf{178623}$$

$$\mathbf{1778} + 1779 + 1780 + \ldots + 6222 + \mathbf{6223} = \mathbf{17786223}$$

$$\mathbf{17778} + 17779 + 17780 + \ldots + 62222 + \mathbf{62223} = \mathbf{1777862223}$$

"She's amazing, isn't she?" said Mom proudly. "But Bethany's math is far too advanced for me."

"Come on, Mom, it's not that hard! The first row says that if you add up the numbers from 2 to 7, you get 27. The second row says that if you add up the numbers from 18 to 63, you get 1863. The third row says that if you add up the numbers from 178 to 623, you get 178,623. This pattern goes on forever."

"I think I actually understand this," said Mom, slowly nodding at the board. "You came up with all this on your own?"

"She sure did," said Dr. Weber, nodding at Mom. "Bethany has more imagination than most of my fourth-year students. She doesn't follow set directions; instead she wants to create new directions. It's like cooking: amateurs follow somebody else's prescriptive recipe, while good cooks don't need the recipe but can't improvise. However, a small number of great chefs are true artists – they throw out the rule book and create new cuisines from scratch. And that's what Bethany is."

"A great chef?" I asked, snickering at the memory of making Kraft Dinner at Grace's house.

"No, an artist," replied Dr. Weber. "A true artist."

"Thank you," I said, beaming.

"Bethany, I am *astonished* by what we discovered today. I'm glad we can work again next week, and do some more number theory together."

Mom pointed to the board. "This is number theory?"

"Yeah," I said. "But it's not always as hard as this. You know, Mom, I just realized you did some number theory in the car today. You said it was a three-hour drive from Sydney to Antigonish, and we wanted to arrive by two o'clock. So that's why you said we had to leave Sydney by eleven. You were calculating in 'modulo 12' when you realized that two minus three is congruent to eleven. That's number theory!"

Dr. Weber turned to Mom. "Bethany has a good point. Maybe you are a mathematician at heart? Like daughter, like mother?"

"I don't think so," said Mom. "Besides, I never went to college or university, so the type of math you and Bethany specialize in doesn't exactly come naturally to me. Had I gone to university, I wouldn't have majored in math. I'm sure I would have picked something else."

"Like what?" I asked, genuinely curious.

Mom paused before answering. "Psychology. Yes, definitely I would have studied psychology. The human brain is amazing – and I would have loved to learn how the brain works, how we're motivated to do certain things, and how and why the brain sometimes doesn't work."

Mom looked at Dr. Weber. "I can tell how much Bethany loves learning. As I watch her, I wish I could be twenty years younger and become a student again."

Dr. Weber turned to face Mom, and I noticed the professor had that gleam in her eyes once again.

"Lucy, what have you been doing for the past two hours?"

"It was my first time on the St. FX campus, so I walked around for a little bit. I picked up these brochures for Bethany, and I spent the rest of the time reading a book in the library. Then I came here."

"Well, I should tell you that the most popular professor at this university teaches an introductory Psychology class. His name is Martin Aucoin and he was this year's winner of the 3M Fellowship, a national award given to the best teaching professors at Canadian universities. As it happens, his Psych 100 class is scheduled on Fridays this semester, from 2–5:00 p.m. Would you like to audit his course?"

Mom looked puzzled. "What does it mean to audit a course?"

"It means you sit in on the lectures, and learn with the rest of the students. You do the readings and participate in class, but don't submit assignments or write the final exam."

"That's a lot different from the auditing that happens at my workplace – tax returns, and all that stuff."

Dr. Weber smiled. "Auditing a course is much more interesting, especially a course taught by Dr. Aucoin. The course wouldn't be for academic credit – it would just be for your personal development."

"You should do it, Mom," I said. "That sounds amazing."

Mom hesitated. She paused and looked at Dr. Weber.

"How much does it cost?"

"Normally the price of auditing a course is one-half the usual tuition fee, and there's a formal admissions process to get you registered as a part-time mature student. But in this specific case, I can get around that, and make a special arrangement for you to audit Psych 100 for free."

"How?"

"I'll just say I have a lot of influence in Dr. Aucoin's day-to-day life."

Mom raised an eyebrow.

"You're his boss? He reports to you?"

She grabbed a frame on her desk and spun it towards us, showing us a picture of six-foot-four Dr. Weber and a handsome grey-haired man nearly as tall, linking arms and smiling at the camera.

Dr. Weber pointed to the wedding ring on the man's finger.

"Yes, he reports to me."

# 42

As usual, the two hours flew by.

But this time, Dr. Weber and I weren't discovering any new results. Instead, we were reviewing the main theorems and deciding what to put in our research paper.

"I can't believe we're going to submit this for publication."

"Yes, Bethany. After just five Fridays together, we've completed all of our research and now it's time to start writing up our paper."

After my first session with Dr. Weber, I called Grace and excitedly told her about my research collaboration. She wanted to get in too, and so we studied the astonishing pairs idea together, treating it like one of our Olympiad practice problems. Of course, we decided to include this research in our three thousand hours, especially as number theory was a common topic on Olympiad contests.

Soon after joining our team, Grace found a much cleaner generating formula, in terms of a single variable $r$.

$$x = \frac{10^n}{2}\left(r + \frac{1}{r} - 2\right) + \frac{r-1}{2r} \quad \text{and} \quad y = \frac{1 + |10^n\left(r - \frac{1}{r}\right) + \frac{1}{r}|}{2}.$$

And then working together on Google Docs while talking to each other on Skype, Grace and I figured out some values of $r$ that produce infinite sequences of astonishing pairs. For example, if $r = \frac{3}{5}$, then we get a simple formula for $x$ and $y$ in terms of $n$.

$$x = \frac{2 \cdot 10^n - 5}{15} \quad \text{and} \quad y = \frac{8 \cdot 10^n - 5}{15}$$

And this produces our first infinite sequence of astonishing pairs – we called this "The Christmas Tree".

(1, 5)
(13, 53)
(133, 533)
(1333, 5333)
etc.

Dr. Weber liked Grace's idea of writing everything in terms of a single variable, and suggested that my friend from Vancouver be included in our collaboration. Especially after Grace and I proved that there were *infinitely* many values of $r$ that generated infinite sequences of astonishing pairs, the professor insisted that the research paper be co-authored by all three of us: Bethany MacDonald, Anna Weber, and Grace Wong – with our last names in alphabetical order.

We had found our "characterization theorem" and we were ready to publish.

"What do you suggest for the title of our paper?" asked Dr. Weber.

"How about *Astonishing Pairs of Numbers*?" I suggested.

"I'm okay with that title, but maybe we can come up with something a little more formal," said Dr. Weber. "Let's talk about it more next week. Also, you should ask Grace what she thinks."

"No problem," I said, and glanced down at my favourite result from our five weeks together.

If $r = \frac{999}{625}$ then we get this messy formula for $x$ and $y$ in terms of $n$:

$$x = \frac{187 \cdot (374 \cdot 10^n + 625)}{624375} \quad \text{and} \quad y = \frac{812 \cdot (374 \cdot 10^n + 625)}{624375},$$

That produces this infinite sequence of astonishing pairs:

(112013, 486388)
(112012813, 486387188)
(112012812813, 486387187188)
(112012812812813, 486387187187188)
(112012812812812813, 486387187187187188)

This sequence was fascinating because we had found a beautiful pattern that seemed to defy explanation: we can recursively generate successive terms by adding 281 just before the end of each $x$ term, and 718 just before the end of each $y$ term; from the above formulas for $x$ and $y$, the ratio of each $y$ to $x$ has to be 812 to 187. And it was a striking coincidence that 281 and 718 involve the same three digits as 812 and 187, respectively.

Furthermore, 281 and 718 wasn't astonishing, but it nearly was:

$$\mathbf{281} + 282 + 283 + \ldots + 717 + \mathbf{718} = \mathbf{218{,}781}$$

For some "astonishing" reason, the number 281 appeared again: if $r = \frac{9999}{15625}$ then we can generate another infinite sequence of astonishing pairs containing the same three digits over and over again:

(10129613, 46136013)
(101296132813, 461360132813)
(1012961328132813, 4613601328132813)
(10129613281328132813, 46136013281328132813)
etc.

In addition to all the results we generated using Dr. Weber's computer program, Grace and I worked through the algebra and formally proved the mathematics behind each of our infinite patterns, explaining why the ratio of each $y$ to $x$ was constant in the case when $r \geq 1$, and that this $y$ to $x$ ratio could be expressed by the simple fraction:

$$\frac{y}{x} = \frac{r+1}{r-1}$$

Dr. Weber told me last week that computers are an essential tool for research in number theory, the same way microscopes are an essential tool for research in molecular biology. The Maple output helps us find the right patterns, and enables us to quickly find counterexamples to show that certain conjectures are false. Naturally, it's much easier to prove something once we know it's actually true.

I turned to Dr. Weber. "So what should I do by next Friday?"

"I'd like you to type up the main results, with clear proofs to all of our theorems. Run it through with Grace, and make sure you two are happy with the initial draft. By then, I'll have written the introduction and conclusion, and then we'll put it together and go through the paper line by line."

"Sounds great," I said. "Grace said she can do a Skype call with us next Friday, since that will be during her lunch hour."

"Excellent," replied Dr. Weber. "I'm looking forward to meeting her – online that is."

The more I collaborated with Dr. Weber, the more I realized I wanted to come here. St. FX was less than three hours from home, and I loved the prospect of learning from Dr. Weber for four more years.

"Maybe next September you'll meet Grace in person," I said. "If she comes to the east coast and also decides to go to St. FX for her undergrad. Then both of us can work with you."

Dr. Weber looked surprised. "But I thought you and Grace were going to the west coast, doing your undergrad in British Columbia. At Quest."

I looked at her in shock.

"How do you know about that?"

There was a knock on the door. Mom poked her head in.

"Mom, you're early," I blurted, still taken aback that Dr. Weber knew about Quest, since I had never mentioned that to her.

Mom smiled. "The second half of Psych 100 was cancelled for me, since the midterm started at 3:30. Since I'm just auditing the course, Dr. Aucoin excused me from the class."

"How was class today, Mom?" I asked, making a mental note to ask Dr. Weber about the comment she just made.

"Today was really deep," replied Mom. "Dr. Aucoin talked about psychological character traits, and how success is determined more by our *grit* than by our IQ or self-esteem or socio-economic status or anything else."

"Ah, yes," replied Dr. Weber, nodding her head. "Grit is Martin's research speciality, so naturally that's his favourite lecture topic."

"Bethany, you've got more grit than anyone I've ever met," said Mom. "Dr. Aucoin gave us a questionnaire, which measures how much we have a passion for something and stick with it, how much we persevere despite facing obstacles and setbacks, and how much we persist at a task until we get it done. I was thinking about you the whole time I was doing the grit survey. You would have scored a perfect five."

"Thanks."

Dr. Weber stood up.

"Yes, Bethany, your grit score is definitely off the charts. Say, do you two have time for a drink before you have to drive back to Sydney? I want to talk to you about something."

"Of course," said Mom, as I thought about having a large hot chocolate to celebrate all the work we had achieved today.

The three of us grabbed our coats and walked from the Annex Building to the Bloomfield Café, a short walk across the main road.

"Thank you again, Anna," said Mom. "I'm learning so much in your husband's class."

"You're welcome," said Dr. Weber, as we sat down at the one empty table. "It's my pleasure."

"Today was quite humbling," said Mom. "Dr. Aucoin was saying that as a culture, we've bought into the 'self-esteem movement', where it's all about feeling good instead of building character. Like many parents, I raised Bethany by trying to protect her from overwhelming challenges, so that she wouldn't get discouraged. But Dr. Aucoin was saying that this doesn't actually work – this type of parenting actually harms the kids."

"Don't worry, Mom," I said, sipping the hot chocolate that Dr. Weber bought for me. "I'm doing fine."

"Not every teenager is like you, Bethany," replied Dr. Weber. "And Lucy, you're being too hard on yourself. However, my husband is right. Sometimes children raised by 'snowplow' parents experience psychological trauma."

"Snowplow parents?" I asked, raising an eyebrow.

"That's right," said Dr. Weber, putting her coffee cup on the table. "It's where well-meaning parents clear obstacles for their kids, and try to make things as easy as possible for them, to set their children up for success. But as my husband has found, this deprives them of opportunities to overcome setbacks, learn from their failures, and develop their grit. So when these young people face real adversity for the first time, they can experience serious psychological trauma. As a university professor, I unfortunately see this far too often among the first-year students."

Mom nodded, and turned to me. "I'm glad you're going for the Math Olympiad, Bethany. As you know, I tried to steer you away from it, many times. But as I was listening to Dr. Aucoin today, I realized how good it was that you went for it anyway."

"Being gritty means being stubborn," I said, smiling at Mom.

"That's true," replied Dr. Weber. "Speaking of the Math Olympiad, I actually wanted to talk to both of you about Bethany's goal."

I felt suddenly nervous. "What's wrong?"

"Bethany, I read your front-page profile in the Cape Breton Post two summers ago."

"You did?" I said, and felt my cheeks turn red. That explained why Dr. Weber knew that I was planning to go to Quest, since that was mentioned in the article.

"Yes," said Dr. Weber. "My husband Martin is an Acadian from Cape Breton. And we were visiting his brother in Cheticamp last summer when that profile came out. For obvious reasons, my brother-in-law thought that I would be interested in reading about the young female mathlete from Nova Scotia. In the article, you 'guaranteed' that you and Grace were going to make the IMO team."

I looked down at the chocolate residue in my empty mug, embarrassed at being reminded of that newspaper article, and that stupid quote of mine. I saw Mom tense up.

"Bethany," said Dr. Weber, "that article came out sixteen months ago, and I know a lot has changed in your life since then. Now that I've met you, and have had the opportunity to work with you over the past five Fridays, I know you understand the difference between grit and overconfidence. I'm not criticizing you. However, I am concerned that you might be treating the IMO as your everything, and that if you don't make it, you're going to be crushed."

"How do you know?" I said, dreading what she was going to say next.

"Because in the article you said that there was nothing you aspired to do other than the IMO, and you felt extra pressure to become an Olympian, since your Mom just missed the Olympics herself. You said 'I'm going to be an Olympian for both of us', or something like that."

Mom looked away. I was reminded of the tension in our home after that article was published, and Mom and I didn't need to go through that again. I cringed.

"Martin's research colleague is a clinical psychologist, and over the years they've conducted a long-term study with dozens of adults who were traumatized due to experiences they had as teenagers. In many of the cases, they or their parents set incredibly high expectations, and they had no safety net to hold them up in case they didn't achieve their dreams."

"That sounds familiar," said Mom quietly.

"I'm so sorry, Lucy," said Dr. Weber. "I know your story."

"How?" asked Mom, her voice barely a whisper.

Dr. Weber turned towards Mom and leaned her head down to avoid eye contact.

"I watched your Olympic Trials on TV. Of course, I never imagined that we'd be meeting in these circumstances, twenty years later."

Mom didn't respond and bit her lip.

Dr. Weber hesitated before slowly continuing: "It was my first year in Nova Scotia, and my first year in Canada, so I remember it well. I was cheering for you, Lucy, because you were from the same province. You were the last skater in the Ladies Free Skate. My heart broke when I saw your reaction after the judges' scores went up, and I'll never forget that. I'm so sorry, Lucy."

I stared at Dr. Weber, surprised she knew more about Mom's figure skating career than me. I looked over at Mom, and saw her take a deep breath, and pause for a few seconds before responding.

"I didn't have a safety net," said Mom. "I had no backup plan. For me, it was the Olympics or nothing, like that hundred-metre runner in *Chariots of Fire* who said he had 'ten lonely seconds to justify his whole existence'. I felt so much pressure from everybody. That's why I tried to discourage Bethany from pursuing the Math Olympiad because I didn't want her to go through what I did."

"We have much in common, Lucy," said Dr. Weber softly. "I also missed the Olympics by one spot."

Mom and I looked at the professor in surprise.

"What was your sport?" asked Mom.

"Basketball?" I guessed.

Dr. Weber shook her head. "Mathematics."

I nearly knocked over the empty hot chocolate mug in front of me.

"You went to the IMO?" I blurted.

"I almost did," said Dr. Weber. "I was the alternate member of the East German team in my last two years of high school. Imagine that, missing the IMO team by one spot, two years in a row."

Dr. Weber paused before continuing.

"In Germany, I grew up in an environment that valued students who excelled in math. There was a lot of media coverage of young mathletes, just like the Canadian media spend a lot of time covering young hockey players. Here in Nova Scotia, many boys grow up dreaming of becoming the next Sidney Crosby, and getting to the World Juniors and the NHL. In Germany and other Eastern European countries, many boys – and girls – grow up dreaming about the IMO.

"There were a lot of us in my year, all competing for the same prize. We were friends, but we were also rivals. And it was heartbreaking to be that close to your dream – twice! – and not touch it. I felt like a failure, that I let down my teachers, my friends, and especially my parents, who invested so much time and money in me."

"How did you cope?" asked Mom.

"Luckily, mathematics is its own safety net. Even though I didn't make the IMO team nearly forty years ago, I still got into the top university in Germany, and was mentored by some of the best math professors in the country. At university, I learned to truly love math for what it is, finally seeing that mathematics is so much more than solving contest questions. I had too narrow a perspective as an eighteen-year-old, and put too much pressure on myself."

"Is that what I'm doing now?" I asked. "Putting too much pressure on myself?"

"Yes, I think so," said Dr. Weber. "You've told me about Grace and how you're training three thousand hours together, even marking down the hours we spend doing research on Fridays in my office. It's as if you're saying you'll make the IMO team if you hit this three thousand-hour target, but that's not how it works. If you have an all-or-nothing attitude, not making the team might make you risk-averse in the future, and for a long time, it may prevent you from committing a hundred percent of your energy and focus towards a goal. That's what happened to me after I didn't make the IMO team. It's a terrible thing for an eighteen-year-old to go through. Or in your case, Bethany, a seventeen-year-old."

As Mom nodded in agreement, I turned to Dr. Weber. "So what should I do?"

"I think you should stop keeping track of your hours. And you should focus your time doing the mathematics you love, not working through hundreds of old Olympiad problems to learn tricks for contests. That's why I think it's so good you and Grace are collaborating with me on this paper, because you're seeing open-ended research mathematics, where correct answers and elegant solutions often don't exist. Of course, it would be wonderful if you and Grace do make the IMO team – but please know that your contribution and value to this world is so much more than some team you make or don't make when you're seventeen years old. What you're training for now is preparing you for the rest of your life."

"I agree with Dr. Weber," added Mom. "For me, the pinnacle was the Olympics. But for you, it's not. What employer wouldn't jump at hiring someone with your grit, your creativity, and your problem-solving ability? There's a lot more demand for someone with your skills, Bethany, than someone like me whose greatest skill was being able to stick a triple Lutz triple toe loop."

I stifled a grin.

"Bethany," said Dr. Weber, "your career options are limitless. It might be becoming a research mathematician like me, it might be working for the government like your mother, it might be teaching at a school like Mr. Marshall, it might be working in the private sector like many people I know, or it might be something completely else. Who knows, someday you might even run for prime minister."

"Yeah, right," I said. "A mathematician for prime minister?"

"Why not?" said Dr. Weber. "Personally, I think many Canadians would gladly vote for a prime minister who makes decisions based on sound logic and data-driven analysis, and forms policy based on rigorous non-ideological scientific evidence. Wouldn't you?"

"I would," said Mom.

"Speaking of which," said Dr. Weber, "have you heard of Angela Merkel?"

"Yes," I said, remembering a discussion we had in Grade 12 Global History last week. "She's the Chancellor of Germany, their prime minister. The only female leader of a G8-country."

Dr. Weber nodded. "Did you know that Angela Merkel has a Ph.D. in Quantum Chemistry?"

"No, I didn't," I said. "She's a scientist by training?"

"Indeed she is," said Dr. Weber. "After she got into politics, one of her first positions was becoming the minister for Environment and Reactor safety. With her scientific background, she was definitely the best person in the government to tackle issues like nuclear waste. Believe it or not, I actually knew Angela Merkel, because we grew up in the same city, and we even went to the same high school. She's older than me, and I looked up to her very much."

Dr. Weber smiled at me.

"Angela Kasner, as she was known back then, was a Math Olympian for East Germany."

"Really?" I said, shocked.

"Indeed," said Dr. Weber. "Keep your career options open, Bethany. Mathematics prepares you for everything. Even politics."

Mom turned to me. "Bethany, if you run for prime minister, I'll vote for you."

"Thanks, Mom."

# 43

Patrick finished the last sentence of his essay, and looked up at all of us. As the class applauded, Patrick sat back down and gave me a high-five.

"Good job," I said.

"Bethany, you're the next speaker," said Mr. Marshall, checking my name off his list.

I stood up and walked towards the front of the classroom, and saw everyone stare back at me. There were just nine of us in Mr. Marshall's Grade 12 Calculus class: me, Patrick, Gillian, and six others.

It was early February, and we had just started the second and final semester of Grade 12. For those of us in the graduating class, university scholarships and admissions would be decided in April, based on our first semester marks and everything before that. But now that the second semester was under way, we could all breathe a sigh of relief. As long as we didn't fail any courses, our scholarships would be honoured.

All of my classes this semester were a joke, since none of the students or teachers cared. The only exception was this course – likely the hardest course in the Nova Scotia curriculum, taught by arguably the most demanding teacher in the province. None of us would be able to slack off in Grade 12 Calculus, and we all knew that.

At the end of Day 1, Mr. Marshall assigned us a 750- to 1,000-word essay for homework, and that we would be reading our essays out loud two days later, in front of the entire class. I had expected some easy essay question where we'd have to research some real-world application of calculus. Instead, Mr. Marshall assigned us a reflective essay, to answer a personal question:

*What have you learned from studying mathematics?*

Five of my classmates had just finished reading their essays, and had touched upon some common themes: improved problem-solving skills; more self-esteem and self-confidence; having a better understanding of how technology like Google actually works; and for those of us who had been taught by Mr. Marshall in the past, understanding the applications of mathematics to social justice and environmental sustainability.

In Patrick's essay, his concluding paragraph was on the fun he got out of learning math with his classmates, and he shared his experiences doing the Nova Scotia Math League, representing Sydney High with the people who became his closest friends. I liked hearing that.

When I was preparing this essay over the past two nights, I decided to take a different approach.

Other than Gillian who was lost in her own thoughts, eight pairs of eyes were staring at me.

"You may start," said Mr. Marshall.

I cleared my throat, and began reading.

> Mathematics is the only subject that claims its theories can be proved, with complete and absolute certainty. (Other subjects, such as biology and economics and psychology, cannot do this.) This is because mathematical theories represent universal truth, not limited to a single person, country, or historical time period. As I've studied mathematics, I've learned how to make clear arguments by replacing feelings and opinions with rigorous logic, helping me improve my ability to *communicate*, both orally and through writing.
>
> Mathematical structure explains patterns in nature, from planetary orbits that trace out a perfect ellipse to fractal-like features in everything from DNA and blood vessels to snowflakes and broccoli. From practicing thousands of contest problems, I've been able to develop my intuition and ability to recognize these patterns. Because these patterns in nature relate to concepts like symmetry and simplicity, I've learned to appreciate the depth and richness of our physical world, as well as in the abstract world where I can explain, for example, why $\pi$ appears in all sorts of unexpected places, such as in the bell curve distribution formula in statistics, or in identities like $\frac{1}{1^2}+\frac{1}{2^2}+\frac{1}{3^2}+\frac{1}{4^2}+\frac{1}{5^2}+\cdots=\frac{\pi^2}{6}$. As I've studied mathematics, I've developed an understanding of *beauty* – not the superficial beauty of magazine covers, but the real thing.

Through the investigation of these patterns in nature, we can unlock the secrets by which this world came to be. As a result, many scientists from modern history, such as Newton and Galileo, believed that the universe was created in the language of mathematics, and viewed mathematical study as sacred and divine, as it gave us the means to understand where we came from and why. As I've studied mathematics, I've learned that I too am engaged in a *sacred activity* – and this is true regardless of whether I believe in the existence of a higher power.

Most importantly, mathematics has taught me *humility*, more than any other subject I've studied, more than anything I've ever experienced. I've learned humility because of what mathematics is.

Just as Plato identified four principal elements (earth, water, air, fire) that function as the building blocks of life, mathematics requires certain axioms which serve as building blocks from which we can rigorously deduce knowledge. For example, without the parallel postulate, Euclid's fifth axiom, we cannot prove that the angles of a triangle add up to 180°. Everything we learned in geometry class is true because we've assumed these five axioms of Euclid. But once we challenge these assumptions, we can develop new geometries – such as non-Euclidean differential geometry, which was used by Einstein nearly a century later to form his theory of general relativity.

Theorems in mathematics are always conditional on our assumptions of various axioms. Furthermore, a logician named Gödel proved that all consistent mathematical systems are incomplete, that there are true statements that cannot be proved within that system. In other words, "truth" is stronger than "provability" – i.e., there exist true statements that cannot be proven, and are beyond the reach of the smartest mathematicians and the fastest super-computers, regardless of how much we evolve, hundreds or even millions of years from now.

That is why we need to be humble when we claim that certain statements are true or false. We need to first be aware of the axioms we have assumed, and also realize that certain knowledge is beyond our physical or intellectual comprehension.

This reminds me of *Flatland*, a book I read last summer. It's the story of a two-dimensional world where all men are squares, pentagons, and other polygons. The narrator, a humble Square, is visited by a three-dimensional person from Spaceland, who can do all sorts of things that are beyond the Square's understanding – such as making himself disappear and re-appear by jumping up and down. To the Square, what's happening is "scientifically impossible", but that's because he's wrongly assuming that the universe operates in only two dimensions.

Using some incredibly complex mathematical techniques, theoretical physicists have been able to show that a three-dimensional or four-dimensional universe is not compatible with both quantum mechanics and general relativity, but that an eleven-dimensional universe is.

In other words, mathematics helps us understand how much we are influenced by our physical boundaries, and proves that certain universal truths, such as the existence or non-existence of an eleven-dimensional God, can *never* be a hundred percent settled. Because I've studied mathematics, I can be humble about questions of faith, knowing that I don't have all the answers, and don't need to.

On the advice of my best friend, I read through *The God Delusion* by Richard Dawkins. The author spent hundreds of pages arguing against the existence of God, using logical scientific reasoning, personal experience, and common sense. But limited to four dimensions in a universe whose primary components are time,

> space, and matter, none of Dawkins' arguments prove the non-existence of an eleven-dimensional timeless/transcendent/spirit God who isn't bounded by our human limitations. So in this way, studying mathematics has helped me think critically, and even find the logical flaws in a bestselling book written by one of the world's most well-respected scientists.
>
> Just as I can't use logical reasoning to fully answer some of my deepest questions about the meaning of life, I also can't always use logical reasoning to explain surprising phenomena in mathematics. For example, between any two rational numbers is an irrational number, between any two irrational numbers is a rational number, yet the set of rational numbers is "countable" but the set of irrational numbers is not.
>
> Such apparent paradoxes, even within mathematics, humble me into realizing how little I know about the subject, and how little I know about life.
>
> This is what I have learned from studying mathematics.
>
> But thanks to these experiences, I want to learn even more, and continue my quest of pursuing beauty and truth.
>
> And through this roller-coaster journey, I've discovered a sense of purpose: that through mathematics, I can and I will, someday, make an extraordinary contribution to society.

I looked up and smiled.

"Well done," said Mr. Marshall, who started the applause.

I beamed and floated back to my seat, returning Patrick's high-five before sitting back down. As I took a deep breath, I realized that public speaking no longer terrified me; I actually kind of enjoyed it.

"Excellent presentation, Bethany. I like how you touched upon the themes of beauty and truth, and how the study of mathematics should inspire humility in all of us."

Mr. Marshall looked down at his clipboard. "Gillian, you're up next."

Gillian stayed in her seat and gazed outside the window.

Mr. Marshall raised his voice: "Gillian, it's now your turn to speak."

Gillian remained in her seat and continued staring at the window, oblivious to the fact that Mr. Marshall was slowly walking towards her. When Mr. Marshall was right in front of her, she turned to face him.

"I didn't do the essay," she said.

"And why is that?" asked Mr. Marshall.

"Because I didn't feel like doing it," she replied. "You got a problem with that?"

I stared at Gillian in shock. Ever since junior kindergarten, she was Miss Goody-Two-Shoes to every teacher she had. Mrs. Ridley was the only teacher to ever get angry at Gillian, and that was for just one hurtful picture of me that she drew over seven years ago. I couldn't believe she would talk defiantly against any teacher, let alone challenge the most intimidating teacher at Sydney High School.

"Gillian, I'd like to talk to you outside. Right now."

Without another word, Gillian stood up and marched out of the classroom, slamming the door behind her. Mr. Marshall calmly instructed the rest of us to open our textbooks and read through the first unit, and went outside to find Gillian.

As soon as the door closed, I turned to Patrick.

"What was that about?" I said, ignoring the textbook in front of me.

"She's under a lot of stress," said Patrick, looking back to make sure the door was closed. He opened up his textbook just in case.

"What kind of stress?" I asked.

"I don't know," he replied. "But I overheard my father on the phone the other day, and he was talking to another teacher about Gillian's mother, and how . . ."

The door opened.

Patrick immediately stopped talking and buried his head in the textbook, pretending to read. Mr. Marshall walked in, with Gillian shuffling behind him, her head down.

I couldn't concentrate for the rest of class, and was just writing down everything Mr. Marshall put on the board without paying attention to what any of it meant. Every now and then, I'd sneak a peek at Gillian, and saw her discreetly wiping her eyes with her finger.

After the bell rang and we all walked out of class, I ran into Bonnie and Breanna by our lockers.

Bonnie showed me the hundred percent mark from her latest quiz. "You should so be in Social Studies with us."

"Yeah," added Breanna. "The class is a joke. Too bad you're stuck in Calculus – with her."

Breanna pointed five metres behind me, where Gillian was by herself, slowly putting her binders back into her locker. Her best friend Vanessa was nowhere to be found – come to think of it, I hadn't seen Gillian with her, or the Chinese twins, in several months.

I stared at Gillian, who had tears streaming down her face.

"Be right back," I said, walking towards Gillian. "Save me a seat in the cafeteria."

"What are you doing?" hissed Breanna, her voice trailing behind me.

As Gillian noticed me walking towards her, she quickly rubbed her face with her arm and sneered at me.

"What do you want?"

I searched for the right words. "I wanted to check you're okay."

"What's it to you?" she said, snorting in disgust. "I bet you're here just to gloat about the essay."

I paused, staring at Gillian who was trying her best to keep herself composed.

"Gillian, let's stop being rivals."

"Don't patronize me, Bethany."

"No, Gillian. I mean it. We've been enemies since we were kids. And I'm sorry."

She clenched her teeth. "Sorry for what?"

"For everything," I said, meaning every word. "It's unfair we ended up at the same schools, and had to compete this whole time. And it's stupid that only one of us could get nominated for the President's Scholarship. We both deserve it."

Gillian's face flushed at the mention of the $32,000 scholarship.

"Why did you have to apply to St. FX, anyway? You said you were going to the west coast, and you screwed me over by being so selfish. I was supposed to win that scholarship – until you took it from me."

"I haven't won anything yet," I said, knowing that the university wouldn't make their decision for at least another month. "And besides, it's not like there's only one scholarship. You've got the best mark in this school, and you probably have the top grade point average in all of Cape Breton. You're easily going to get one of those $24,000 scholarships to St. FX. That's six grand a year – more than enough for tuition and…"

"That's not good enough," she snapped.

"For who?"

"For my family. Because of you, they think I'm a failure."

"Your family doesn't think you're a failure," I said.

"How would you know?" said Gillian, raising her voice. "Have you ever met my mother?"

I paused and took a deep breath. "If your Mom thinks you're a failure, that's not my fault or your fault. It's your mother's fault. Who cares what she thinks, anyway?"

"It doesn't work that way, Bethany."

"Sure it does," I replied. "Your mother doesn't control you. It's your life, not hers. Why should you have to live somebody else's dream?"

A thought suddenly came to me.

"Hey, do you even want to go to St. FX?"

Gillian looked away, and folded her arms.

"I thought so," I said. "Why are you accusing me of taking something away from you when you don't want it yourself? Why don't *you* go to a different school?"

"Well, that's easy for you to say," said Gillian, folding her arms and glaring at me. "Life's a bit harder when both your older sisters, and both your parents, and all four of your grandparents, all graduated 'summa cum laude' from the same university. And when your mother and sisters then go on to win the Rhodes Scholarship, there are things expected of you. You're lucky you have options in life. I don't."

"You do, Gillian," I said. "You've got options. You've got so many options!"

Gillian snorted. She marched off, not bothering to acknowledge my last comment. I stared at her as she turned the corner and headed towards the cafeteria, tears streaming down her face.

I closed my eyes and sighed. I didn't like Gillian, and I probably never would, but at least I was finally beginning to understand her.

"You've got options," I whispered. "You really do."

# 44

Just as we had done every Friday afternoon for the past five months, we arrived at the St. FX campus at 1:50, and drove into the small parking lot off Notre Dame Avenue.

Before getting out of the car, we put on our jacket, gloves, scarf, and toque. Even though it was early March, it was still freezing outside.

"Have a good class, Mom."

"Thanks, you too!" said Mom, as she walked west towards Nicholson Hall and I headed south towards Dr. Weber's office in the Annex Building.

Mom had finished the introductory Psychology 100 course in December, and this semester was once again auditing a course from Dr. Aucoin. This time, it was a second-year course on Developmental Psychology, conveniently offered during the same 2–5 time slot on Friday afternoons.

For her Christmas gift to herself, Mom bought a new laptop, and had spent most evenings in January and February reading a thick stack of papers and typing up various documents that she said was related to Developmental Psychology. I was shocked Mom was working this hard for a course she was just auditing for fun. Mom would work late into the evenings, and the majority of the time, the TV never came on.

After a quick washroom stop, I walked up to the second floor and knocked on Dr. Weber's door.

"Come in!"

"Hi, Dr. Weber," I said, opening the door.

She looked excited, and greeted me with a big smile. Not quite as excited and animated like Mr. Collins, but close.

"Bethany, have you heard the news?"

"No, what news?" I asked.

"Come and see!" she said, pointing at her laptop and motioning for me to sit next to her.

Without bothering to take off my jacket, I sat down and stared at the screen of her inbox, showing a message from the Editor-in-Chief of *Crux Mathematicorum*, the Canadian problem-solving journal.

I stared at the screen in amazement.

Dear Ms. MacDonald, Dr. Weber, and Ms. Wong,

I am pleased to announce that your paper, "Astonishing Pairs of Integers", has been accepted for publication in *Crux Mathematicorum*, and will appear in the upcoming September issue. In the attached PDF file, you will find the reviewers' comments and suggestions.

Please submit your revised paper directly to me by May 1. I look forward to receiving it. If you have any further questions, don't hesitate to contact me.

Sincerely,
Jeffrey Sato
Editor-in-Chief

"Congratulations, Bethany! It's your first paper."

"Wow. We got accepted. We actually got accepted."

"It's a tremendous achievement," said Dr. Weber, with a big smile on her face. "I got my first publication right after defending my Ph.D. thesis. But you and Grace have already gotten your first publication – and both of you are still in high school!"

"But we only got published because you taught us all this hard number theory and showed us how to write up a paper. If it weren't for you, this wouldn't have happened."

"And I'd argue that if it weren't for *you*, none of this would have happened. Don't forget it was you who came up with the original idea of the astonishing pairs, and that it was you and Grace, who worked together over many evenings to develop the theory and create all this original mathematics. Well done to both of you."

"Thank you," I said, taking off my jacket. "Grace and I now have one publication. We've got a long way before we catch up to you. Don't you have over seventy?"

"Something like that," said Dr. Weber, shrugging modestly. "But don't forget that I'm almost forty years older than you."

"Good point," I said. "So do you get a raise for doing this?"

Dr. Weber seemed puzzled by the question. "What do you mean?"

"You know, more money for every paper you publish?"

"Not at all," she replied. "I'm a full professor, with tenure, so any paper I publish doesn't change my salary. I can publish as much I want or as little as I want. I may not be as prolific as some of my colleagues in the Canadian math community, but I still maintain a strong and active research program."

"So whether you publish zero papers or seventy papers, your salary remains the same?"

"That's right," said Dr. Weber. "Technically, I could have stopped doing research the day I got tenure."

"Why didn't you?" I asked. "I mean, what's in it for you?"

"All this," she said, pointing to the big whiteboard which contained the ideas from our session the previous Friday. "Creating new knowledge. Working with talented students like yourself and mentoring the next generation. Learning from students and from my research collaborators, creating synergy, knowing I'm making a difference in the lives of others. Bethany, how could I possibly give this up?"

She opened up another e-mail from her inbox, and pointed to it.

"And I have more exciting news to share. A few days ago, I got a letter from NSERC, the federal government agency for scientific research. They told me that my grant got renewed for another five years. So this means I can build a better research lab to take advantage of recent advances in scientific computing; more importantly, I can use the grant money to hire summer students and take them to conferences. What this means is that you and Grace, if you choose to, can work with me full-time for twelve weeks next summer – and you'll get paid for it too!"

"That would be awesome," I said, thrilled by the prospect. "And congratulations on getting your research grant!"

I paused, suddenly remembering something important.

"But what if Grace and I end up going to Quest? Can we still be your summer students?"

"Of course," replied Dr. Weber. "Many universities have collaborative partnerships, where they actively encourage their students to go elsewhere for a summer and work with other researchers. So whether you come go to St. FX, or to Quest, or to another university, I'd be delighted to take you on as research assistants – for one summer, or for multiple summers."

"So there's no catch?"

"None whatsoever. As I've told you many times, Bethany, our collaboration has been about much more than me trying to recruit you to my university, and I've gotten just as much out of this as you have, if not more. Personally, I think it's a good thing that you and Grace aren't rushing your decision on where to spend the next four years of your life. It's a big commitment, with many factors to consider."

I nodded, relieved that Dr. Weber wasn't pressuring me to go to St. FX and stay on the east coast, since Grace was still subtly pressuring me to join her on the west coast.

"Yeah, we're still not sure whether we should do a broad liberal arts and sciences degree, or specialize and do a math major. We're going to make our decision once we learn where we get in, and whether we win anything. I can't afford either school without a scholarship."

"Both St. FX and Quest would be excellent choices for your undergraduate degree, as would many other schools," said Dr. Weber. "Just so you know, I know all the math 'tutors' at Quest."

"Really?" I said, having a flashback to my visit there two summers ago. "Grace and I met Robert Cooper during a campus visit."

"Oh, yes," said Dr. Weber, nodding. "He's quite the handsome young man. I'm sure you and Grace were extremely impressed."

"Not really," I said. "He looks exactly like my ex-boyfriend."

Dr. Weber smiled, somewhat apologetically.

"It's no big deal," I replied. "We broke up a long time ago."

Dr. Weber quickly changed the subject, and told me about another of the Quest math tutors, who she met when he was a graduate student in Nova Scotia. She said that her friend did a lot of math contests in high school and university, and from that, learned the ability to solve hard problems by re-converting them into equivalent simpler problems. This skill came in handy when he later did his Ph.D. thesis, successfully solving some open problems in algebra and combinatorics using Graph Theory, creating new techniques that connected different topics in pure mathematics.

But then, Dr. Weber continued, this colleague of hers got really discouraged, since so few people had bothered to read any of the seven publications that came out of his thesis. He felt all that work had no impact, and he became cynical about academic research in pure abstract mathematics, which required one to specialize in a narrow area so that the results were only relevant and meaningful to just a handful of other mathematicians. He became so convinced that theoretical math research was pointless that he took a job as an applied mathematician in the Government of Canada.

"Just like Mom," I said. "She works for the government too."

Dr. Weber explained that her colleague from Quest switched from being a mathematical specialist to becoming a mathematical generalist, and he used computer science and statistics to improve border security and efficiency. The last project he did was applying a technique called "integer programming" to design a more efficient staffing schedule for customs officers to cut down wait times at border crossings. He and his staff showed how a centralized math-based scheduling system would increase efficiency at every border crossing and airport in Canada.

While the higher-ups loved the idea in principle, they were slow to implement it, instead choosing to deploy his scheduling program at only a handful of airports, one or two each year. Dr. Weber explained that her colleague got discouraged again, wondering why they were paying him so much money for work that didn't lead to meaningful change for Canadians.

"Yeah, that's government," I said. "Mom always complains about how government is so slow to change, that even the simplest of projects take years to get done. She says that I've developed a lot of perseverance from doing math contests, but that I'd develop even more if I joined the public service."

Dr. Weber laughed and continued her story: "A few years later, my friend moved to Japan and found a position working at a research lab in Tokyo. One day, riding the train into downtown Tokyo, he was looking at the regular-season schedule for the professional baseball team in his hometown, and noticed how inefficient the schedule was. He realized that each team could save a lot of time and money by re-ordering the sequence of games to maximize multiple-game road trips against different opponents whose home stadiums were located in the same geographical area. He realized it was a complicated scheduling problem, with similarities to his final government project on reducing border wait times.

"While thinking about baseball scheduling on the train that day, he popped open his cell phone and used the little applet that told him how to get to his destination in the quickest way. Because the Tokyo subway system is just one gigantic graph, with train stations as vertices and train lines as edges, my friend knew that this applet was an application of a well-known shortest-path algorithm, a method in Graph Theory that explains how to find the shortest path from point A to point B."

"Dijkstra's Algorithm," I said, remembering a lesson from Mr. Collins years ago.

"Yes, that's right!" said Dr. Weber. "And then my colleague from Quest had a crazy thought, that this complex problem in baseball scheduling could be converted into a simpler equivalent problem of finding the 'shortest path' from the start of the season to the end of the season. It took him about a month to work through all of the details to come up with a formal proof of why the two problems were equivalent, but because he had so much theoretical training in pure mathematics, he was able to figure it out. By solving the scheduling problem using Graph Theory, he showed how the Japanese baseball league could save hundreds of thousands of dollars in travel costs each year, while simultaneously reducing their greenhouse gas emissions."

"Did the baseball league hear about this?" I asked.

"Yes," she replied. "It took a long time for them to agree to meet my friend, but eventually they brought him in as the scheduling consultant."

"Mr. Marshall would love this – he says mathematics is the foundation for economic/environmental win-wins."

Dr. Weber nodded. "So why do I mention this story? It's because you've asked me about the value of pure research. And even in the life of one mathematician, you see how seemingly 'useless' research from a barely-read Ph.D. thesis, combined with the disappointment of a government project that never got fully implemented, was exactly what my colleague needed to make the connection between the simple train connector applet on his cell phone and the complex scheduling problem for Japan's billion-dollar professional baseball league."

"That's like your story of the British mathematician Hardy," I said. "You know, the applications comes later, like number theory turning into the basis for security on the internet?"

"Not exactly," replied Dr. Weber. "In the case of my colleague at Quest, his Olympiad training and his graduate school research didn't lead to any practical real-world applications, but it was what he learned from that process that led to all these unexpected benefits fifteen years later. That's how mathematicians often do their research – choosing problems for the pursuit of truth and beauty, rather than problems they think might be 'useful' someday. Then, all these unexpected benefits arise.

"Unlike other scientific disciplines that require millions or billions of dollars in advanced equipment, mathematical research hardly costs anything, and yet you'd be hard-pressed to name a twenty-first century innovation that doesn't have pure mathematics as its foundation. That's why, in this age of economic and environmental uncertainty, we need more funding for mathematical research, to develop young minds like yourself to become specialists or generalists, and drive the innovation for Canada's future."

"Should I become a specialist or a generalist? What would you recommend?"

"That's completely up to you," replied Dr. Weber. "And that is a decision you can make many years from now. What we're doing is looking at something specialized and abstract, exploring solutions to Diophantine Equations in number theory. But look at how we've done that, using Maple – and in the process, you've acquired key skills in computer programming. By the time you finish your schooling, you'll have the training to become a specialist or a generalist, or possibly both!

"If you're interested in applications, you'll be interested to know that Canada is among the world leaders in some key research clusters emerging in the twenty-first century: information processing, computational algorithms, wireless technology, digital media technology, speech recognition, image recognition, fuel cell technology, and more. Of course, mathematics is at the heart of each one. Bethany, I can easily see you becoming a world leader in one of these areas."

"Really?" I asked. "I mean, you really think so?"

As Dr. Weber nodded and smiled, I turned to face her. There was something I needed to say.

"If I hadn't met you, I would have never known that math is so much bigger than the IMO. Thank you."

"You're welcome," said Dr. Weber. "Of course, the IMO is an important goal, and I'm so glad you're pursuing it. But now you've learned that there are all these unexpected benefits from the training you've put in over the past few years, and that someday, society will reap the rewards. And as you've discovered, there's so much to your life beyond the Olympiad."

She paused and looked at the calendar on her wall. "Speaking of which, isn't your big contest this month?"

"That's right," I replied. "I'm writing the Canadian Math Olympiad in two weeks. I'll be nervous, but Grace and I have been training every day – doing contest prep on our own and original research with you – and we're ready."

"Yes, Bethany, you're definitely ready."

I gazed at my professor in admiration.

"Ever since I stepped foot on this campus, I feel like my perspective has changed so much – and because of these Friday visits, my life is now heading in a completely new direction."

Dr. Weber smiled and stood up from her chair, and walked towards the window. She invited me to come up and stand next to her, where she pointed her finger towards Nicholson Hall, where the Psychology department held their classes. Dr. Weber turned towards me.

"I know someone else who feels the same way you do."

# 45

I put my pen down and checked the two sheets of paper. Yes, every sentence was identical.

Carefully folding each page, I inserted each letter into a separate envelope. I opened the door to my room, and stepped out into the hallway.

"Mom, I'm ready."

"Great," she said, looking up from her *Developmental Psychology* textbook and inviting me to sit next to her on the couch. We huddled close to the fireplace, which we used every evening throughout the winter. Even though it was already mid-March, it was still cold outside, and we both loved the smell of the warm fireplace.

Tomorrow was the big day: the Canadian Mathematical Olympiad.

Six years of training had come down to this – five questions, three hours, one dream.

I knew I wouldn't be getting much sleep tonight.

Over the past week, I'd been thinking about so many things, and Mom encouraged me to write my thoughts down on paper. I had just spent the past two hours collecting my scattered sheets of reflections, and synthesizing everything onto a single page. My English essay could wait until tomorrow.

I read and re-read my letter and felt that I had said everything I wanted, knowing that I wouldn't be tossing and turning tonight, wondering if there was something I had forgotten to say.

*9:00 p.m.*

I glanced at the clock, realizing that the contest would begin in exactly twelve hours.

Mr. Marshall had graciously arranged for me to write the CMO in the office of the Cape Breton Regional School Board, just a few minutes away by car. They had a boardroom that wasn't being used tomorrow morning, and Mr. Marshall thought that I would benefit from writing the contest in my own space, away from any distractions. After that experience with Gillian in the school library, I was grateful.

"Mom, do you know how to get to the School Board office?"

"Of course," she replied. "I drive past there every day. I'll get you there tomorrow in plenty of time."

Mom looked ecstatic, and was beaming throughout dinner. I had no idea what she was so happy about, but I'm sure it had something to do with the white envelope she was holding close to her chest, some important "news" she promised she would tell me after I had finished writing my letter.

"Would you like some tea, Bethany?"

"No, I'm fine."

I took a seat next to her and handed her one of my two envelopes, while holding the other in my hand.

"One for you, one for me."

"Thank you," replied Mom, opening her envelope to reveal my letter. "Would you read it to me?"

I smiled and took the letter from her hand, and began reading the words I had so carefully prepared.

> Tomorrow, I'm writing the Canadian Mathematical Olympiad. If I can solve all five problems within the three-hour time limit, I'll fulfill my dream of becoming a Math Olympian.
>
> Over the past six years, so many people have helped me get to where I am today. Regardless of what happens tomorrow, I'm dedicating this contest to each of them.
>
> I'm writing for Mr. Collins: for working with me every Saturday for three years, for introducing me to the wonder of math and inspiring me to love the subject, and for showing me that math is applicable and relevant to everything in this world.
>
> I'm writing for Miss Carvery: for encouraging me after I came in second place at the Pinecrest Spelling Bee, for celebrating with me after I got perfect on the Gauss Contest, and for being the most amazing vice-principal anybody could ask for.
>
> I'm writing for Rachel Mullen: for inspiring me to follow in her footsteps and join her on this roller-coaster journey, for encouraging me to pursue a life of courage rather than a life of comfort, and for showing me that a small-town girl can indeed achieve her wildest dreams.

I'm writing for Marlene Thomas: for inviting me to the Canada Math Camp and changing the direction of my life, for opening my eyes to the importance of struggling in something to master it, and for believing in my potential which inspired me to believe in my potential too.

I'm writing for Mr. Marshall: for challenging me to set high expectations, for demonstrating unwavering integrity, and for showing me that math can be used to solve real-life problems to address social justice issues and create an environmentally-sustainable world.

I'm writing for Mr. Campbell: for choosing me as the captain of the Cross-Country team, for believing that I am a leader, for teaching me that grit and desire are more important than natural talent, and for celebrating with me when I ran my personal best at the provincial finals.

I'm writing for Patrick and all the people on the Sydney High School Math League team: for showing me that math is so much more fun when it's done in cooperation rather than in competition, for developing my love of teaching, and for being great teammates and even better friends.

I'm writing for Bonnie and Breanna: for being so supportive of me all these years and including me in everything, for keeping me as their friend when I was selfish and stuck in my own world, and for encouraging me to pursue this dream, no matter how unrealistic it seemed at times.

I'm writing for Raymond Wong: for being such an amazing dad to Grace and supporting both of us in Vancouver, for teaching me that math enables us to think deeply about the big questions of life, and for helping me understand that there is no contradiction between faith and science.

I'm writing for Dr. Weber: for mentoring me every Friday and teaching me the joy of creating original mathematics, for showing me that abstract theoretical math is worth doing for its own sake, and for opening my eyes to the exciting possibilities for life beyond the Olympiad.

I'm writing for Grace: for the memories of our time in Ottawa and Vancouver where we developed an unbreakable bond, for all the hours we spent training together and supporting each other's dreams, and for being the most amazing friend anybody could ever ask or imagine.

I'm writing for Sydney High School, the Cape Breton School Board, and the entire province of Nova Scotia. I know that no Nova Scotian, boy or girl, has ever made the Canadian team to the International Mathematical Olympiad. Tomorrow I'll do my best to change that.

Most importantly, I'm writing tomorrow's contest for Mom.

You've always loved me, even when I was too stubborn to realize how much you cared.

You've always sacrificed for me, even becoming Ella's figure-skating coach, to grant me hope and opportunity.

You've always been there for me, not for 3,000 hours but for 150,000 – every hour I've been alive.

Mom, you mean so much to me. I'm going to do my best for you tomorrow and will honour you by giving it my all.

Thank you, Mom. For everything.

I love you.

When I looked up, Mom had tears in her eyes.

"Thank you, Bethany."

"You're welcome," I said, returning her copy of the letter. "That's yours to keep."

"I appreciate that," she replied. "In fact, I know just what to do with this."

Mom carefully folded the letter and put it back in the envelope. She then got off the couch, took a few steps forward, and calmly dropped my masterpiece into the fireplace.

"What are you doing?" I shouted, jumping off the couch.

I stared as the white envelope slowly turned bright orange and then disintegrated into tiny black ashes.

After the final corner of the letter had melted into the fireplace, I clenched my jaw and glared at Mom, who gently touched my arms and motioned for me to sit back down on the couch with her.

"Bethany . . ."

"Mom, I can't believe you just did that!"

"I want you . . ."

"You had no right to do that!"

Mom grabbed my hands and looked into my eyes. I saw an intensity in her eyes that matched the fire in mine, and was taken aback. Mom never showed that kind of emotion.

"Bethany, listen to me!"

I folded my arms and leaned back on the couch. I shook my head in disbelief, shocked that Mom would do something so hurtful, especially after I wrote how much I loved her.

I turned my head away and blinked back a tear. Eyes red, I turned back to face Mom.

"Okay, fine. I'm listening."

Mom turned her body towards me so that we were face-to-face.

"Bethany, you're not writing tomorrow's contest *for* anyone. You're not writing it for me, or for Grace, or for Mr. Collins, or for Rachel, or for anybody else you mentioned. And you're certainly not writing it for the entire province of Nova Scotia!"

"But, Mom . . ."

"Think about how much pressure you're putting on yourself, feeling the weight of the province on your shoulders, thinking about how so many people have invested in your life over all these years, and how you want to perform your very best so that you won't let any of them down."

My face flushed. Mom was right.

All I had been thinking about was *not* bombing tomorrow's contest, and hoping somehow that my performance would be able to live up to the expectations of those in my life.

Mom touched my hands and spoke quietly.

"At the Olympic Trials, I had a flawless short program; it was the best skate of my life. I was in first place heading into the long program, by a wide margin. All I had to do was skate my four-minute long program, which I had practiced thousands of times, and I'd make the Olympic team. I couldn't sleep that night. All I could do was think about the people who had helped me get to that point – my coaches, my teachers, my friends, the locals around Cape Breton, and especially my parents.

"You know that my mother, your grandmother, died of cancer just a couple of years before you were born. She died just two weeks after the Olympic Trials. And while she was in the hospital, I put so much pressure on myself, to try to have the perfect skate – for her. To thank her for all the sacrifices she made to let me pursue my dream.

"I put the wrong type of pressure on myself, not the healthy pressure that expects success, but the unhealthy pressure that fears failure. We were talking about healthy and unhealthy pressure in Dr. Aucoin's class the other day – it was tough to listen because I've experienced everything he's talking about. For most of the students in the class, it was just a bunch of content they need to know to pass a course. But for me, I lived it."

"What happened?" I whispered, knowing I had been waiting my whole life to hear this story.

"The long program was a disaster. I fell on my final three jumps. I even messed up my closing spin, and even the least experienced amateurs don't do that. On the most important skate of my life, I choked. I absolutely choked."

I reached over and held Mom's hand.

"The first three minutes were perfect, but then I fell on my fifth jump. And even though I was easily in first place, I started to feel panic, thinking to myself that I better not mess up the next jump. And then I under-rotated my double axel, and I fell again. My confidence was shot. I began thinking of my Mom, and how I couldn't afford to let her down. I was determined to nail the final jump for her, that somehow if I could stick the landing, I'd make the team and her cancer would disappear. I jumped way too high and couldn't control the landing. And I fell again. I got up and barely finished. And that was it. In just sixty seconds, I went from first to third, and my Olympic dream was over."

I bit my lip and tried to hold it together.

"That's all it took, Bethany. One minute. And because of that one terrible minute, I missed the Olympic team by one-tenth of a point. I was that close."

A tear fell out of my eye.

"Bethany, look at me."

As I turned to face her, Mom stared lovingly into my eyes.

"You're not writing tomorrow's contest to please anybody. You're not doing it to satisfy the expectations of others. You're doing it because math gives you joy."

I nodded.

"Okay," I whispered. "Just me and the five problems. That's it."

"Exactly," said Mom, standing up. "And this way you won't spend the next twenty years of your life regretting what happens tomorrow. Because you're pursuing this for the right reason."

I stood up too.

Mom looked at me. "Now close your eyes and open your hands."

I did, and noticed that Mom put my copy of the letter into my cupped hands. Mom lowered her voice so that she was almost whispering.

"Now think of all the people on your list – and think of all the pressure you feel they're putting on you, all the expectations you feel they have of you to win the Olympiad tomorrow. Do that for each person on your list."

"Okay," I said, feeling overwhelmed by all the pressure building up inside my body. I went through the list one by one: Mr. Collins, Miss Carvery, Rachel, Marlene, Mr. Marshall, Mr. Campbell, Patrick, the Math League team, Bonnie, Breanna, Ray, Dr. Weber, Grace, the Cape Breton community, and Mom.

After about three, four, five minutes, my body began shivering.

"Now take a deep breath, and exhale all that pressure inside of you, and put it into that envelope you're holding."

"Okay."

"Bethany, take all the pressure you feel, all of it, and put it into that envelope. And when you've finished doing that, open your eyes."

When I opened my eyes, the envelope felt heavy in my hands, and it shook from side to side as I trembled. Mom looked at me.

"You know what the next step is."

Without any hesitation, I marched up to the fireplace and tossed the heavy envelope onto the burning coals. I stared silently as the flames burned up six years of built-up pressure welling inside of me. As the final corner of the letter melted away into nothingness, I took a deep breath, feeling the weight of the world lift off my shoulders.

Mom walked up to me, and as we stared at the fireplace together, she put her arm around me.

"Doesn't that feel great?"

"Yeah, it does," I softly replied. "I feel so much lighter."

She turned to face me.

"You got to spend the last six years pursuing a dream that filled your life with energy and passion. You got to spend several thousand hours reaching for a goal that gave you purpose and meaning. Regardless of what happens tomorrow, whether you solve zero problems or all five, you'll have no regrets."

"Yeah," I nodded. "No regrets."

Mom motioned towards the couch. "Here, let's sit back down."

"You should become a professional psychologist," I said. "You're a natural."

Mom laughed. "Thanks. That's really flattering."

"No, I mean it. You're learning all this psychology stuff from Dr. Aucoin, and you'd be amazing at it."

"Well, thank you, Bethany. I appreciate that."

Mom grabbed the one remaining unburned envelope, the letter that she kept around her for the past few hours, and looked into my eyes.

"I know you want to watch *Soul Surfer* before you go to bed tonight, and I'm looking forward to seeing the movie with you. But before we watch the movie, there's something I want you to know."

I stared at her, wondering what could possibly be in that envelope. She smiled.

"I'm quitting my job."

My jaw dropped. Of all the things I thought she'd say, this was the last thing I imagined I'd hear. Mom despised her job as an administrative assistant at the Canada Revenue Agency, but always said she never could, or ever would, walk away from the perks of a permanent job in the federal government: the salary, the security, the benefits, the pension.

I was at a loss for words.

After a few seconds, I pointed to the thick white envelope in her hand.

"You got a new job?" I asked.

"Yes," she said, handing the envelope to me. "I most certainly did."

I opened the letter and before I finished reading the first sentence, I could feel the tears welling up.

As I finished the letter, I put my hand up to my mouth, in disbelief, and started to sob.

I threw my arms around Mom.

"Thank you, Bethany," she whispered. "For everything. I love you."

## The Canadian Mathematical Olympiad, Problem #5

**Let $m$ be a positive integer.**

**Define the sequence $x_0, x_1, x_2, \ldots$ by $x_0 = 0$, $x_1 = m$, and $x_{n+1} = m^2 x_n - x_{n-1}$ for $n = 1, 2, 3 \ldots$.**

**Prove that an ordered pair $(a, b)$ of non-negative integers, with $a \leq b$, gives a solution to the equation**

$$\frac{a^2 + b^2}{ab + 1} = m^2$$

**if and only if $(a, b)$ is of the form $(x_n, x_{n+1})$ for some $n \geq 0$.**

# Solution to Problem #5

I review what I've discovered so far, hoping for a breakthrough.

For the case $m = 2$, I have the sequence of integers $x_0, x_1, x_2 \ldots$ whose initial values are $x_0 = 0$ and $x_1 = 2$. All other terms are defined by the recurrence relation $x_{n+1} = 4x_n - x_{n-1}$.

For a general value of $m$, I have the sequence of integers $x_0, x_1, x_2 \ldots$ whose initial values are $x_0 = 0$ and $x_1 = m$. All other terms are defined by the recurrence relation $x_{n+1} = m^2 x_n - x_{n-1}$.

Substituting $n = 1$, I get $x_2 = m^2 x_1 - x_0 = m^2 \times m - 0 = m^3$.

Substituting $n = 2$, I get $x_3 = m^2 x_2 - x_1 = m^2 \times m^3 - m = m^5 - m$.

And so on.

I need to show that each pair $(a, b) = (x_n, x_{n+1})$ of consecutive terms in my sequence satisfies the equation $\frac{a^2+b^2}{ab+1} = m^2$. And I know that I can do this with just a few lines of algebra, using the same technique of mathematical induction that I did for the specific case $m = 2$.

Glancing at the clock, I realize that I have no time to spare. I begin writing frantically, and a few minutes later I have a proof for the first half, the *sufficiency* direction. The hard part is the other direction, to prove that this condition is also *necessary*:

If $(a, b)$ satisfies $\frac{a^2+b^2}{ab+1} = m^2$, then $(a, b)$ must be of the form $(x_n, x_{n+1})$ for some value of $n \geq 0$.

*11:45 a.m.*

Just fifteen minutes left. I'm so close.

The most natural approach is to try Proof by Contradiction, the technique of assuming that the desired statement is not true, and showing this assumption leads to a logical contradiction. With so little time left, I have to try something.

I assume there exists some solution $(A, B)$, with $0 \leq A \leq B$, that is *not* of the form $(x_n, x_{n+1})$ for any value of $n$. Somehow I need to show this assumption leads to an impossible statement.

This "bad pair" satisfies the equation $\frac{A^2+B^2}{AB+1} = m^2$, which is equivalent to $A^2 + B^2 = m^2(AB + 1)$, or $B^2 - (m^2 A)B + (A^2 - m^2) = 0$.

Great! I have a quadratic equation in terms of the variable $B$.

Because there are too many variables to deal with, I just re-consider my simple case from earlier, $m = 2$, and hope that I can find some kind of general pattern that works for all values of $m$.

Letting $m = 2$, I have $B^2 - (4A)B + (A^2 - 4) = 0$. I choose to re-write this quadratic equation in a more familiar form, using the variable $x$:

$$x^2 - 4Ax + (A^2 - 4) = 0$$

If $A$ is fixed, this quadratic equation *must* have $B$ as a root, as well as some other value $C$. For example, if $A$=8, then this quadratic equation becomes $x^2 - 32x + 60 = 0$, and from my earlier calculations, I know that there are two solutions: $B = 30$ and $C = 2$.

I nearly jump out of my seat, seeing a critical insight that eluded me a few minutes earlier.

*Yes, I think this works!*

Given some pair $(a, b)$ of non-negative integers, with $0 \leq a \leq b$, satisfying the equation $\frac{a^2+b^2}{ab+1} = m^2$, we can automatically produce another pair of integers that satisfies the same equation. I use some algebra to show that this pair must be of the form $(c, a)$, with $0 \leq c \leq a$.

For example, if we start with $(A, B) = (8, 30)$, this satisfies the equation $\frac{A^2+B^2}{AB+1} = 4$. From this, we solve the resulting quadratic equation to show that the smaller root is $C = 2$, which produces another solution, (2, 8).

If $A = x_n$ for some value of $n$, I know that $B$ must equal $x_{n+1}$ from my algebraic induction work earlier: there are only two solutions to the quadratic equation for $A = x_n$, with the larger solution $B = x_{n+1}$ and the smaller solution $C = x_{n-1}$.

*11:48 a.m.*

I can feel my heart pounding.

Starting with any pair $(A, B)$ satisfying the equation $\frac{A^2+B^2}{AB+1} = 4$, we can generate a *smaller* pair $(C, A)$ also satisfying this equation.

I see this from above: if $(A, B) = (x_n, x_{n+1})$ satisfies the equation, then so must $(C, A) = (x_{n-1}, x_n)$.

If $(A, B)$ is a good pair, then we can always generate a smaller good pair $(C, A)$. It's like induction, only backwards.

But what if $(A, B)$ is a bad pair, i.e., one that is *not* of the form $(x_n, x_{n+1})$ for any value of $n$. Does that mean that $(C, A)$ is also a bad pair?

A quick case analysis shows that yes indeed, that's true: if $(A, B)$ is a bad pair, then $(C, A)$ must also be a bad pair. Furthermore, $(C, A)$ is a *smaller* bad pair, since $C < A < B$.

Great. But what can I do with this?

*Oh my God.*

The solution is staring me in the face.

1) I need to prove that no bad pairs exist.
2) In other words, I need to prove there cannot exist any solution $(a, b)$ in non-negative integers, with $a \leq b$ where $(a, b)$ is not of the required form $(x_n, x_{n+1})$.
3) I assume that the conclusion is false – there exists at least one "bad pair".
4) Because we're dealing with non-negative integers, there must exist a *smallest* bad pair $(A, B)$. Specifically, among all bad pairs, there exists a pair where the sum of the two terms is minimized.
5) I use this assumption to prove the existence of a bad pair $(C, A)$, which is smaller than this bad pair $(A, B)$. Because we've explicitly said that $(A, B)$ is the smallest bad pair, we've obtained a contradiction.

*Yes!*

I pick up my pen and start writing as fast as I can, knowing the contest ends in a few minutes. I start writing up the second half of my solution, and ignore the sound of the boardroom door opening.

If $(A, B)$ is the smallest bad pair, I use a few lines of algebra to show that there must exist another solution $(C, A)$ to the initial equation. To complete the proof, I need to show that $(C, A)$ is a smaller bad pair. I first justify that $C$ must be an integer, and that this integer $C$ is less than $A$. There are three cases to consider: either $C$ is negative, $C$ is zero, or $C$ is positive.

*11:57 a.m.*

If $C < 0$, I prove that this produces an impossible equation where the left side is less than 0 and the right side is greater than 0. This is a contradiction. Check.

*11:58 a.m.*

If $C = 0$, I explain that this contradicts our hypothesis that $(A, B)$ is a bad pair. Check.

"One minute left, Bethany," says a loud voice, momentarily breaking my concentration.

If $C > 0$, that means $(C, A)$ is a bad pair – and I've already shown that it is a smaller bad pair, so that's my final contradiction. I have proven that no bad pairs exist.

I write furiously, ignoring the searing pain in my shoulder, and finish writing my final sentence.

"Time's up, Bethany. Please put your pen down."

I look up and start hyperventilating.

*12:00 p.m.*

Oh my God, I did it. I solved all five problems.

*Yes, yes, yes, yes, yes!*

**Problem Number:** 5
**Contestant Name:** Bethany MacDonald

We will prove that $(a,b) = (x_n, x_{n+1})$ is a necessary and sufficient condition for the pair $(a,b)$ of non-negative integers, with $0 \leq a \leq b$, to satisfy the equation $\frac{a^2+b^2}{ab+1} = m^2$.

To prove sufficiency, we proceed by induction. The case $n = 0$ is clear, since $(a,b) = (x_0, x_1) = (0, m)$, which satisfies the desired equation.

By definition, we have $x_{k+1} = m^2 x_k - x_{k-1}$, and so

$$\frac{{x_{k-1}}^2 + {x_k}^2}{x_{k-1}x_k + 1} = m^2 \Rightarrow {x_{k-1}}^2 + {x_k}^2 = m^2(x_{k-1}x_k + 1)$$

$$\Rightarrow {x_{k-1}}^2 + {x_k}^2 + (m^4 x_k^2 - 2m^2 x_k x_{k-1}) = m^2(x_{k-1}x_k + 1) + (m^4 x_k^2 - 2m^2 x_k x_{k-1})$$

$$\Rightarrow {x_k}^2 + (m^2 x_k - x_{k-1})^2 = m^2(x_k(m^2 x_k - x_{k-1}) + 1)$$

$$\Rightarrow \quad {x_k}^2 + {x_{k+1}}^2 = m^2(x_k x_{k+1} + 1) \quad \Rightarrow \quad \frac{{x_k}^2 + {x_{k+1}}^2}{x_k x_{k+1} + 1} = m^2$$

Thus, we have shown that if $(x_{k-1}, x_k)$ satisfies the equation $\frac{a^2+b^2}{ab+1} = m^2$, then so must $(x_k, x_{k+1})$. This completes the induction, and we conclude that $(a,b) = (x_n, x_{n+1})$ is a solution for all integers $n \geq 0$.

We now prove that this condition is also necessary, that there are no other pairs $(a,b)$ of non-negative integers for which $\frac{a^2+b^2}{ab+1} = m^2$.

Suppose on the contrary that at least one "bad pair" exists, i.e., there is a pair $(a,b)$ of non-negative integers with $a \leq b$ for which $(a,b) \neq (x_n, x_{n+1})$ for every $n \geq 0$. Consider the bad pair $(A,B)$, with $0 \leq A \leq B$, for which the sum $A + B$ is *minimized* over the set of bad pairs.

Then $\frac{A^2+B^2}{AB+1} = m^2$, which implies that $A^2 + B^2 = m^2(AB+1)$, or $B^2 - (m^2 A)B + (A^2 - m^2) = 0$.

If $A = 0$, then $B^2 = m^2$, implying that $B = m$. Then, $(A,B) = (0,m) = (x_0, x_1)$, contradicting the definition of a bad pair.

So we can assume that $A \neq 0$ and that $0 < A \leq B$. Let $A$ be fixed. Then the quadratic equation $x^2 - (m^2A)x + (A^2 - m^2) = 0$ has $B$ as a root, as well as some other value $C$.

Since $x^2 - (m^2A)x + (A^2 - m^2) = (x - B)(x - C)$, we have $B + C = m^2A$ and $BC = A^2 - m^2$.

Since $0 < A \leq B$, this implies that $C = \frac{A^2 - m^2}{B} \leq \frac{A^2 - m^2}{A} = A - \frac{m^2}{A} < A$. Note that $C$ is an integer, since $C = m^2A - B$, and $A, B, m$ are all integers by definition. We have three cases to consider:

Case 1: If $C < 0$, then $A^2 - m^2 = BC = C(m^2A - C)$, which simplifies to $m^2(AC + 1) = A^2 + C^2$. This is a contradiction, since the left side can't be positive (since $AC \leq -1$), but the right side is.

Case 2: If $C = 0$, then $A^2 = m^2$. Since $A$ is non-negative, this forces $A = m$ and $B = m^2A - C = m^3$. This implies that $(A, B) = (m, m^3) = (x_1, x_2)$, contradicting the definition of a bad pair.

Case 3: If $C > 0$, then $(C, A)$ is a bad pair with $0 < C < A$, whose terms sum to $C + A$. Since $C + A < A + B$, this is a contradiction as we have specified that $(A, B)$ is the bad pair with *minimal* sum!

We have therefore proven that bad pairs do not exist, i.e., the only solutions to $\frac{a^2+b^2}{ab+1} = m^2$ are precisely the pairs $(a, b)$ that equal $(x_n, x_{n+1})$ for some $n \geq 0$.

Our proof is complete!

## Press Release

The results of the CMO were officially announced one month later.

OTTAWA, Ontario — The Canadian Mathematical Society (CMS) is pleased to announce the winners of this year's annual Canadian Mathematical Olympiad (CMO) competition.

1st Prize: Bethany MacDonald, Sydney High School, Sydney, NS
(tie) Albert Suzuki, J.G. Diefenbaker Academy, Toronto, ON

2nd Prize: Raju Gupta, Metcalfe Secondary School, Ottawa, ON

3rd Prize: Grace Wong, Vancouver Independent School, Vancouver, BC

"I would like to congratulate the winners of this year's CMO on their outstanding performance," said J. William Graham, the executive director of the CMS. "To obtain such excellent standings on an advanced national mathematics competition is a testament to the dedication and hard work they put into preparing for this competition."

The CMO is Canada's premier national advanced mathematics competition and is staged by the Canadian Mathematical Society (CMS). Students who participate in the CMO are invited based on their scores in the Canadian Open Math Challenge (COMC) or their scores in other recognized competitions. Students invited to participate in the CMO must be of Canadian citizenship or be permanent residents of Canada.

This year's 1st Prize Winners, Bethany MacDonald and Albert Suzuki, both received a perfect score of 50 out of 50.

"The CMO is an advanced competition that deeply challenges our students. Fifty students studied diligently to prepare for this year's CMO," said Dr. Graham. "These types of competitions are essential to education because they inspire students to develop their problem-solving ability – a skill that is useful no matter what career path they decide to pursue."

# Epilogue

It's the first Sunday in September.

I take my fifth and final photograph, turn it around with my fingers, and stick an adhesive strip on each of the four corners. I carefully press the photo onto the freshly-painted wall, in the exact position I want it, just above my new desk. It's perfect.

"All done."

"Let me see," says Grace, walking across the imaginary line that separates our new dorm room, and surveys my artistic creation.

There are five photographs on the wall, with the upper-right corner of each photo touching the bottom-left corner of the photo above.

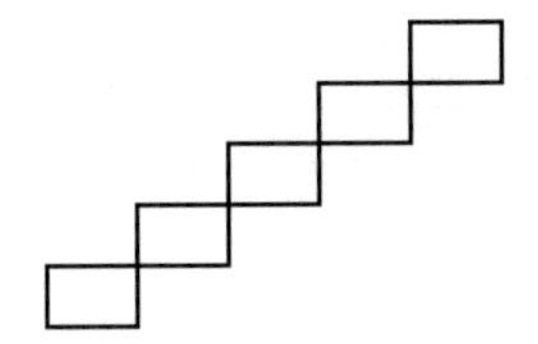

"It's a staircase," says Grace, smiling in approval. "Of course."

"Naturally," I reply.

Photo #1 is a picture of Mr. Collins, crossing the finish line at his most recent marathon, back in April. The photographer captured the moment perfectly: the big display marked the finishing time of 3:59:59, just underneath the even bigger sign containing two iconic words: "Boston Marathon".

On the week of his sixty-ninth birthday, Mr. Collins desperately wanted to finish Boston in under four hours, and sprinted to the end so that he'd reach his target. He didn't realize until after the race was over that his actual finishing time was nearly five minutes faster, since he was behind thousands of other runners at the starting line. So not only did Mr. Collins attain his goal, he also beat his personal best. In Boston.

Photo #2 is a picture of Grace and me from three months ago at the Waterloo Math Seminar, posing with Rachel and Marlene. Having tied with Albert for first overall on the Canadian Mathematical Olympiad, I finally received my invitation to Waterloo. The lunching ladies turned into the dining dames, as Grace and I sat with our mentors from the Canada Math Camp, at the CMO Awards Banquet.

Rachel had solved a long-standing open problem in Group Theory for her Masters' thesis, and was now halfway into her Ph.D. program at the University of Waterloo. While Rachel could walk to the banquet hall from her office, Marlene ended up driving all the way from Ottawa to be in attendance that evening. Rachel and Marlene were delighted that the top four finishers were all alumni of the Canada Math Camp. There were so many pictures of that memorable evening, but none was more special than this one.

Photo #3 is a picture of five people: Mr. Marshall in the middle, with Bonnie and Breanna on the left, and Patrick and me on the right. We're all smiling, wearing formal clothes, celebrating Graduation Day at Sydney High School. Mr. Marshall had a special pin on his lapel, which he received from our country's leader the day he flew to Ottawa to accept the Prime Minister's Award for Teaching Excellence.

Bonnie, Breanna, and Patrick all got in to St. FX, and are in the same residence together this year, with Bonnie and Breanna sharing a room on the second floor and Patrick just one floor up.

Against the advice of her mother, Gillian decided to enroll at the Nova Scotia College of Art and Design, where she won a scholarship to pursue a degree in Fine Arts. A few weeks ago, I wrote Gillian a long e-mail where I asked her to forgive and forget all the junk from our past, and wished her the very best in her new life in Halifax. I apologized for our rivalry, which brought out the worst in ourselves. I asked if we could start over.

Gillian didn't reply to my e-mail, but the next day, she sent me an invitation to become friends on Facebook. I accepted the invitation. One day, I hope we'll be friends for real.

Photo #4 is a picture of Grace and her father Ray, taken at Grace's high school graduation. I fondly remember that life-changing week in Vancouver and recall how and why I got interested in spirituality.

During orientation yesterday, Grace and I picked up a dozen brochures of various clubs on campus, including one from the multi-faith fellowship group. I introduced myself to the guy behind the booth and asked him whether it was okay if I could join, even though I'm not a "believer" of any particular faith or religion. He responded enthusiastically. Thanks to Ray's influence, I'm excited to learn about his faith, and explore whether it can provide answers to my deepest questions.

Photo #5 is a picture of Dr. Weber and me, eating dinner at Gabrieau's in Antigonish. During our four-hour meal of Nova Scotian lamb and cedar-roasted Atlantic salmon, I announced to Dr. Weber what university Grace and I would be attending. Dr. Weber congratulated us on our choice, and was thrilled that each of us had won the university's largest entrance scholarship.

Grace and I found a perfect solution to our geographical dilemma: we'll do our undergraduate degrees on one coast, and pursue our graduate degrees on the other coast. We'll spend every Christmas with Ray in British Columbia, and every summer in Nova Scotia as Dr. Weber's research assistants.

After that, we'll see where life takes us.

I look over at my best friend, and smile.

We no longer have to live four time zones away. We're now at the same university and are roommates. One chapter of our lives had recently ended, and the first of many would begin tomorrow.

"I've met so many amazing people," I say, pointing to the five pictures on the wall. "And I can see them every day for the next four years."

"It's great," says Grace. "Especially because I'm the only person in two pictures."

I roll my eyes and laugh.

"Okay, I'm ready for the capstone."

Grace nods and takes out a sealed package from the bottom of her clothes drawer. She carefully cuts the tape with scissors, and opens the package to reveal two felt-wrapped items, and lays them carefully on my desk.

I know what's inside, but I haven't seen the finished product. Grace has, and she says it's amazing.

With Grace on the left and me on the right, we simultaneously open up the felt-wrapped item in front of us. The small circular object shines, and as I hold it in my hand and rotate it, I'm stunned by what I see.

"It's beautiful."

"Which one do you want?" asked Grace, pointing at the two disks.

"Does it matter?" I ask. "Aren't they identical?"

"Good point," responds Grace, nodding in agreement. "The guy at the jewellery store said this was the strangest request he'd heard in thirty years. He thought it was a stupid idea, that it would destroy the value of each piece. He said that one-half plus one-half would add up to zero."

"He's wrong," I say. "In this case, one-half plus one-half equals two."

"Exactly," says Grace, motioning towards the wall and inviting me to complete my artistic mural.

I take my adhesive strips and stick them on the back of my metallic disk, firmly pressing the object onto the wall so that it forms the sixth and final rung of my staircase.

We stand back and see the words INTERNATIONAL MATHEMATICAL OLYMPIAD staring back at us, with the left hemisphere made out of sterling silver, and the right hemisphere made out of pure gold.

The two halves of our Olympic medals are perfectly soldered together, and as I stare at the shiny object, I realize that I've spent six years, and several thousand hours, working for something that weighs less than half a pound. And yet, I know that I'll never ever get bored of looking at this piece of metal that will forever bind Grace and me.

Grace holds her Olympic medal in her hand, with the two halves switched, gold on the left and silver on the right. With her free hand, she points to my wall.

"It's the perfect capstone."

"I like that word," I respond. "The top stone of a structure. The final touch."

"It's more than that," says Grace. "The capstone is the crowning achievement, the culmination of all your effort – and when you look back, you see that all those dreams, all that work, was worth it."

I stare at the five photos on the wall, seeing the five major chapters of my life unfold before me: Mr. Collins, the Canada Math Camp, Sydney High School, the week in Vancouver, and my Friday afternoons with Dr. Weber.

I see how each chapter formed a staircase, one on top of the other, and how each experience was instrumental to my perfect score on the Canadian Mathematical Olympiad, leading to the half-silver half-gold medal I won for Canada, six weeks ago at the IMO.

But something's not right. Instantly, I know.

"It's the wrong capstone."

"Huh?" says Grace, raising an eyebrow.

I reach into my bookshelf and pull out a photo album, containing several hundred pictures. I leaf through the album, and on the final page, I find the picture I'm looking for.

I take it out and apply the adhesives one final time, pressing the photo onto the wall to form the seventh and final rung of my staircase.

"That's the capstone," I say.

Grace and I stare at a picture of Mom, looking radiant in a summer dress, with the biggest smile I've ever seen. She's pointing to a neon sign with her left hand, waving a white envelope with her right hand.

"You're right," says Grace. "That's the capstone."

Mom is re-taking Psych 100 this semester, enrolled in the same class as Patrick, Bonnie, and Breanna. I smile at the irony of my three Sydney High School friends being "tutored" by Mom; but as Mom says, everything is so much easier when you're doing it for the second time.

Grace and I lean closer and we see the words on the neon sign.

WELCOME TO SAINT FRANCIS XAVIER UNIVERSITY.

The white envelope contains Mom's acceptance letter from the St. FX Admissions Office, now proudly framed in the living room of her new one-bedroom apartment. She's living just a few hundred metres away from the St. FX Psychology building, Mom's academic home for the next four years.

Tomorrow, Mom starts Frosh Week.

I chuckle at the thought, and I'm so excited for her new beginning.

I close my eyes, and am overcome with joy, knowing that our roller-coaster journey has only just begun.

# Acknowledgements

First and foremost, I am grateful to my wife Karen, my best friend and life partner. For supporting this improbable project in every way and believing in Bethany's story despite my feeble initial efforts at creative writing, I thank you. I am blessed to have your love and encouragement, and for being "all my reasons".

Like Bethany, I have had numerous mentors over the years who inspired me with a love of learning and provided opportunities beyond my wildest dreams. I am deeply indebted to: Kayoko and Osami Hoshino, my parents; Jean Collins, my high school math teacher; Edward Barbeau, J.P. Grossman, Ravi Vakil, and all my former Math Olympiad coaches; Ian VanderBurgh and his remarkable team at the University of Waterloo's Centre for Education in Mathematics and Computing; Graham Wright from the Canadian Mathematical Society; Tom and Marlene Griffiths from the CMS National Math Camp; Jason Brown, Karl Dilcher, Jeannette Janssen, Richard Nowakowski, Dorette Pronk, and Keith Taylor from Dalhousie University; Cathy Beehan from Action Canada; Mark Winston from the Simon Fraser University Centre for Dialogue; Alain Jolicoeur and Diane Keller from the Canada Border Services Agency; Graham Flack and Adam Hendriks from the federal government's Recruitment of Policy Leaders Program; Mary Anne Moser and Jay Ingram from the Banff Science Communications Program; Ken-ichi Kawarbayashi from the National Institute of Informatics in Tokyo; and the entire community of Canadian mathematics educators, especially Peter Taylor from Queen's University and John McLoughlin from the University of New Brunswick.

I'm grateful for the numerous friends who read my original manuscript, whose frank advice and constructive feedback made the final novel so much more polished. In particular, I'd like to acknowledge the contributions of Sarah Aldous, Andrew Curran, Leonid Chindelevitch, Graham Duke, Matt Finlayson, Sharonna Greenberg, Chris Harder, Zdena Harder, Jack Koenka, Christopher Kowalchuk, Steve LaRocque, Derek Lemieux, Stephanie Mitchell, Anatole Papadopoulous, David Sandomierski, Naoki Sato, Adrian Tang, Ilya Volnyansky, and Wai Ling Yee.

After completing the first few chapters of this book, I posted what I had written to the Art of Problem Solving forum. I received valuable comments from members of the AoPS community, including high school students, undergraduate students, university professors, and the parent of a former Math Olympian. Thank you to Alex Chen, Jing Jing Li, Jonathan Love, Fedja Nazarov, Mary O'Keeffe, and Luyi Zhang.

Since February 2013, I have been a Mathematics Tutor at Quest University Canada in Squamish, British Columbia. I am grateful to my colleagues on the Quest faculty for deepening my love for teaching and reminding me of the privilege we have of serving the next generation. Thanks also to the students at Quest, including Rebecca Gross, Anna Harvey-Vieira, Anna Marie Obermeier, and Ariel van Brummelen, who read the manuscript and provided feedback on character voice and plot development. I am particularly grateful for the contributions of Sophia Matthew and Deanna Kronenberg, two remarkable students who worked with me every week for nine months, as they went through the text line by line, showing me how to make the story more compelling, and being the best editors any first-time writer could ask for.

Thank you to the entire team at FriesenPress, whose professionalism and talent taught me so much about the value of assisted self-publishing. I acknowledge the work of Dana Mills and her entire team for turning my manuscript into a finished novel.

Finally, I dedicate this book to all the Bethany MacDonalds out there, young students who are interested in mathematics but wonder if there is something more to the subject for them. I wrote this book with you in mind, to share the message that through relentless perseverance and hard work, anyone can succeed in mathematics and develop the confidence, creativity, and critical-thinking skills so essential in life. Like Bethany, I hope that studying mathematics will provide you with a deep clarity of purpose, and through this roller-coaster journey, you will discover how to serve society and live life to the full.

# Q&A with the Author

**Q: Why write a math novel?**

**A:** In March 2010, I moved to Japan with my wife Karen after she landed an amazing job teaching English at a top-notch Japanese university. As an unemployed house-husband living in a new country, I wondered how I could best apply my passion and experience to contribute to society. Having a love for expressing myself through writing, as well as possessing a Ph.D. in mathematics, I felt inspired to write a "math novel". Nearly five years later, I published *The Math Olympian*, the story of an insecure teenager who commits herself to pursuing the crazy and unrealistic goal of representing her country at the International Mathematical Olympiad, and thanks to the support of innovative mentors, combined with her own relentless perseverance, discovers meaning, purpose, and joy.

**Q: But why a novel, and not a textbook?**

**A:** I loved how *Sophie's World* reached millions of readers with no background in philosophy. The author did a masterful job of making philosophy accessible and enjoyable. Similarly, my goal is to reach a wider audience with *The Math Olympian*, sharing beautiful Olympiad-level math with the general public, and revealing the surprising and unexpected applications of mathematics to everything in this world. While my target audience is high school students (particularly female students who have an interest in math), I hope that many others will enjoy the book too!

**Q: Why did you feel qualified to do this?**

**A:** I felt qualified to attempt this ambitious project, given my experiences as a former Math Olympian for Canada, as a coach and trainer for Canada's Math Olympiad team, as the founder of two math outreach programs that have reached thousands of high school students in Nova Scotia, and as a mathematician for the Government of Canada who spent four years developing math-based solutions to improve the security and efficiency of the Canadian border. Given my experiences, as well as the mentorship I have received from so many world-class teachers, I felt that I could write a novel that would reveal creative ways to teach and learn mathematics, and show how the subject develops problem-solving skills, daring, critical-thinking and imagination – the types of skills Canadians require if we're going to be at the forefront of innovation in the twenty-first century.

**Q: How much of your novel is based on real-life experiences?**

**A**: All of the characters are fictional, but they are hybrids of real people who have influenced my life. For example, Bethany meets several mentors throughout the course of the book; all of her mentors are inspired by actual mentors I've had over the past twenty years, who modeled innovative teaching techniques and showed me that mathematics isn't about memorizing formulas or rules (in high school) or about memorizing theorems and proofs (in university). Also the settings of various scenes parallel some of my own personal experiences. For example, all of Chapter 2 is set in Ottawa, at a week-long event called the "Canada Math Camp" (formerly known as the National Math Camp). As I was the co-director of the National Math Camp for seven years, I could draw upon my experiences as a coach and trainer to create each scene and show how difficult problems could be solved in a myriad of elegant ways.

**Q: How did you come up with the five Olympiad problems that form the core of the book?**

**A**: I wanted to present actual Olympiad problems, not watered-down versions of the real thing. As a result, I selected five actual problems that have previously appeared on the Canadian Mathematical Olympiad. I chose these five problems as they are five problems that I know really well, based on my own experiences of attempting to make the Canadian IMO team nearly twenty years ago. Bethany's solutions to these problems are completely (or nearly) identical to my solutions on these contests, which I still vividly remember.

**Q: You were really driven to publish this book. Why?**

**A**: In 2009, I applied for a job that would have given me the platform to make a significant contribution to the Canadian math community by becoming a champion for pure and applied math research, and an advocate for improving math education at all levels, as well as reaching the target audience I was most passionate about – high school students and their teachers. And after I didn't get that job, I felt I had lost my one chance, my one platform, to make that significant contribution to the community that had given me so much. A mentor gently suggested that I didn't need to have *that* job in order to have that kind of impact. Perhaps it was this conversation that inspired *The Math Olympian*, realizing that through writing and publishing this book, I could make the contribution I had longed for.

**Q: How did you choose the name Bethany MacDonald?**

**A**: Many names have a special meaning, such as Amy (beloved), Karen (pure), and Sophia (wise). Based on the experiences the main character goes through, I wanted a female name that meant something like "One who perseveres through adversity and becomes a champion". But I couldn't find the right name. I eventually settled on the name Bethany because it was a name I've always liked, despite not having any meaning other than "House of Figs" in Hebrew. However, I did find one website claiming that Bethany means "New Beginnings", and I decided that was a good name for this character, especially given the parallel plot line of Bethany's mother.

Over a year after I started writing the book, I heard the remarkable story of a young woman who had achieved her dream of becoming a world-champion professional surfer despite losing her left arm as a teenager in a shark attack. This inspiring story became the hit movie *Soul Surfer*, which I saw in Toronto in the summer of 2011. When I learned that the surfer's first name was Bethany, I knew that wasn't a coincidence – I definitely had the right name for my main character! As for Bethany's last name, I went with MacDonald, the most common last name in Cape Breton, and also the last name of the celebrated Cape Breton family in *No Great Mischief*, Alistair MacLeod's masterpiece of loyalty, family, and discovering your identity.

**Q: Why is your main character a girl from Cape Breton?**

**A**: I want to challenge the common stereotype that mathematics can only be done by boys, nerds, and Asians (i.e., people like myself). I want *The Math Olympian* to reveal how with inspired mentorship, anyone can succeed in mathematics and develop the confidence, creativity, and critical-thinking skills so essential in life. Through my involvement with math outreach programs at Dalhousie University, I met young women like Bethany all throughout Nova Scotia. It is my hope that Bethany's story will inspire high school students, girls especially, to participate in math contests and math outreach activities, take mathematics courses in college or university, and pursue a future career in mathematics to tackle the complex challenges of the 21st century.

To support these goals and "pay it forward", I will donate 10% of my author royalties to the Canadian Mathematical Society (CMS), an organization that promotes the advancement, discovery, learning and application of mathematics in Canada. Half of this donation will go to the CMS Math Camps Program, and the other half will go to support the initiatives of the CMS Women in Mathematics Committee.

**Q: What are the key messages of your novel?**

**A**: There are three key messages that I tried to convey in *The Math Olympian*, all of which are connected. The first is that our dreams are worth pursuing, no matter how unrealistic, because they motivate us to reach our fullest potential and maximize our contribution to society. The second is that by choosing a roller-coaster life, that is, a life of courage rather than a life of comfort, we inspire those around us to do the same. And finally, in searching for truth with all of our heart and relentlessly pursuing the calling of our life, we'll be given opportunities beyond anything we could ever ask or imagine.

# ABOUT THE AUTHOR

Richard Hoshino, a first-time novelist, teaches mathematics at Quest University Canada in Squamish, British Columbia. After completing his Ph.D. at Nova Scotia's Dalhousie University, he worked for four years as a Research Scientist for the Government of Canada, using mathematics and statistics to improve the security and efficiency of the Canadian border. He later accepted a Post-Doctoral Fellowship at the National Institute of Informatics in Tokyo, and helped the Japanese Professional Baseball League design a season schedule that cut travel costs and reduced greenhouse gas emissions. Richard is a former Math Olympian, and won a silver medal for Canada at the 1996 International Mathematical Olympiad in Mumbai, India.

To get in touch with Richard, please send him an e-mail at richard.hoshino@gmail.com or visit www.richardhoshino.com.